...acent Lands &c: from Hockley-Bridge to Mr. Toney's Ga...
...ich and its Houses, with the lower part of Money-Bag Hill;
...ts of View.

Secondly, as it relates to my Interest.

forward with the hopes of my Descendants being Opulent and respectable Manufacturers, at Soho, to the Third and Fourth Generation, rather than dependant Courtiers —

I should live in peace with my Neighbours, and steer clear of Lawsuits & Quarrels —

Every Shilling I then laid out in Building &c: would be for the benifit of my own Successor, and my property would always retain its full Value for Sale or otherwise —

Compare the probable Value of Soho after my Death, if under the circumstances I have stated, of Not Buying, with the probable Value of it, if under the circumstance of being a Freehold —

If these Blessings and Comforts are procurable by laying out any Sum in Land, so as to make it yeild, for certain, Five ℏCent, and with a fair prospect of its yeilding in a few years Ten ℏCent —

I ask, if under all these circumstances I ought to Buy? —

But before that Question is answered, Examine the Account hereunto annexed of Cost, with profit & Loss —

MATTHEW BOULTON

MATTHEW BOULTON

Selling what all the world desires

edited by
Shena Mason

Birmingham City Council
in association with
Yale University Press, New Haven and London

This publication accompanies the Matthew Boulton Bicentenary Exhibition
Selling what all the world desires
organized by Birmingham Museum & Art Gallery
at the Gas Hall Exhibition Gallery from May to September 2009

Copyright © 2009 by Birmingham City Council

All rights reserved. This book may not be reproduced, in whole or in part, including illustrations, in any form (beyond that copying permitted by Sections 107 and 108 of the U.S. Copyright Law and except by reviewers for the public press), without written permission from the publishers.

Designed by Catherine Bowe and Sally Salvesen
Printed in Italy by Conti Tipocolor S.p.A., Florence

Library of Congress Cataloging-in-Publication Data
Matthew Boulton : selling what all the world desires / edited by Shena Mason.
    p.   cm.
  Includes bibliographical references and index.
  ISBN 978-0-300-14358-4 (alk. paper)
  1. Boulton, Matthew, 1728-1809. 2. Industrialists--Great Britain--Biography. 3. Mechanical engineers--Great Britain--Biography. I. Mason, Shena.
  TS140.B68M38 2009
  739.2'3092--dc22
  [B]
                            2008035075

Illustrations:
Half title: Silver medal issued by Edward Thomason to commemorate the death of Matthew Boulton.
Engraved by P. Wyon after P. Rouw. Diameter 113mm.
In his memoirs, Sir Edward wrote that he had been 'one of Mr Boulton's pupils from the age of sixteen years to that of twenty-one', and that after Boulton's death he commissioned Wyon to engrave the die, which was 'the deepest cut die then extant'. The medal was thought to be the largest in Europe, and of the highest relief. This claim was challenged by the Russian Ambassador who produced a larger diameter Russian medal, but taking the elevation of the Boulton medal into account, a mathematician calculated that its surface area was greater than the Russian one. The medal is on show at Boulton's home, Soho House.

Frontispiece: Lemuel Francis Abbott, Portrait of Matthew Boulton, oil on canvas, c.1798–1801 (see Cat. 1).

Contents

Foreword — vii
Acknowledgments — viii
Lenders to the Exhibition — ix
Contributors — xi
Chronology — xii
Boulton family tree — xiii

Introduction — 1
Rita McLean

1. **Matthew Boulton and the Lunar Society** — 7
   Jenny Uglow

2. **'The Hôtel d'amitié sur Handsworth Heath': Soho House and the Boultons** — 14
   Shena Mason

3. **Picturing Soho: Images of Matthew Boulton's Manufactory** — 22
   Val Loggie

4. **The Context of Neo-Classicism** — 31
   Nicholas Goodison

5. **Matthew Boulton's Silver and Sheffield Plate** — 41
   Kenneth Quickenden

6. **'I am very desirous of being a great silversmith': Matthew Boulton and the Birmingham Assay Office** — 47
   Sally Baggott

7. **Ormolu Ornaments** — 55
   Nicholas Goodison

8. **The Soho Steam Engine Business** — 63
   Jim Andrew

9. **'I had L[or]ds and Ladys to wait on yesterday…': Visitors to the Soho Manufactory** — 71
   Peter Jones

10. **Matthew Boulton's Mints: Copper to Customer** — 80
    Sue Tungate

11.	'Bringing to Perfection the Art of Coining': What did they make at the Soho Mint? David Symons	89
12.	A Walking Tour of the Three Sohos George Demidowicz	99
13.	How do we know what we know? The Archives of Soho Fiona Tait	108
	Catalogue entries 1–404	117
	Notes and references	237
	Bibliography	249
	Picture credits	250
	Index	251

Foreword

In August 1809, an estimated 10,000 people lined the roads to watch the funeral procession of a man who had done more than anyone before him to put his home town of Birmingham on the map. That outpouring of affection and respect was for Matthew Boulton, the pioneer industrialist whose inventive mind, ability to recognise potential, social flair and sheer energy had helped to establish Birmingham as one of England's greatest manufacturing centres, and England as the industrial leader of the known world.

Two centuries later, in the age of globalisation, the political, economic and social maps are dramatically different. The City of Birmingham, however, is still renowned for nurturing innovation, aspiration and industry, and our local spirit still embraces those same characteristics which made Matthew Boulton such a successful and well-loved figure.

It is, therefore, entirely fitting that in 2009 Birmingham City Council and our Museum & Art Gallery, in collaboration with a number of key institutions within and beyond the city, are acknowledging and celebrating Matthew Boulton's life and work. This is being done through the major Boulton Bicentenary exhibition, and through the pages of this book, which brings together contributions from a number of authors working in the field of Boulton studies and an unsurpassed collection of Boulton-related images (many of them published here for the first time).

Boulton's range of interests (as engineer, scientist and manufacturer) seems incredible for one lifetime. The exhibition looks at all aspects of his life, revealing something of the enquiring mind and personality of a man who moved easily through the social strata of his day, as welcome at the court of King George III as he was at the firesides of his friends.

If the life of Matthew Boulton can teach us anything, surely it is that openness to new ideas, and determination to pursue a vision, are what move our world forward. This book, and the exhibition, are a fitting demonstration of that lesson from a man who was proud to be known as 'the father of Birmingham'.

Mike Whitby
Leader, Birmingham City Council

Acknowledgments

The project to create the Matthew Boulton Bicentenary Exhibition and this book, *Matthew Boulton: Selling What all the World Desires*, as well as the related programme of events, would not have been possible without the help of many institutions and individuals. Funding support for the project has come from a number of sources. These include: Birmingham City Council, The Friends of Birmingham Museums & Art Gallery, Birmingham Museum & Art Gallery Development Trust (The Soho House Appeal), Advantage West Midlands, The Heritage Lottery Fund, The Leche Trust, the Edward Cadbury Trust, The Bryant Trust, and the Sir Robert Gooch Trust. We are also grateful for the collaboration of the Birmingham Assay Office.

That we have been able to bring together probably the largest selection of Boulton-related objects ever exhibited is due to the co-operation of the owners of many public and private collections, who are listed opposite. Birmingham Museum & Art Gallery's own Boulton-related collections are much the richer for the help of the Heritage Lottery Fund, The Art Fund, MLA/V&A Purchase Grant Fund, and the Friends of Birmingham Museums & Art Gallery, who have contributed to the acquisition of so many objects over the years.

Special mention should be made of one or two people including Dr David Symons, who with Laura Cox had the mammoth task of identifying and sourcing the objects for this many-stranded exhibition. Fiona Tait, Siân Roberts and the other staff of Birmingham Archives & Heritage have been endlessly patient in helping to find and evaluate documents, and Sir Nicholas Goodison has been a source of much good advice. Finally, many other people who are not named here have given willingly of their time, advice, knowledge and enthusiasm, in particular the members of the Boulton 2009 Advisory Board, which has met regularly over the long planning phase.

It is hoped that this book will be a valuable reference source for Boulton scholars for many years to come. Its publication has been generously part-funded by the Paul Mellon Centre for Studies in British Art and by The Goldsmiths' Company, home of Assay Office London. Thanks are also due to the writers who have contributed to the book, who are listed elsewhere, and to Shena Mason who has seen the book through from its inception to publication. That it is so fully and so beautifully illustrated is due in part to the helpfulness of lenders in providing images of loaned objects, but in the main to the intensive work of Birmingham Museum & Art Gallery's photographer, David Rowan, picture librarian, Tom Heaven, and Richard Albutt of Birmingham Archives and Heritage.

Our warmest thanks are due to them all.

Rita McLean
Head of Birmingham Museums & Art Gallery
March 2009

Lenders to the Exhibition

The exhibition which this book accompanies has drawn extensively on Birmingham Museums & Art Gallery's own Boulton collections and those of Birmingham Archives & Heritage (Birmingham Reference Library). It has also brought together Boulton-related objects from many other sources, including the following:

The Birmingham Assay Office
The British Museum
Christie's
Derby Museum & Art Gallery
City of Leeds Art Gallery, Temple Newsam
The Masonic Grand Lodge
The National Gallery
The National Maritime Museum
The National Portrait Gallery
The National Trust
The National Waterways Museum
The Royal Collections Trust
The Science Museum
Tate Britain
Thinktank
Tyne & Wear Museum Service
The University of Birmingham
University College London
The Victoria & Albert Museum
The Weston Park Foundation
Wolverhampton Museum & Art Gallery
and
A number of private collectors

The Contributors

Jim Andrew is an engineer who has worked with Birmingham's industrial collections since 1974 with responsibilities including the oldest working Watt engine in the world and the excavation of its original engine house in Smethwick in 1984. In 1991 he was awarded his PhD by Birmingham University for research on Boulton & Watt's water supply engines including those used by canal companies. Following retirement he continues as Collections Advisor to ThinkTank, the Birmingham Science Museum.

Sally Baggott. Formerly a chemical analyst with Severn Trent, Sally Baggott went on to study for a BA in English Literature and Cultural Studies at the University of Birmingham; her subsequent PhD, funded by the AHRC, was on late nineteenth-century writing on the lower classes of London. She has worked at the Museum of the Jewellery Quarter and ThinkTank, and is now librarian and curator at Birmingham Assay Office.

George Demidowicz is head of the Conservation and Archaeology team at Coventry City Council. His research interests are mainly in building and landscape history, and industrial archaeology. As a result of many years' research a volume on the origins, development and layout of the three Soho sites is close to completion, to be published in 2009.

Nicholas Goodison is the author of *Matthew Boulton: Ormolu* (2002), a revised edition of his *Ormolu: the Work of Matthew Boulton*, first published in 1974. He has published a number of other papers and articles about Boulton, including 'Matthew Boulton's Trafalgar Medal' (Birmingham Museum and Art Gallery, 2007). He was formerly Chairman of the London Stock Exchange, the Courtauld Institute of Art, the Burlington Magazine and the National Art Collections Fund, and is President of the Furniture History Society.

Peter Jones is Professor of French History at the University of Birmingham. He published an article on visitors to the Soho Manufactory in the July 2008 issue of the journal *Midland History*. His book *Industrial Enlightenment: Science, Technology and Culture in Birmingham and the West Midlands, 1760–1820*, was published by Manchester University Press in 2009.

Val Loggie is the former curator of Soho House and was part of the project development team that restored the house. She has also worked at Erasmus Darwin House in Lichfield, Pickford House in Derby and Tamworth Castle. She is currently working on an AHRC-funded collaborative PhD at the University of Birmingham and Birmingham Museum and Art Gallery.

Shena Mason was administrator of the Soho House and Archives Appeal, and part of the Soho House project development team. She has also helped with some of the cataloguing of the Matthew Boulton Papers. A graduate of the University of Warwick, she is the author of *The Hardware Man's Daughter: Matthew Boulton and his 'Dear Girl'* (2005) and *Jewellery Making in Birmingham, 1750–1995* (1998).

Rita McLean is Head of Birmingham Museums and Art Gallery (BMAG). In a former role within Birmingham Museums Service, she led the restoration and refurbishment of Matthew Boulton's home – Soho House, and oversaw its opening to the public as a museum in 1995.

Kenneth Quickenden is the Jewellery Heritage Professor at Birmingham City University and has been researching the silver and Sheffield plate of Matthew Boulton since the 1970s.

David Symons has a BA in Ancient History and Archaeology and a PhD on Anglo-Saxon numismatics, both from the University of Birmingham. He has worked at Birmingham Museum and Art Gallery since 1977 and is now Curator of Antiquities and Numismatics. He has served on the Council of the British Numismatic Society and from 2000 to 2007 was Editor of the Society's journal, the *British Numismatic Journal*.

Fiona Tait has worked in Birmingham Archives & Heritage for over twenty years. She has a wide-ranging knowledge of the Archives of Soho and has always enjoyed assisting researchers. She was seconded to catalogue the papers of James Watt & family for the Archives of Soho project.

Sue Tungate is a PhD student using an AHRC Collaborative Award to study 'The coins, medals and tokens of Matthew Boulton's mints', as part of the joint University of Birmingham/Birmingham Museums and Art Gallery project, 'Matthew Boulton: Visual and Material Cultures of Industrialisation'. Previously she was a science teacher in Birmingham. She has a B.Sc. in Biological Sciences and an M.A. in Local History.

Jenny Uglow is a writer and publisher: her books include the story of the Lunar Society of Birmingham, *The Lunar Men: The Friends Who Made the Future* (2002), as well as biographies of, among others, Elizabeth Gaskell, William Hogarth, and Thomas Bewick.

Descriptions of the objects in the exhibition have been written by the following people: Jim Andrew, Sally Baggott, Laura Cox (Curator-Manager/ Deputy Curator-Manager, Museum of the Jewellery Quarter, BMAG), Sylvia Crawley (Curator, Applied Art, BMAG), Gordon Crosskey (Principal Lecturer and Fellow, Royal Northern College of Music, also researching the 18th century plated trade), Jo-Ann Curtis (Curator of History, BMAG), George Demidowicz, Martin Ellis (Curator, Applied Art, BMAG), Brendan Flynn (Curator, Fine Art, BMAG), Nicholas Goodison, Peter Jones, Jack Kirby (Collections Interpretation Manager, Thinktank), Henrietta Lockhart (Curator of History, BMAG), Val Loggie, Pamela Magrill (freelance curator and lecturer), Shena Mason, Clare Parsons (Curator-Manager, Soho House, BMAG), David Powell (Administrative Officer, BMAG), Victoria Osborne (Curator, Fine Art, BMAG), Chris Rice (Head of Community Museums, BMAG), Fiona Slattery (Curator, Applied Art, BMAG), David Symons, Sue Tungate and Fiona Tait.

Chronology

1728	Matthew Boulton born in Birmingham, 3 September. Parents: Matthew Boulton senior (a button and bucklemaker) and Christiana Peers.
1745	Leaves school; joins the family business.
1749	Marries Mary Robinson of Lichfield.
1759	Death of wife Mary and father, Matthew Boulton senior.
1760	Marries Ann Robinson of Lichfield (sister of Mary).
1761	Leases 13 acres of land at Handsworth, including Soho House (built c.1757).
1762–4	Builds Soho Manufactory. Now exports jewellery, silver and decorative wares throughout Europe with partner John Fothergill.
1766	Lunar Society begins, with Dr. Erasmus Darwin, Matthew Boulton and Dr. William Small. Boulton and his second wife move into Soho House.
1768	Daughter, Anne Boulton, born, 29 January. Matthew Boulton on the General Hospital committee organising the first of the Birmingham Music Festivals.
1770	Son, Matthew Robinson Boulton, born, 8 August.
1772	Boulton supplies 'green glass earrings' as bartering goods for Captain James Cook's second circumnavigation of the globe.
1773	Succeeds in campaign to establish Birmingham Assay Office.
1775	The Scottish engineer James Watt joins Boulton in partnership at Soho, as Boulton & Watt.
1785	Boulton elected a Fellow of the Royal Society.
1790	Soho House remodelled by James and Samuel Wyatt. Matthew Robinson Boulton joins the business.
1793	Matthew Boulton Chairman of the Birmingham Theatre Proprietors' Committee
1794	Serves as High Sherriff of Staffordshire. Also Chairman of the Governors of the Birmingham Dispensary, which provides medicines and medical care to the poor.
1797	Awarded Royal Mint contract to strike copper coinage at Soho. Boulton & Watt steam engines and minting machinery now exported worldwide.
1809	Matthew Boulton dies at Soho House, 17 August.

Family Tree

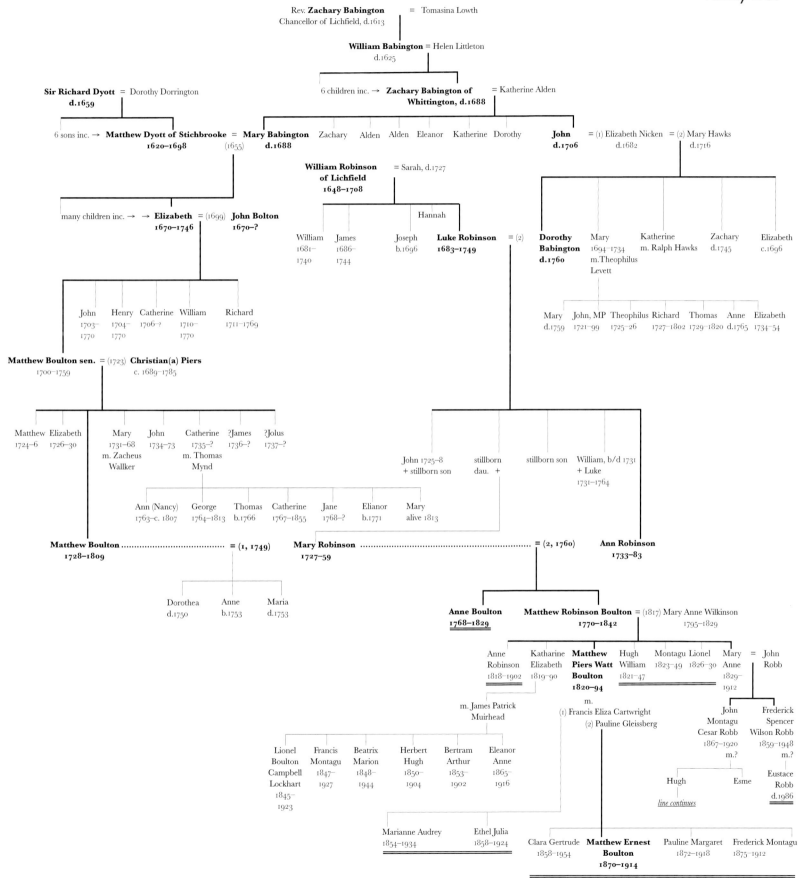

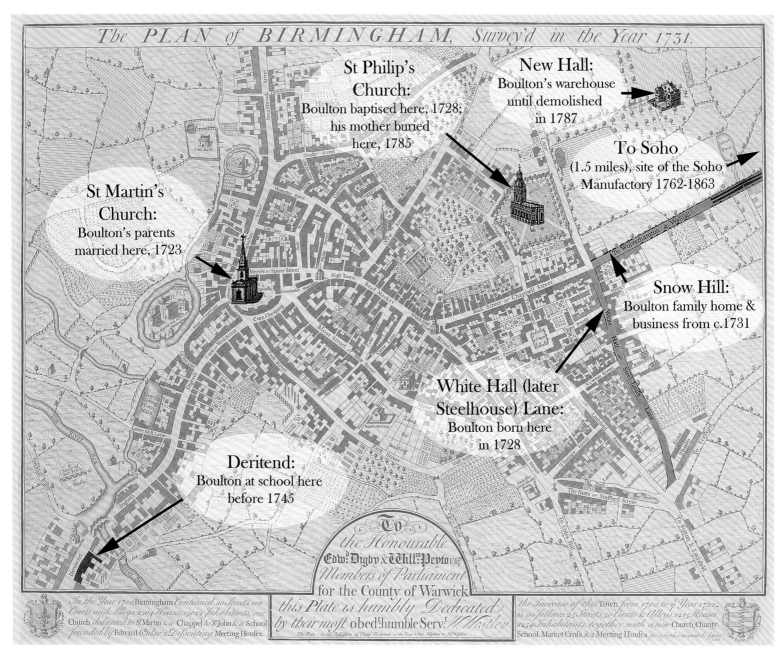

This map, based on William Westley's 1731 Plan of Birmingham (Cat. No. 17), shows the locations of significant places in the life of Matthew Boulton. Westley's orientation has West at the top and North at the right. Soho, where Boulton built his Manufactory from 1762, and where he lived from 1766, lies about one-and-a-half miles to the northwest beyond the northern edge of the map. Boulton is buried in Handsworth Parish Church, not far from Soho (Map: Birmingham Museum & Art Gallery; treatment: J.S. Mason).

Introduction: Matthew Boulton, 1728–1809

Rita McLean

Matthew Boulton, the bicentenary of whose death this volume commemorates, was undoubtedly one of the great manufacturers and entrepreneurs of his age. His activities contributed to some of the most important technological advances made anywhere during the second half of the eighteenth century. Boulton has perhaps been most widely known for his steam-engine partnership with James Watt. His significance, however, is by no means confined to his exploits in this field, for he was pre-eminent, both as a manufacturer of useful and ornamental metalwares, and because of his achievements in the improved production of coinage. While these were the three major branches of his commercial enterprises, Boulton's career – which spanned more than sixty-five years – was also characterised by his constant venturing into numerous other business concerns. His life was interwoven with an extraordinary range of personal interests, and he had a fascinating and extensive network of eminent friends, acquaintances, and collaborators.

Though Boulton was a figure of renown to his contemporaries, the detail of his life and the breadth and significance of his work have been less widely appreciated in recent times. Boulton's businesses have not continued to within living memory, unlike for instance the pottery concern of his friend Josiah Wedgwood. Matthew Boulton's best-known metalwares – his silver and plated wares, ormolu, and coins and medals – are preserved in museums and in important private collections around the world. Their historical importance, artistic merit and intrinsic value have ensured that they have remained expensive articles, and thus not widely known or attainable. Boulton's cheaper products, the masses of buttons, buckles, chains and other items were less durable, discarded more readily as fashions changed, and are in any case largely anonymous amongst the articles of this type that have survived to the present time. Inevitably, the steam engines produced by the firm of Boulton & Watt (and its successor firms) became obsolete. The few surviving items remain preserved either as museum pieces or at industrial heritage sites.

Thanks to the existence of Boulton's personal papers and business records (housed at Birmingham Archives & Heritage, and outlined by Fiona Tait in Chapter 13, pp. 108–15), we have an exceptionally detailed picture of Matthew Boulton's endeavours, and the scale and diversity of his interests and activities, yet the sheer magnitude of the archive has perhaps inhibited the writing of a comprehensive account of Boulton's life and work. Only two biographies of Matthew Boulton have been produced. The first of these, by the Scottish author and reformer Samuel Smiles, was published in 1865 and looked jointly at the lives and work of Boulton and Watt.[1] The only work that has so far been exclusively devoted to Boulton was written in 1936 by H. W. Dickinson, who himself acknowledged that 'the aspects of Boulton's life were so many that to do justice to them no single author could hope to command the qualifications that are necessary to enable him to do so.' Dickinson presented his own work 'only as a step in the right direction', and consigned the task of producing an exhaustive and fully documented account to a future writer.[2] Many decades later, no such account has yet emerged, though modern studies of discrete aspects of Boulton's activities have been produced.

To an extent, that pattern is continued in this book. However, by bringing together the knowledge of a group of people currently working in the field of Boulton studies, it is hoped to present an introduction to Boulton's life and work, and provide a context for the associated Matthew Boulton Bicentenary Exhibition in Birmingham. A further aim is to stimulate future research and publications on the exploits and enterprises of this remarkable figure – who is undoubtedly deserving of greater recognition.

Early life and the Lunar Society

Matthew Boulton was born in Birmingham on 3 September 1728. The Boulton family's roots were in Lichfield, and the Boulton ancestry included the old Staffordshire county families of Babington and Dyott, both substantial landowners.[3] Matthew Boulton's father, Matthew Boulton senior, had come to Birmingham as a youth to serve an apprenticeship.[4] In June 1723, at the age of twenty-three, he married Christiana Piers, said to be the daughter of Daniel Piers of Chester, at St Martin's, Birmingham's Parish Church in the Bullring.[5] The Matthew Boulton who forms the subject of this volume was their third child. The register of St Philip's Church (now Birmingham Cathedral) records his baptism on 18 September 1728.[6] At the time of his birth, the family was living and working at Whitehalls Lane (later renamed Steelhouse Lane); when young Matthew was about three years old they moved to a house in Snow Hill.[7]

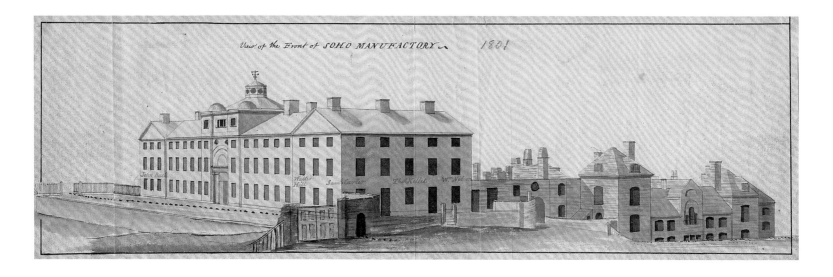

1 *The Soho Manufactory*, a drawing done in 1801 as evidence in the prosecution case against robbers who had attempted a wages snatch in December 1800. (Birmingham Archives & Heritage, MS 3069)

Matthew Boulton senior was a manufacturer of metal 'toys' (not children's playthings but small personal accessories, such as shoe buckles and snuffboxes). Evidence of the scale and exact nature of Boulton senior's business is very limited, but toy-making was important in Birmingham: by the middle of the eighteenth century, some of the town's leading manufacturers reported to a Select Committee of the House of Commons that twenty thousand people were employed in the toy trades in Birmingham and the surrounding districts, and toys to the value of about £600,000 per year were produced, of which £500,000 in value were exported.[8]

Samuel Smiles records that young Boulton was educated at a private academy in Birmingham run by the Reverend John 'Ansted' (correctly, Hausted), who was the chaplain of St John's Chapel, Deritend – although it must be said that Smiles' source for this information is not known.[9] *Memoirs of Matthew Boulton, Esq., F.R.S., Late of SOHO, Handsworth, Staffordshire* (1809) records that he was taught 'drawing under Worledge, and mathematics under Cooper', but again the source of the information is not known (Maria Edgeworth's copy of this pamphlet is in the Beinecke Library at Yale University). After leaving school in about 1745 the youth went to work for his father. Their firm produced fashionable goods and exported much of its output to France, from where it was often re-imported to Britain as the latest 'French' novelty.[10] In 1749, twenty-one-year-old Matthew Boulton married Mary Robinson, daughter of a prosperous Lichfield mercer, Luke Robinson. The couple were distantly related, for they shared Babington ancestors.

Boulton's personal and business papers chart his continual inclination to broaden his knowledge and technical abilities, and by his late twenties he could count among his friends such able and scientifically minded individuals as Erasmus Darwin, John Whitehurst and Benjamin Franklin (see Cat. 37). By the late 1760s this small circle of 'philosophical' friends had enlarged to form the Lunar Society, which Jenny Uglow describes in Chapter 1, pp. 7–13. This group has frequently been acknowledged as the eighteenth century's foremost provincial philosophical society.[11] Almost all of the members became Fellows of the Royal Society of London. Indeed Boulton, his partner James Watt and William Withering were all elected to this body on 24 November 1785, Whitehurst writing to Boulton later, 'I congratulate you on your election into the Royal Society and particularly as there was not a negative ball against you.'[12] The group provided considerable practical assistance to Boulton in a range of his manufacturing projects.

The 'most complete manufacturer in England in metal'
From the late 1750s onwards, Boulton's manufacturing ventures are documented in considerable detail. It may well have been in the aftermath of his father's death in 1759 that Boulton felt free to develop the business on a larger scale. His first wife had also died in 1759, shortly before his father; less than a year later, somewhat controversially, Boulton married her sister, Ann Robinson of Lichfield, and Shena Mason gives some account of the Boultons' family life and their home, Soho House, in Chapter 2, pp. 14–21. Though the addition of Ann's capital was no doubt useful, Boulton's letters to her, both before and after their marriage, portray considerable affection. Nevertheless the property and capital gained through this second alliance with the Robinson family would certainly have helped finance the expansion of his business. Ann's inheritance of estates worth over £20,000 in 1764, on the death of her brother Luke, also added to Boulton's security and his ability to raise capital. But although the inheritances obtained from the Robinson family estates were of undoubted importance, Boulton's business activities were predominantly financed by a staggering array of loans which he was to juggle with for the greater part of his career.[13]

In 1761, the year after his second marriage, Boulton leased thirteen acres of land at Soho, in Handsworth. Before long he was immersed in an ambitious programme to build the Soho Manufactory, around which George Demidowicz guides us in Chapter 12, pp. 99–107. Even before the completion of the principal building, Matthew Boulton stated that he was employing four hundred people at the Soho Manufactory,[14] and after visiting it in 1767, Josiah Wedgwood described Boulton as 'the first or most complete manufacturer in England in metal'.[15] In 1768 Boulton himself claimed that he had established 'the largest Hardware Manufactury in the world'.[16] The grandiose buildings (Fig. 1) were depicted by contemporary artists in a variety of styles and for a variety of purposes, discussed by Val Loggie in Chapter 3, pp. 22–30. The site attracted numerous visitors. Some were simply tourists with money to spend in the associated showroom and tea room; others came with more covert agendas, and Peter Jones has given some account of them in Chapter 9, pp. 71–9.

The toy trade

Although Boulton began in business as a toy maker, no comprehensive study of his enterprise in this field has yet been attempted, and thus it has been somewhat overshadowed by his other business concerns. However, the production of these small articles represented a significant aspect of the Soho Manufactory's trade until the late 1790s. Boulton's toy making activities were conducted through a series of business partnerships, the first of these being a twenty-year alliance with John Fothergill, from 1762 to 1782 (Fig. 2). Fothergill had served an apprenticeship with a merchant house in Königsberg, where his family originated.[17] Immediately before joining Boulton, he had worked for the Birmingham toy makers Ingram and Duncomb, probably as a traveller engaged in selling their wares on the continent. Fothergill therefore brought to the partnership with Boulton his expertise and experience of the continental toy trade, and no doubt the valuable contacts he had with merchants and customers abroad.[18] He also invested heavily in the toy making enterprise, putting just over £5,000 into the business at the start of the partnership, alongside Boulton's own investment of £6,000, to help fund the programme of building work that Boulton had embarked upon at Soho.[19] Fothergill travelled extensively on the Continent to seek orders, in one year travelling from Hamburg to Lübeck and Königsberg, then through Denmark, Sweden and on to St Petersburg. His return journey took him through Narva, Metteau, Riga, and back to Königsberg.

Boulton's chief aims were to produce high-quality and fashionable articles, which would serve to overturn Birmingham's reputation for producing cheap and shoddy goods derogatorily known as 'Brummagem wares'; and to produce these items by employing advanced manufacturing methods. Despite Boulton's evident success in winning the custom of the nobility and aristocracy, and in presenting his Manufactory to the outside world as a hive of industry leading the way in production technology and the superiority of its products, in reality Boulton and Fothergill's toy business was highly unprofitable. While there is

2 J. S. C. Schaak, *John Fothergill*, oil on canvas, 75 × 61.5 cm. (Dr Alastair Brown)

no doubt that the Manufactory was technologically advanced, and capable of producing excellent and eminently marketable articles, the partners failed to make the business pay. In his detailed financial analysis of Matthew Boulton's business activities, J. E. Cule discerned the causes that lay behind this state of affairs. In short, the overall manufacturing operations at Soho were run in what can only be described as a spectacularly disorganised manner. Insufficient attention was given to overseeing the workforce, to ensuring that products were priced at a level which would make them profitable, to monitoring and controlling stock, and to recovering payments owed from sales – particularly from customers abroad. Matthew Boulton's frequent absences from Soho to court prospective customers, and his propensity to veer off into launching new business enterprises, also both caused and exacerbated the failings of the manufacturing operations. Above all, the partners were saddled with huge debts arising from the building of the Soho Manufactory, which forced them to borrow heavily to keep their enterprise going. By 1773 Fothergill considered their financial situation to be serious enough for him to suggest that they should cease trading. For Boulton, a man with huge ambition and unswerving determination, such a course of action was unthinkable; his response was to raise further loans to keep the toy business afloat, support

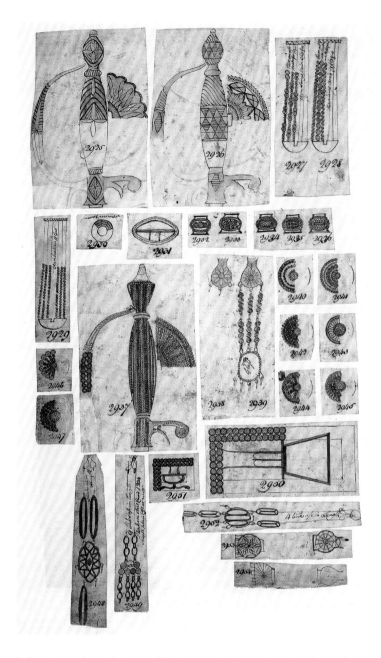

3 Sample page from a Boulton and Scale pattern book, showing designs for jewellery, buttons and sword hilts. (Birmingham Archives & Heritage)

his diversification into manufacturing silver and ormolu wares, and allow investment in a new commercial venture, namely the steam-engine enterprise with Watt.[20] Indeed after Fothergill's death in 1782, Boulton formed a new partnership with John Scale to continue the manufacture of buttons, steel chains and sword hilts (and he also established other partnerships to pursue this area of business at Soho after Scale's death).

As stated above, to date, no comprehensive study of this aspect of Boulton's business activities has been undertaken. However, it is worth noting that the records that have survived relating to this part of the business at Soho include a pattern book – probably dating from c. 1785–90 during the period of his business partnership with John Scale (Fig. 3). This single volume, almost certainly illustrating only a part of the firm's output, contains over 1,400 designs for cut-steel[21] buttons, buckles, chatelaines, sword hilts, watch chains and other articles.

Silver and plated wares

Despite the financial and management problems of the toy enterprise, within a few years of his arrival at Soho, Boulton embarked on a plan to branch out into manufacturing larger-scale and more expensive articles in silver and Sheffield plate, described by Kenneth Quickenden in Chapter 5 (pp. 41–6).

The absence of an Assay Office in the town was perhaps the greatest hindrance to Boulton's ambitions in developing this line of trade, and in Chapter 6, pp. 47–54, Sally Baggott writes about his lengthy and, in 1773, ultimately successful campaign to establish the Birmingham Assay Office, and about the Assay Office's fine collection of Birmingham – and especially Boulton – silver.

As well as the hallmarked silver, Boulton produced lower-priced Sheffield plate wares at the Soho Manufactory and was one of the earliest people outside Sheffield to do so. Fothergill was a reluctant participant in this enterprise – with some justification, for Boulton was again overreaching himself in branching out, at the expense of concentrating efforts on ensuring their toy trade was established on a secure and profitable footing.

Nevertheless, from the late 1760s to the early 1780s, an elegant and important body of work in silver and Sheffield plate was produced at Soho. Once again, Boulton's aims were to produce high-quality and fashionable products in great quantities, and at more competitive prices, through the application of technologically advanced production methods. But despite the notable quality of much of the silverware produced at Soho, this line of trade remained unsound for many years. Correspondence with Boulton & Fothergill reveals the great difficulty the firm found in executing orders within a reasonable length of time, and the frequent challenges presented by satisfying the capricious demands of their customers, many of whom had a tendency to expect modifications to designs to suit their tastes and requirements. On many occasions the firm trod a fine line between giving customers what they wanted and deterring them from straying too far from the standard range of forms and ornaments which went into creating the silverwares. Too many variations required the costly production of new dies, stamps and other tools, and were therefore not an economic proposition. Consequently, although it did ultimately become a profitable enterprise, Soho's silver business was as unprofitable as the toy concern throughout the course of the Boulton & Fothergill partnership, and for largely the same reasons: Boulton's captivation with courting and pleasing his aristocratic customers above costing his silverwares accurately, his distraction with other business concerns, and his lengthy absences from Soho, all militated against achieving financial success.

4 John Russell, *Sir Joseph Banks*, pastel, 1788, 73.7 × 58.4 cm. (Private collection). Banks is shown holding a copy of Russell's *Face of the Moon* (see Cat. 55).

5 Unknown artist, *Dr William Small*, miniature on paper, 60 × 45 mm. (Private collection). James Millar is known to have painted Small's portrait in 1779, four years after his death.

Ormolu
The manufacture of ormolu ornaments was another significant metalware venture which Boulton launched within a few years of the Soho Manufactory's establishment, and Nicholas Goodison discusses both this business and the prevailing design climate in Chapters 4 and 7, pp. 31–40 and pp. 55–62. During the period 1768–82, Boulton & Fothergill became the largest and preeminent producers of these costly decorative items in England, and commissions from George III and Queen Charlotte must surely have given Boulton an enormous sense of achievement.[22] As with his toys and silverwares, Boulton's ormolu ornaments were marketed to customers in Britain and overseas. Again, despite his achievement in producing high-quality articles that found a place in some of the wealthiest and most important palaces and households both at home and abroad, the venture brought no direct financial reward, losses being made rather than profits accrued. However, during the course of conducting this business Boulton made immensely important contacts. His name became known and respected in the highest circles, and in practical terms his dealings with members of parliament, ambassadors, and other influential people in the commercial and political world played an invaluable part in ensuring that his other endeavours in steam engineering and minting were ultimately successful.

Steam engines
Although there is a considerable contrast between the manufacture of decorative metalwares and steam engineering, it was Boulton's activities in metalworking that drew him into this very different field of business, the development of which is described by Jim Andrew in Chapter 8, pp. 63–70.

Soon after his arrival at Soho in 1761, Boulton was faced with resolving the problem of the inadequate water supply at the site that affected the efficiency of his water mill, which was crucial to powering the machinery used to grind, scour and polish the metal components of the toys he produced.

It is evident that Boulton explored the idea of using a steam engine to provide power, but it is not clear whether he envisaged utilising a pumping engine to circulate an adequate supply of water from the Hockley Brook to his water wheel, or an engine that directly powered the metal-working machinery in the mill. From as early as 1765, he had exchanged ideas with Benjamin Franklin on the subject of steam engines. It is also clear from his correspondence with Franklin and others that Boulton constructed an experimental model of a steam engine which he sent to London while Franklin was there, for demonstration – quite possibly to the Royal Society.[23] By 1768, it seems that Boulton had constructed some kind of steam engine at Soho, for the botanist and explorer Joseph Banks (Fig. 4) who visited Boulton in the January of that year noted seeing a 'fire engine' which he described as working with a wheel instead of a beam.[24]

During these early years of steam-engine experiments, Boulton's work was carried out in collaboration with Dr William Small, one of the founder-members of the Lunar Society (Fig. 5). Originally from Scotland, Small had served as a professor of natural philosophy at the College of William and Mary in Virginia from 1758 to 1764, when he returned to England. In the following year he ventured to Birmingham in pursuit of a medical post. He soon became not only Boulton's family physician, but also a close friend and collaborator. Before long he was deeply engrossed in Boulton's steam-engine experiments. However, after the two became aware of James Watt's progress in developing an improved steam engine, Small was highly instrumental in encouraging Watt's work. Boulton, too, clearly recognised the superiority of Watt's engine and its suitability for his own immediate needs, but more significantly he perceived its wider application and thus its commercial potential.

Boulton and Watt's first orders for engines were received in 1775, the year that their partnership began, and sold to customers both in Britain and overseas. Initially the firm designed pumping or blowing engines. The majority of the pumping engines were used for mine drainage, and they were supplied predominantly to Cornish copper and tin mines, while others were supplied to collieries, to canal companies for lifting water at canal locks and to waterworks.[25] A number of the early blowing engines were supplied to the ironmaster John Wilkinson for the blast furnaces at his various ironworks in Shropshire, Staffordshire and Lancashire.[26]

Once the partners had designed engines capable of producing rotary motion which could thus directly power machinery, the application of Boulton & Watt engines to industry was considerably broadened. The firm's rotative engines were applied to 'working forge hammers, to winding coal and ores in mines, to supplying power in breweries, starch manufactories, and bleach works, to driving oil mills, corn mills, metal-rolling mills, glass grinding machinery, fulling mills, woollen mills, and cotton-spinning mills'.[27]

Boulton's own technical contribution to the development of Watt's steam engine is too often discounted. In many general historical accounts of the Boulton & Watt steam-engine partnership, Boulton has been and continues to be simplistically depicted as merely the financier of the enterprise, and Watt is portrayed as the technological genius. More accurate analyses of their roles have been presented by a number of specialist historians.[28]

Minting

During the last two decades of the eighteenth century Boulton developed the first steam-powered mint in the world at Soho (see Sue Tungate's account in Chapter 10, pp. 80–88). The impetus which led him into this sphere of business activity arose from a combination of factors: his distaste for shoddy goods – the poor quality of coins in circulation at that time and the scale of counterfeit coinage in circulation – much of which emanated from Birmingham and which created for him and other employers the practical problem of paying their workforces with legal tender. Boulton felt that by taking up minting himself he could bring about the improvements necessary to resolve the poor state of the nation's coinage, but also profit from the commercial rewards this would reap as well as the potential prestige to be gained. Characteristically, he envisaged applying advanced technology to minting. As with so many of his projects, his ambitions in this field took many years of relentless lobbying and negotiating, but ultimately in 1797 he was awarded a government contract to mint a new copper coinage, a contract he had finally succeeded in winning by being able to demonstrate beyond doubt that he possessed the most advanced minting technology in existence. And in addition to the famous 'cartwheel' copper pennies, halfpence and twopence pieces, Soho's skilled die engravers produced a succession of beautiful commemorative medals, and other numismatic products described in detail by David Symons in Chapter 11, pp. 89–98.

All in all, Matthew Boulton – the 'ingenious proprietor' of Soho – whose varied and skilfully juggled exploits are so well described in the chapters that follow, comes across as a person of boundless energy and ambition. His character is summed up in a remark that is echoed in our title: *Matthew Boulton: Selling What All the World Desires* which is based on James Boswell's account of his visit to Boulton's Soho Manufactory in 1776. Boswell writes that he saw:

> … the great works of Mr Bolton, at a place which he has called Soho, about two miles from Birmingham, which the very ingenious proprietor showed me himself to the very best advantage. I wish that Johnson had been with us; for it was a scene which I should have been glad to contemplate by his light. The vastness and the contrivance of some of the machinery would have 'matched his mighty mind'. I shall never forget Mr Bolton's expression to me: 'I sell here, Sir, what all the world desires to have – POWER.'[29]

1: Matthew Boulton and the Lunar Society

Jenny Uglow

Matthew Boulton was central to the story of the Lunar Society of Birmingham from its misty beginnings in the late 1750s until the end of the century, and it was largely his energy and enthusiasm that kept the group going for so many years. The Lunar Society was a small club of pioneering natural philosophers, doctors and manufacturers, whose members have been called the fathers of the industrial revolution. It was an informal body with no membership lists, officers or minutes (although Boulton did try to lay down a firmer framework in 1775). All over Britain, clubs met monthly near the full moon – hence the name – providing light to ride home by, but the importance of this particular Society stems from its pioneering work in experimental chemistry, physics, engineering and medicine, combined with leadership in manufacturing and commerce, and with political and social ideals. Its members were brilliant representatives of the informal scientific web which cut across class, blending the inherited skills of craftsmen with the theoretical advances of scholars, a key factor in Britain's leap ahead of the rest of Europe.

The Society never numbered more than fourteen. As well as Boulton himself, the leading members were the doctor, poet and inventor Erasmus Darwin (grandfather of Charles Darwin), Boulton's partner James Watt, the potter Josiah Wedgwood (although there is some debate about whether he was ever formally a member) and the chemist and leader of Radical Dissenters, Joseph Priestley.

Most had been entranced by mechanics in childhood in the 1730s and 1740s, when itinerant lecturers toured the country displaying electrical and mechanical marvels. The first step in the group's creation came when Darwin and Boulton met in Lichfield in the late 1750s: Darwin was then a young doctor and his patients included the Robinsons, the family of Boulton's wife, Mary. Darwin had studied medicine in Cambridge and Edinburgh, imbibing the ethos of the Scottish Enlightenment, and had just sent his first paper to the Royal Society. Boulton, by contrast, left school at seventeen, became his father's partner and would gain a fortune by two successive marriages, the second of which was controversial. Darwin's theoretical knowledge and social skills complemented Boulton's technical expertise and entrepreneurial drive.

Boulton was thus at the centre of the Society from the very beginning. It has been argued that his chief interest in the group was getting ideas and increasing profits, but there was more to it than that. As James Keir wrote after his death:

> Mr B. is proof of how much scientific knowledge may be acquired without much regular study, by means of a quick & just apprehension, much practical application, and nice mechanical feelings. He had very correct notions of the several branches of natural philosophy, was master of every metallic art, & possessed all the chemistry that had any relations to the objects of his various manufactures. Electricity and astronomy were at one time among his favourite amusements.[1]

And like Watt in Glasgow and Wedgwood in the Potteries, he learned most of all through friendship, as Keir acknowledged: 'It cannot be doubted that he was indebted for much of his knowledge to the best preceptor, the conversation of eminent men.'

Science, as we now loosely call it, was a real passion of Boulton's youth. In 1755, writing notes on electricity, he plumped for material rather than mystical interpretations. However subtle it was, this crackling form of energy had nothing to do with the soul: 'We know tis matter & there is wrong to call it Spirit.' Thanks to recent work, he thought:

> we are much better enabled to say what Elect^y is, to know its uses & understand its Laws & propertys than the Philosophers of any preceding Age for we can both hear it see it smell it & feel it… We have it as much in our power as any of the other Elements we are acquainted with to experiment upon therfor let us consider it just as it appears to our senses.[2]

In the same year Boulton made a note of books bought to set up the beginnings of a study, modelled on a gentleman's library. Among them were several works on electricity including '3 Vols of Franklin's' and treatises by Freke and Benjamin Wilson, lectures by Abbé Nollet, the French expert and 'Gowin Knight on Attraction etc'.[3] He also recorded in his notebook experiments on precipitation, temperatures of different liquids, the freezing and boiling point of mercury, the expansion of different types of cords, flax and wire. And his interests were various: he jotted down entries on people's pulse rates at different ages, on sunbeams, on the movements of the planets, on how to make phosphorus and sealing wax and even 'To write in a secret manner' – a recipe for disappearing ink.[4]

That last entry reminds us how important keeping secrets was to a manufacturer, and, much later, the concerns of the Lunar Society would turn to protecting their inventions and experiments from industrial spies, as much as disseminating them to the public. Even at this early stage, many of Boulton's notes were concerned with his own trade, as seen in entries on alloying copper and silver and hardening tin. He could see how such experiments might push his craft further. Thus in 1757, when he placed an order with Benjamin Huntsman, the Sheffield inventor of 'fine crucible steel', he added: 'I hope thy Philosophic Spirit still laboureth within thee, and may it soon bring forth Fruit useful to mankind, but particularly to thyself, is the sincere wish of Thy Obliged Friend.'[5] He was also searching for accurate measurements of heat, testing the thermometers of Reamer, Fahrenheit and Celsius, and attempting to make his own. In 1762 Darwin joked that their friend Dr Petit, 'desires I would use my Interest with your Worship, to procure him a Thermometer or two – now why won't you sell these Thermometers, for I want one also myself.'[6]

Boulton and Darwin were natural allies. Darwin bubbled with ideas, such as a scheme for a 'fiery chariot', a steam-driven car, in the 1760s.[7] Another later wild, but far-seeing, Darwin idea was the invention of a speaking machine, for which Boulton, ever the entrepreneur, offered him a contract and a thousand pounds if he could get it to say the Lord's Prayer – a good bet either way.

Both men also knew the Derby clock-maker John Whitehurst, and the trio were soon corresponding about experiments, instruments and problems of heat. In 1758 Whitehurst told Boulton that he had finally built a pyrometer which 'has all the perfection I could wish for, and will, I think, ascertain the expansion of Metals with more exactness than any machine extant.' He would come over to Birmingham soon, 'and hope to spend one day with you in trying all necessary experiments'.[8]

Boulton and Darwin also shared a hero in Benjamin Franklin, who already knew Whitehurst. And when Franklin visited relations in the Midlands in 1758, he came with a note from another of Darwin's friends, John Michell, the expert on magnetism, introducing him to Boulton.[9] The connections were beginning to intertwine, and Boulton swept Franklin round his friends.

At the beginning of the next decade, Boulton's interest in experiment slowed slightly, while he plunged into setting up his new Manufactory at Soho; in 1763, Darwin wrote to him wryly, 'As you are now become a sober plodding Man of Business, I scarcely dare trouble you to do me a favour in the nicknachatory, alias philosophical way'.[10]

But although he was so preoccupied, the impact of Soho in fact made Boulton even more eager to get 'philosophical' and technical expertise, and the Lunar Society was soon to expand, and would partly fill this need. In 1765 Franklin introduced the young Scottish doctor and mathematician William Small, who was seeking a post in Birmingham. Small had previously taught in Williamsburg, Virginia (where his pupils included Thomas Jefferson, who later credited him with having 'fixed the destinies' of his life and remembered him as 'a man profound in most of the useful branches of science, with a happy talent of communication, correct and gentlemanly manners and an enlarged and liberal mind'[11]). Small's natural diplomacy welded the group together.

The strength of the group was already evident: varying expertise created a broad knowledge base, helping them to solve problems and explore different avenues. The application of steam power and the development of James Watt's improved steam engine was a case in point. In 1767, on his way back from canal business in London, Watt stopped in Birmingham. Boulton was away, but Small showed him round Soho, where he was overwhelmed by the ingenuity of the machines. On his way north, Watt stayed at Lichfield with Darwin, and was soon exchanging letters about carts and steering. 'The Plan of your Steam-Improvements I have religiously kept secret, but begin myself to see some difficulties in the Execution that did not strike me when you was here', Darwin wrote to him later.[12] Over the next few years, largely through the constant encouragement of William Small, Watt kept in touch with the group through detailed letters. Their shared interests were far from steam alone, covering cobalts as a semi-metal, the resin of tropical trees used as varnish, lenses and clocks, alkalis and acids. This shared endeavour proved extremely useful. When Boulton was considering making earthenware, for example, Watt – who was also experimenting with pottery – sent him a heavy package of potential ingredients.

The interest in earthenware was spurred partly by Boulton's urge to compete with the Lunar Society's second great manufacturer, Josiah Wedgwood, who had been in regular contact with the group from the time when he was lobbying for the building of the Trent-Mersey canal in 1765, and had asked Darwin to assist with the prospectus. Wedgwood shared the Lunar Society's scientific and commercial interests: his experiment book on the chemistry of ceramic bodies and glazes contains nearly 5,000 carefully recorded trials. He also had an eye for design and a talent for marketing, and took Boulton's Soho works as a model for his own new factory, opened in 1769 (it was Darwin who suggested the name, 'Etruria'). Wedgwood was a Unitarian, and through his partner Thomas Bentley, who was involved with the Warrington Academy, he became an early supporter of Joseph Priestley. Over the next few years the whole Society followed Priestley's experiments with electricity, optics and chemistry, especially his discovery of pure 'dephlogisticated air' (oxygen) and other gases.

During these early years the friends joined each other when they could, to try experiments and talk about new findings, rather than establishing regular monthly meetings. A more radical note had already been introduced in the mid-1760s by two idealistic followers of Rousseau: the voluble, charming, Irish inventor Richard Lovell Edgeworth and the eccentric and wealthy Thomas Day, author of *Sandford and Merton* (1783). From 1766 Boulton found another new ally when James Keir, Darwin's fellow student in Edinburgh, settled in the Midlands. A shrewd, humorous man and a talented chemist, Keir later became a glass-maker and innovatory soap manufacturer. He was the 'rock', to whom all – particularly Boulton – would turn in times of stress. In the 1760s, Keir was working on a new translation of Macquer's *Dictionary of Chemistry* (1766), and in preparing a booklist for Darwin, he included three works on metallurgy, pointing to studies of the method of assaying ores, tables of affinities, experiments on the density of metallic alloys, and accounts of smelting and forging. This was, of course, of great interest to Boulton, not only for his ornamental business, but because of his interest in steam engines, whose cylinders needed metals that could be cast and worked easily but would also resist corrosion and heavy wear.

Boulton was always hunting for unusual metallic ores and materials, and asked his friends and foreign agents to send samples 'for philosophical purposes'. In June 1768 Darwin reported that 'Mr Boulton has got a new metal which rivals Silver both in Lustre, Whiteness, and endures the Air with as little tarnish. Capt. Keir is endeavouring to unravel this metal.'[13] In 1769 Boulton, Small and Keir were all working on cobalt, nickel and manganese and were borrowing books from the London metallurgist Peter Woulfe who offered to do chemical trials. Boulton called on Keir constantly to analyse samples and make equipment. In 1772, for example, Keir was promising Boulton that his order for 'chemical glasses' would be 'executed as speedily as possible, especially those for your own Experiments as I well know the Impatience of my fellow-schemers and I should also be sorry to check by delay your present hobbyhorsicality for chemistry'.[14] Another little-known member of the Lunar Society believed to have had an interest in chemistry was Robert Augustus Johnson, whose wife was the sister of the 6th Lord Craven. They lived for a time at Kenilworth, not far from the Craven seat at Combe Abbey, but at the time of his death Johnson was Rector of Hamstall Ridware in Staffordshire.

For twenty-five years from the start of the partnership in 1774 Boulton & Watt installed engines from Cornwall to the Clyde, and it was partly the usefulness of Boulton's philosophic friends as a think-tank, and partly the realisation that he might have to spend much time away from home, especially in Cornwall, that made him try to establish the group on a firmer footing. The spur, however, was the death of William Small, who had been the key communicator between the busy individuals, in the summer of 1775. The following New Year's Eve saw the start of the Lunar Society proper. The plan was that they would meet each month for a few hours, beginning with dinner at two in the afternoon and carrying on until at least eight in the evening. They would

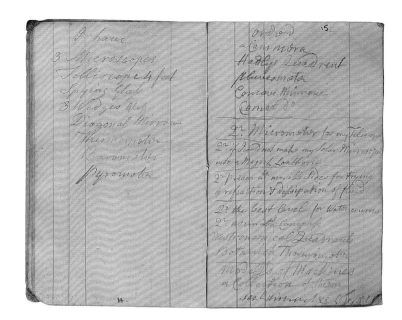

6 List of scientific instruments already owned or on order, from Matthew Boulton's Notebook for 1772. (Birmingham Archives & Heritage, MS3782/12/108/7) (See Cat. 42)

meet on the Sunday nearest to the full moon, and agreed that they could bring interested guests on occasion: many of these meetings took place at Boulton's Soho House. On 3 March 1776 he told Keir and Watt that he was planning to 'make many motions to the members', concerning laws and regulations 'such as will tend to prevent the decline of a Society which I hope will be lasting'. For the next meeting he ordered an electrical machine, noting 'as this is for a philosophical society beg everything be most accurately fitted'.[15]

This was a small, high-powered experimental group. That summer, for example, they met to test the Comte de Buffon's theories that heat affected weight, setting up experiments using Boulton's fine balances to see if there was any difference between a cold ball of iron and a heated one: some members and guests wrote papers on these experiments. Lunar interests were kaleidoscopic, ranging from optics and astronomy, chemistry and mechanics, hydraulics and minerals to meteorology and magnetism, ballooning and ballistics (Fig. 8). The practical and theoretical merged. All were interested in minerals for example, Boulton for metallurgy, Wedgwood for ceramics, Watt, Darwin, Keir, Withering and Priestley for chemical examination. Boulton was undoubtedly the boldest exploiter of the group's findings, using the precious Derbyshire fluorite 'blue john' in the early 1770s, not for metallurgy or experiment, but as bodies for his ormolu vases, joining Wedgwood in supplying the taste for classical vases. And when Boulton was contemplating the production of Argand oil-lamps, Darwin, Watt, Wedgwood and Edgeworth were all working busily on lamp problems.

As with their work on vases, several of the members were as fascinated by aesthetics and design as they were by scientific research and application. Their interests were also linked to

7 Frontispiece from William Withering's *An Account of the Foxglove* (1785). (Birmingham Archives & Heritage)

significant rift. (Less damagingly, Withering also argued with the young doctor and botanist Jonathan Stokes, a member of the Society from 1783 to 1788.)

Despite such disputes, the group's real aim was collaboration and the 1770s and 1780s were their most productive years. Some schemes were exotic, like Darwin's proposal for towing icebergs to the equator to cool the tropics: others were futuristic – using diving bells to explore the ocean bed and balloons to cross continents. The mood is conveyed in Darwin's apology to Boulton in 1778 regretting that the 'infernal Divinities' who bring diseases

> should have prevented my seeing all you great Men at Soho today – Lord! What inventions, what wit, what rhetoric, metaphysical, mechanical and pyrotechnical, will be on the wing, bandy'd like a shuttlecock from one to another of your troop of philosophers! While I, poor I, I by myself I, imprison'd in a post chaise, am jogged, and jostl'd and bump'd and bruis'd along the King's high road, to make war upon a pox or a fever!'[16]

As this note suggests, it was hard to maintain the group's cohesion with so many different preoccupations and busy working lives. In the same year, Boulton sold his telescope, which had lain outside suffering from rain and neglect, to the astronomer Alexander Aubert.[17]

It was not surprising that the telescope lay unused. Although Soho appeared to be flourishing, its financial situation was always precarious and Boulton turned unashamedly to the group for help. From Thomas Day he borrowed considerable sums of money, managing to avoid repaying him for many years. From James Keir he sought practical help: Keir effectively took over the running of Soho while Boulton was occupied in Cornwall, getting little thanks and even less financial reward. In spite of this, Keir retained his affection for Boulton, later recalling admiringly, 'To understand the character of Mr Boulton's mind, it is necessary to recollect that whatever he did or attempted, his successes and his failures were all on a grand scale.'[18]

Then in 1780 Darwin moved to Derby, where he set up his own philosophical group, although he remained in close correspondence with the Birmingham 'Lunaticks'. As Darwin departed, however, Joseph Priestley arrived. He was appointed minister of the Unitarian New Meeting House in Birmingham and immediately joined the Society. For years he had kept in touch with its work. In 1775 the American War of Independence had caused rifts between the more radical and the conservative spirits, the latter group including Boulton, who was anxious to keep on good terms with his landed and government contacts. But late that year he sent Priestley a box of Derbyshire spar, and Priestley replied in delight, hoping they would meet again soon. 'It will be a great pleasure to me to see your improvements on fire engines and all your other valuable improvements in mechanics and the arts… I shall not quarrel with you on account of our different sentiments in Politiks.'[19]

wider concerns. John Whitehurst's work on the volcanic rocks of the Peak District, published in his *Inquiry into the Original State and Formation of the Earth* (1778), fuelled the intense debates of the period about the creation of the earth. Another shared interest, although not so close to Boulton's heart, was in botany. After William Small's early death his place was taken by William Withering, who became Boulton's physician and whose botanical interests, especially the new Linnaean taxonomy, provoked Darwin's exuberant rivalry. Darwin's Linnaean dictionaries, his impulsive use of digitalis on patients and publication of its effects, through his augmented version of his late son's dissertation, before Withering published his carefully gathered results in *An Account of the Foxglove* (1785) (Fig. 7), would create the group's only

Priestley's arrival in Birmingham galvanised the group, and all Boulton's old interest in chemistry was revived. Early in 1781 he wrote to Logan Henderson from Cornwall, where he was working with his manager, another brilliant Scottish inventor William Murdock. Using Keir's earlier image, he admitted that 'Chemistry has for some time been my hobby horse'.[20] He had made great progress recently, he said,

> and am almost an adept in metallurgical moist chemistry… I have annihilated Mr Murdock's bed-chamber, having taken away the floor, and made the chicken-kitchen into one high-room with shelves and these I have filled with chemical apparatus. I have likewise set up a Priestleyan water-tub, and likewise a mercurial tub for experiments on gases, vapours etc, and next year I shall annex these to a laboratory with furnaces of all sorts, and other utensils for dry chemistry.

The laboratory was never built, but Boulton's enthusiasm was undoubtedly genuine. In 1784, when the Lunar Society, like the rest of the nation, was enthralled by the new vogue for balloons, he showed his huge enjoyment in experiment for its own sake, however dangerous (Fig. 8). He and Watt designed an experiment 'to determine whether the growling of thunder is owing to echoes, or to successive explosions'.[21] To do this they built an 'explosive balloon', made of thin paper, about five feet in diameter, coated with varnish and filled with a mixture of one part 'common air' and two parts 'inflammable air' (hydrogen). Then Boulton tied a firework, 'a common squib' with a two-foot fuse, into the neck of the balloon. The night was ideal, calm and dark, and the balloon rose beautifully, but the fuse was so slow that instead of exploding overhead it floated a couple of miles before igniting, with a sound so loud that people 'took the balloon for a meteor, and the explosion for real thunder'. There was no way that they could time the echoes, mixed up as they were with the roars of the crowd. But, wrote Watt, 'it exhibited a fine fireworks for a few seconds', so if little was learned about thunder, a good time was had by all.

Priestley, too, was energetically involved in balloon experiments, as well as his investigations into gases and the nature of air. The Lunar Society members, including a new recruit from 1781, the Quaker gun manufacturer Samuel Galton, funded Priestley's research and provided equipment. Watt worked both in collaboration and in competition with him in the race against Henry Cavendish and Antoine Lavoisier to discover the composition of water. Less happily, the group also backed Priestley's outdated phlogiston theory against Lavoisier's new French chemistry with its different theory of combustion and its new nomenclature. Darwin was the first to be won over to Lavoisier's theory, incorporating novel words like 'oxygen' and 'hydrogen' in his poems *The Botanic Garden*, and *The Economy of Vegetation* in 1791–2 (the first published uses of these words in the English language).[22]

With his natural bonhomie and bumptious, at times ruthless,

8 Frontispiece from *Descriptions des expériences de la Machine Aérostatique de MM. de Montgolfier*, by Faujas de St Fond, who visited Soho in 1784. This is Matthew Boulton's own copy (Cat. 54). (Birmingham Museum & Art Gallery, 1997 F 764.1–3)

charm, Boulton skilfully managed to avoid taking sides in all these arguments, remaining close friends with both Darwin and Withering, and with Priestley and Watt. And although interested, he also stayed carefully detached when the group carried their experiments into the social realm, such as in developing new methods of education (seen, for example in Darwin's radical prospectus, *A Plan for the Conduct of Female Education in Boarding Schools* (1797) and Richard Lovell and Maria Edgeworth's *Practical Education*, 1798). He did, however, join the rest of the Lunar Society in supporting the anti-slavery movement, for which Wedgwood produced a medallion, showing a chained slave with the motto 'Am I not a Man and a Brother?', and formed part of the deputation, with Galton and Priestley, to welcome Olaudah Equiano to Birmingham in 1789, when the freed slave came to speak about the suffering of his compatriots.

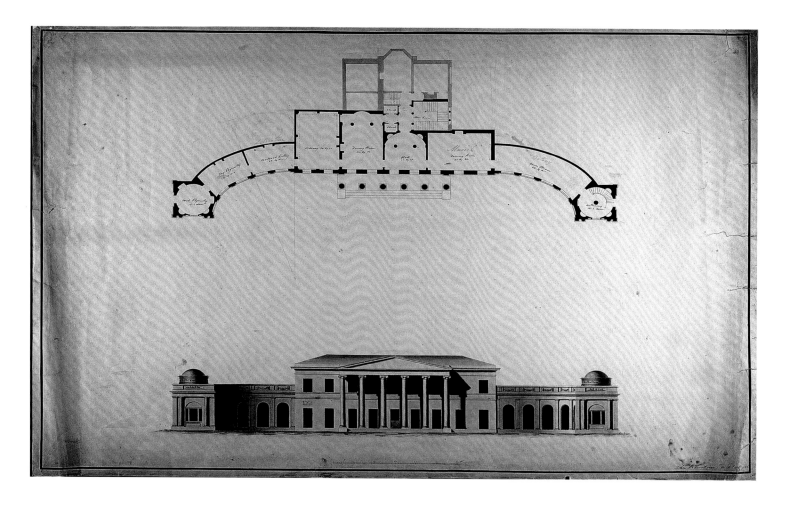

In politics Boulton remained a pragmatic, rather than ideological, Tory and Anglican. With him, business came first, and he was keenly involved in the 1780s when the group came together to work out ways of preventing industrial espionage (although his own desire to show off his machines, and ventures like the Albion Mills, often laid him open to the charge of carelessness). Similarly, with Wedgwood and Watt he actively backed the furious attempts of Arkwright to protect the patent for his carding machine, however dubiously this had been acquired, and however much he disliked him personally.

Boulton's connections still made him vulnerable however, sometimes to his exasperation. Through their links with Priestley all the members of the Lunar Society became involved in the battle of Radical Dissenters for repeal of the Test Acts, which banned Dissenters from Oxford and Cambridge and from holding any public office; and though Withering and Boulton shared their horror at the execution of Louis XVI in February 1793,[23] the group as a whole had welcomed the French Revolution, seeing in it an act of liberation. Their perceived radicalism was their downfall. In July 1791, a dinner was held by some Dissenters to celebrate the second anniversary of the Fall of the Bastille. Priestley was not present, but Birmingham rioters chanting 'Church and King!' razed his New Meeting House and sacked and burnt his house and laboratory, and those of several other prominent citizens. But after five days of rioting in the town, Soho remained unharmed and Boulton wrote with cool sang-froid to a French friend:

> The Town of Birmgm's quite unharmonied party spirit and Rancour tears all good neighbourhood to pieces but I am happy in living alone in the Country & am almost silent upon this dissonant Subject. By minding my own business I live peaceably & securely amidst the Flames, Rapin, plunder, anarchy & confusion of these Unitarians, Trinitarians, pre-destinarians and tarians of all sorts.[24]

In 1794 Priestley emigrated to America, dedicating his last work published in Britain, *Experiments on the Generation of Air from Water* (1793), to the Lunar Society. After this, the Lunar spirit faded, although occasional meetings continued until around 1800. Wedgwood died in 1795; Darwin retreated to write his monumental poems and books, and to develop his theory of biological evolution before he, too, died in 1802; far away in Northumberland, Pennsylvania, Priestley died in 1804, still mourning the loss of his Birmingham philosophical circle. Boulton, Watt and Keir flourished as manufacturers. Britain was swept up in the wars against Napoleon. Regretting the loss of Johnson, who had left Kenilworth, Withering noted in 1798 that

9 John Rawst(h)orne, Plan and elevation of the proposed alterations to Soho House, 1788. Rawstorne's proposals were not carried out (Cat. 41). (Birmingham Archives & Heritage)

10 William Pether, print after Joseph Wright's *A Philosopher Giving a Lecture on the Orrery*, 1768, print on paper, 48.4 x 58.2 cm (Cat. 44). (Derby Museum & Art Gallery.) Attending 'philosophical' lectures, often given by travelling lecturers, became a fashionable pursuit in the eighteenth century and whetted many people's interest in science.

they missed him at their Lunar meetings, which had never flourished since the departure of Priestley. 'The Members are to a Man either too busy, too idle, or too much indisposed to do anything; and the interest wch everyone feels in the state of public affairs draws the conversation out of its proper course,' he wrote.[25]

The Lunar Society passionately believed that their discoveries would make the world a better place. They were the optimistic, and idealistic, forebears of a new class, the nonconformist industrialists and reformers who would dominate nineteenth-century Britain and America. Part of their strength came from their long and deep friendships, providing support in times of trouble. Although Boulton drew on these friends to enable Soho to expand, and to forward all his projects, from vases to steam engines, and from lamps to minting coins, his real delight was in the knowledge they shared. He did see himself as an experimental 'philosopher', as well as a great man of industry and commerce. The Lunar Society was integral to this vision. In 1787 he had commissioned Samuel Wyatt to remodel Soho House. But a year after the Wyatt plans an alternative set, never used, were drawn by John Rawst(h)orne, with rooms in the wings marked for 'Wet Chymistry', 'Dry Chymistry', 'Natural History', 'Botany – Green House', 'Astronomy'.[26] The pursuits and passions of the Lunar Society were, literally, built into Boulton's dreams (Fig. 9).

11 Francis Eginton junior, *Soho Park*, showing Soho House on top of the hill among the trees, engraving for Stebbing Shaw's *History and Antiquities of the County of Staffordshire* (1798). (Birmingham Museum & Art Gallery)

2: 'The Hôtel d'amitié sur Handsworth Heath'
Soho House and the Boultons

Shena Mason

Soho House stands in Handsworth, about two miles northwest of Birmingham city centre. Today's visitors find it a small, tranquil oasis surrounded by dense housing and commercial development, but in the eighteenth century it commanded distant open views across country. At the time the house was built, high on the windy expanse of Handsworth Heath, around 1757, Matthew Boulton had no thought that it would one day be his home. Twenty-nine years old that year, he was married to his first wife, Mary, and they were perhaps living at his parents' home near his father's workshop in Snow Hill, on the northern edge of Birmingham town centre.[1] The energy, drive and ambition, which would later take him out of town to Handsworth, then just inside Staffordshire, were fully employed in the family business. Matthew Boulton the pioneering industrialist was still to make his appearance on the wider stage. Yet this was the man who later, with his business partner, the engineer James Watt, would put Birmingham – and Britain – firmly at the heart of the brave new industrial world of the late eighteenth century. In that world, huge Boulton & Watt steam engines would power Lancashire cotton mills, pump out Cornish tin mines, drive the great rumbling stone mill wheels in the world's largest flour mill in London, crush sugar cane in the heat of the Caribbean, stamp coins in chilly St Petersburg, and, later still, power steamboats on the Danube and around Britain's coasts. But at the time the bricklayers were at work on the shell of Soho House, all this was still in the future and Matthew Boulton's manufacturing skills and business acumen were being honed in a much smaller-scale environment: his father's workshop, where the output was tiny buckle components, buttons and decorative trinkets.

In 1759 Boulton suffered a double blow. In August that year Mary died. He buried her in the Robinson family vault at Whittington, near Lichfield, placing a poem he had written to her in the coffin (see Cat. 92).[2] A month later his father, Matthew Boulton senior, also died. Boulton now took control of the family business. Not only that, but within a matter of months he was writing ardent love letters to Mary's sister, Ann Robinson, whom he married in June 1760.

Even as he courted his second wife, Boulton's business ambitions were growing far beyond what his father had entertained, and beyond what he would ever be able to accommodate in the cramped workshops at Snow Hill. With the benefit of two wives' capital behind him, Boulton began looking for a site to build a much bigger factory. When in 1761 he bought the lease on a rolling mill and thirteen acres of land at Soho,* Soho House already stood on the hilltop above the rolling mill; the house accounted for £300 of the £1,000 Boulton paid for the lease.[3]

Soho House was not the initial focus of Boulton's attention, though he had some unfinished jobs there completed and moved his mother and one of his sisters, probably Mary, into it; he and Ann meanwhile remained at the Snow Hill house. All of his energies and resources went into replacing the small rolling mill with what was to become the world-famous Soho Manufactory. In 1762, Boulton moved his mother and sister back into Birmingham so that his new business partner, John Fothergill, could live at Soho House.

Once the Manufactory was in production, however, and with Fothergill travelling extensively on the Continent, Boulton felt the need to live nearby, and at last turned his attention to Soho House, insisting on Fothergill vacating it for a house near their warehouse in Birmingham town centre. Fothergill was greatly put out by this but after some temporary acrimony the partnership survived.[4] Meanwhile Boulton and Ann finally moved into Soho House in 1766. From then until his death on 17 August 1809 Soho House was Matthew Boulton's home. The two children of his second marriage, Anne (Fig. 12) and Matthew Robinson Boulton (Fig. 13) were born there.

Soho House at the time the Boultons took up residence looked very different from how it came to look after Boulton had it remodelled in the 1790s, and how it looks today, though it was evidently already regarded as an attractive house. When Lord and Lady Shelburne visited Soho the year the Boultons moved in, Lady Shelburne wrote in her diary, 'His house is a very pretty one about a mile out of the town, and his workshops newly built at the end of his garden where they take up a large piece of ground.'[5] The complex architectural history of the house is still

---

* The name 'Soho' is believed to come, like that of its London counterpart, from a hunting cry; there are contemporary newspaper accounts of a nearby inn sign showing a huntsman blowing a horn, from the mouth of which issued the word 'Soho'.

12 Attributed to Tilly Kettle, *Miss Anne Boulton*, c. 1778, oil on canvas. (Private collection).

not fully understood, though the area occupied by the present hall, dining room, breakfast room and drawing room appears to be the oldest part of the building, and it may have remained within this footprint or with only minor additions for some years. An archaeological investigation carried out in 1990 concluded that alterations and extensions were made to the house in several phases between 1761 and 1788,[6] but there is little documentary evidence of such work from the first twenty years or so of the Boultons' occupancy. In any case, money for house alterations would have been tight in the first decades, due to the immense cost of building the Soho Manufactory, which led to much juggling of debts for many years.

Notwithstanding the relatively modest size of Soho House at the time they occupied it, the Boultons soon made themselves a comfortable and stylish home there, suitable for entertaining business contacts and friends, including the 'philosophical' friends who made up the Lunar Society. The Boultons' visitors would have been assured of a good dinner, perhaps (judging from household accounts) including salmon or lobster or venison, and sometimes pineapple if one had arrived from London, or some other dessert (bills show that sugar consumption – something of a marker of prosperity in the eighteenth century – was high in the Boulton household, with predictable though not then understood consequences for the children's teeth). The food was washed down with fine wines and the meal finished off with a good port or cognac, all carefully stored in casks or on the slate racks down in the cool cellar and regularly inspected and listed by the head of the household.[7] The servants (up to nine including butler, footman, housekeeper, cook, coachman, groom, gardener and maids) also had to be provided with food and drink.

Boulton was now a family man, a busy manufacturer, and a leading member of the Midlands' growing intellectual and social circles. He was beginning to play a role in public life, too, as a member of the committee of the General Hospital, as well as campaigning for Birmingham to be allowed to have its own Assay Office, and lobbying for a better theatre. All this might have been enough for most people, but Boulton seems to have been an individual of boundless energy, endless curiosity and perceptive vision. He was quick to see the potential in the engineer James Watt's development work on the steam engine, and offered him a bed at 'The Hôtel d'amitié sur Handsworth Heath'[8] whenever he should need it on his travels. When Watt accepted Boulton's invitation to move down from Scotland and join him in the business at Soho in 1774 and then form a partnership the following year, the future path of the enterprise was set, and with it, that of much of Britain's industrial history.

While the captain of industry steered his ship through sometimes decidedly rocky channels, domestic life hummed along at Soho House, interrupted regularly by the sound of what Boulton called 'the Golden Knocker', the rap on the front door announcing the arrival of yet another visitor.[9] In addition to the housekeeping bills which detail food, candles and consumables, other surviving bills, notebooks and letters give an indication of how the family dressed and amused themselves; Mrs Boulton in particular had a taste for dancing, concerts and games (a letter from her husband reproaches her for playing the 'dangerous' game of blind man's buff, which had resulted in a sprained ankle).[10] Trying to mollify her for his frequent absences from home, mainly in London, Boulton regularly bought lengths of silk and lace, jewellery, caps and gloves for his wife and small gifts for the children, and from time to time wrote plaintively asking her for a word or two of family news, for he was constantly anxious about the children's health. Whether or not she acceded to these requests it is not possible to know – there are no letters from Mrs Boulton in the Archives. Their daughter Anne in particular Boulton regarded as delicate, for she had been born with a slight disability (possibly a club foot).[11] In reality, judging from her travels and activities, Anne seems to have been a sturdy enough child, but as she went through the common but dangerous childhood illnesses of measles and whooping cough, her father was tormented with worry. This anxiety about the children's welfare did not prevent him from sending them to boarding schools, however, though the reason that schools in or near London were chosen initially for both children was perhaps because Boulton was there so frequently that it was easy for him to visit them, which he did, often, always reporting back to their mother on how he had found them. If at times the children had to be taken out of school because of an infectious illness, there were also willing and affectionate surrogate parents in London, in the persons of Boulton's bankers, William and Charlotte Matthews, who took care of the Boulton children from time to time.

Having the children at boarding school also meant that Mrs Boulton was free to accompany her husband on business trips, but following a 1778 visit to Cornwall (shortly after ten-year-old Anne and eight-year-old Matt had first gone away to school) she seems to have been reluctant to leave the comfort of Soho House again for the rigours of long journeys, and dug her heels in whenever the subject of a trip was raised, her husband sending disconsolate letters home pleading with her to reconsider joining him. Both Boulton and Watt spent months at a time in Cornwall, where they had substantial orders for steam engines. Neither of their wives liked Cornwall (which Boulton himself described as 'this Siberia-like country'[12]), even after they took a house near Truro to make the long visits more comfortable, and Mrs Boulton seems to have gone to Cornwall only once more, in 1782.

In the summer of 1783 Mrs Boulton was found dead in a pool in the grounds of Soho House. There was some gossip locally about suicide, but medical opinion at the inquest was that she had not drowned, but had fallen into the water as a result of a stroke or fit of some kind while walking near the edge of the pool, where she had been seen shortly beforehand apparently in good spirits. William and Charlotte Matthews came and fetched the children back to stay with them in London while Boulton adjusted to being a widower again. He wrote to his son and daughter regularly as they grew up, and seems to have been especially close to Anne, if their correspondence is anything to go by. His letters to Matt are affectionate enough, but often rather more concerned with the boy's character and academic progress, though Boulton was soft-hearted enough to worry about his son feeling the cold and to send him warm waistcoats and money for firewood at his lodgings, when sixteen-year-old Matt went off to Versailles to continue his education.[13] During a visit to Holland many years earlier when Matt was still a little boy, Boulton had speculated that the university at Leiden might be a good place for a young man to study,[14] but in the event Matt did not go to university. Instead, after Versailles, he spent a while in Paris, and later near Frankfurt with a tutor who had also instructed James Watt junior and the sons of one or two other Birmingham manufacturers. Matt became proficient in French and German, and studied subjects such as mathematics and engineering drawing. His future was mapped out.

Each time the children came home from school for the holidays, they would have found changes at Soho House, especially in the gardens. At the time Boulton had acquired the house, the surrounding land had been unpromising: exposed, sandy heath with a typical heathland vegetation of heather and gorse, inhabited by little more than rabbits. One of his first decisions in the

13 Jean Etienne Liotard, *Matthew Robinson Boulton aged three*, pastel on paper laid on copper plate, 28 × 23 cm. (Birmingham Museum & Art Gallery) (See Cat. 94)

1760s had been to plant 'above 2,000 firs' to form a protective windbreak for the house.[15] He had also instituted a massive soil improvement programme, and many cartloads of marl and manure were brought in and thickly spread on the land to improve water retention and fertility. From the early 1770s onwards, Boulton gave much thought to the development of his grounds, which gradually took on the aspect of a small landscaped park (Fig. 11), with all the features to be found in much bigger, statelier establishments, such as statuary, cascades, ornamental buildings (Fig. 14), and colourful flowerbeds and borders.[16] It is probably no accident that the one known portrait of Miss Anne Boulton, painted when she was probably about ten years old, shows her with an apron full of summer flowers which she is placing in a flower stand. A year or two later, Anne would be instructed in botany by Lunar Society member Dr William Withering.

Towards the end of the 1780s Boulton began to give serious consideration to major alterations to Soho House itself. His friend the astronomer William Herschel described Soho House as 'your Elysium' and 'that blissful Mansion',[17] but it was about to undergo a transformation. In 1794 Boulton bought the free-

hold of the entire estate (Fig. 15). Not long after making this decision, he acknowledged 'It's true my House wants many Conveniences, and now it's my own I shall take an early opportunity of consulting Mr Wyatt in some necessary alterations'.[18]

The brothers James and Samuel Wyatt were family friends and both now rising stars of the smart architectural world. Their planned changes included extending the house with a new west wing (Fig. 16). James Wyatt also drew up designs for a new and imposing neo-classical front block, about which Boulton wrote to him, 'Your Ionick front with columns accords with my taste the best & yet it may be said to be an old fashion Front, as such were built 2000 years ago, but at the same time it may be said to have stood the test of Criticks & Time without amendment.'[19] The scheme never materialised in the way James Wyatt envisaged, however. The planned new west wing *was* built, housing a larger library, billiard room, kitchen and other domestic quarters, but the projected new front block was never built and Boulton contented himself with Samuel Wyatt's modified and more modest front elevation to the original house, with its pilasters and portico which survive today. The external brick walls were clad with Penrhyn slate and painted with a sand-and-paint mixture to give something of the impression of ashlar-built walls, a technique Samuel Wyatt had already employed at Thomas Anson's great Staffordshire mansion, Shugborough. The curtailment of the building project was due, perhaps, to a combination of declining health, rising cost and waning enthusiasm on Boulton's part for having builders in, for he grumbled to James Wyatt:

> I have already paid a very large sum of Money to bring my dwelling house into the most uncomfortable state possible as the Winds Rain & Snow drives into it; & for want of the main Stack of Chimneys being built up to the top of the House it is constantly filld with Smoak by which my Books are spoiled, my daughters health much injurd & my servants obliged to live out of Doors.[20]

Having abandoned a substantial proportion of the Wyatts' proposed scheme, Boulton and his son and daughter (now both in their twenties) went to town on the interiors, with redecoration and new furniture throughout. The well-known interior designer and theatrical scene painter Cornelius Dixon, who had worked on the new British Museum, and the smart London furniture maker James Newton, were brought in to transform the house into a fashionable gentleman's residence. Here Boulton could entertain distinguished guests in style and impress them with up-to-the-minute modern facilities, including warm-air-ducted

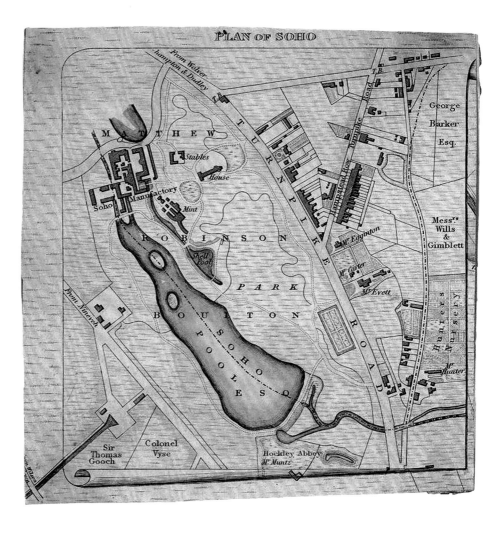

14 John Phillp, *Temple of Flora, Soho*, watercolour on paper, 1794, 8 x 11.4 cm. (Birmingham Museum & Art Gallery, 2003.0031.8)

15 John Arrowsmith, *Plan of Soho*, 1834, showing the extent of Boulton's land and the relationship between Soho House, the Manufactory and the Mint, 11.4 x 11.5 cm (Cat. 100). (Birmingham Museum & Art Gallery, 1970 V 1082)

central heating (possibly the earliest surviving domestic central-heating system, apart from those installed by the Romans), the huge steam-heated bath (its unlikely dimensions are given in one of Boulton's notebooks as 8ft. 3ins. long x 5ft. 9ins. wide x 5ft. 6ins. deep,[21]) and upstairs and downstairs Bramah patent flushing water closets. It was typical of him to want to try out the most up-to-date ideas in his house, and when his Lunar Society friend, the metallurgist James Keir came up with a new alloy of copper, iron and zinc, which he called 'Eldorado', Boulton had the glazing bars of the bow window in the dining room made of this metal. Being very strong, it allowed the frames to be thinner, thus increasing the area of glass and the amount of light admitted. The central and right-hand windows of this bow still have their 'Eldorado' glazing bars.

It took several years, but eventually the refurbished Soho House exuded an air of warmth and comfort, leading James Watt's younger son, Gregory, to write to his father in 1798, 'Soho is nearly compleat and shines refulgent in all the grandeur of painting, gilding, Red Morocco & buff curtains'.[22]

Here in comfort Boulton spent his declining years. Until the early years of the nineteenth century he remained vigorous and involved with the business, and frustrated when painful kidney problems prevented him from getting out and about. In 1802, when he was ill, friends in London, the Dumergues, arranged with Mrs Fitzherbert for him to have the use of the Prince of Wales' wheelchair while the Prince was out of the country for a while.[23] The conveyance was crated up and sent to Soho, and enabled Boulton to be wheeled once more about his beloved garden. The Dumergue household included Charles Dumergue (dentist to the Boultons and also to the royal family), his nephew Charles junior (also a dentist), Charles senior's daughter Sophia, his ward Charlotte Carpenter (later Mrs Walter Scott), and their friend-cum-housekeeper Sarah (Sally) Nicholson. They all arrived at Soho from time to time with gifts for Boulton, Anne and Matt, and tit-bits for Dash the dog. Other visitors also continued to arrive at the house – the opera singer Elizabeth Billington came and sang to Boulton, and the actress Sarah Siddons came to dinner and wrote to him. Both ladies sent numerous kisses via the Dumergues. Boulton may have been temporarily cheered, but nothing made up for a loss of involvement in the business, and James Watt junior observed in a letter to a friend, 'It is not in his nature to be penned up in his arm chair and resign the fasces of imperial sway to his successor. This he will not do whilst he is able to wield a pen or issue a mandate…'[24] On 17 August 1809, after months of pain and fitful sleep, and with his consciousness dulled by laudanum, Matthew

16 *James Wyatt's rear elevation for the extended Soho House, 1796. (Birmingham Archives & Heritage, MS 1682/6/5)*

Boulton died in his room at Soho House, with his son and daughter at his bedside. John Furnell Tuffen, a Quaker wine merchant and banker who was friendly with the Watts and the Boultons, attended the funeral. He wrote to James Watt (who was in Scotland with his wife at the time of his old friend and partner's death) that 'at least 10,000 persons' had lined the route of the funeral procession to Handsworth Parish Church, and added 'Thus my dear Sir has the Grave closed on one of our oldest & dearest friends, whose like, take him for all in all, we shall not see again.'[25]

After his father's death, Matthew Robinson Boulton continued to carry out various improvements to Soho House and its grounds. In 1815 he bought a country estate in Oxfordshire, initially for recreational use. In 1817, at the age of forty-seven, Matt Boulton married and he and his young wife, Mary Anne Wilkinson,[26] continued to make their home at Soho House, and to raise their seven children there (Mary Anne was to die in 1829 following the birth of their last child, a daughter). Matt's sister, Anne Boulton, remained single, and after the birth of her brother's first child moved out of Soho House to set up her own household, not far away at Thornhill House, where she died a few months after her sister-in-law, in November 1829, at the age of sixty-one.

The Boulton family stopped using Soho House in the mid-nineteenth century. From that time onwards, the house was let to a succession of tenants for some 150 years, and went through several changes of use. Over this period Soho House became, by turn, a vicarage, a girls' school, a hotel, a hall of residence for General Electric Company (GEC) apprentices and a hostel for single policemen in the West Midlands Constabulary. Over the same period the estate itself gradually shrank in size, as leases for building plots were sold off and most of Boulton's park disappeared under high-density housing, leaving just a small garden of less than an acre around the house. The builder of many of these houses, James Wilson, lived in and enjoyed Soho House himself for a time. The Soho Manufactory buildings were demolished in the early 1860s, having stood for just a century (and just long enough to be photographed). The lake, having been used for some years as a public boating pool, was drained in 1868, but Joseph Chamberlain's approach to have the area designated as a public park was turned down by the Boulton family. The land the lake had occupied became for a while a public tip and then a coal yard, before being sold to the London and North Western Railway Company as a goods yard.[27] Today part of it is occupied by a small industrial estate. The ornamental Shell Pool nearer the house was re-landscaped and opened as a garden to subscribers only, most of whom would have been the residents of the upmarket new villas being built on part of the surrounding land. This, too, has long since disappeared under later redevelopment.

After the West Midlands Constabulary moved out and Birmingham Museums & Art Gallery embarked on the project to restore and develop Soho House as a Museum in 1990, it was decided to use the 1790s alterations as the reference point for the restoration. Archive evidence for this period was much more plentiful and informative than for the earlier part of the house's history. In the intervening years the Wyatts' west wing had been

17 Front elevation of Soho House, home of Matthew Boulton from 1766 to 1809. (Birmingham Museum & Art Gallery)

demolished, to be replaced when the house was a hotel in the 1920s with an unsympathetic brick extension. During the time that the house was occupied by West Midlands Constabulary, a new brick block had been built, attached by a short corridor to the drawing room bow window on the east side of the house. As part of the restoration and development project, the brick extension on the west side of the house was demolished, but the separate block on the east side was retained, though without its link which spoiled the integrity of the bow window. This block, refurbished to be more in sympathy with the old house, now forms the Visitor Centre through which visitors enter the site.

Soho House itself was carefully restored, retaining the many surviving original features, returning the room plan to its late eighteenth-century layout, and carrying out thoughtful and painstaking reproduction of lost features, such as some of the fireplaces. Archive evidence was available to inform all of this work. Decorative schemes for the interiors were likewise based on both archive research and chemical analysis of surviving eighteenth-century plasters and pigments in the house. To furnish it, the City of Birmingham had already taken advantage of the opportunity afforded by a Christie's sale in 1987 to buy some items of Matthew Boulton's own furniture. Of these pieces, perhaps the most evocative is the dining table, which was made for the Soho House dining room by Benjamin Wyatt. Some notable visitors had sat around this table, including the Lunar Society members, Count Woronzow the Russian Ambassador, the botanist Sir Joseph Banks, the astronomer Sir William Herschel, the actress Sarah Siddons, the opera singer Elizabeth Billington, and possibly even Nelson, who visited Soho House in 1802 when a sick Boulton was obliged to receive him in his bedchamber.

The furniture acquired in 1987 set the tone for the other pieces needed to present fully furnished rooms. Where it proved possible to obtain items with a Boulton provenance, this was done. In the absence of Boulton's own furniture, items from the same London maker, James Newton, were sought. And failing that, pieces were chosen which appeared to match closely Newton's own detailed descriptions of the furniture he had supplied for Soho House (for example, the 1780s gothic-backed dining chairs by Gillows, which are in the dining room today and are probably very like those supplied by James Newton).[28] The Archives also provided useful detailed information from which to order curtains and carpets. The result is a house where the furnished rooms give a good impression of the style in which Matthew Boulton and his family lived from the 1790s onwards. From time to time further items are acquired as they emerge onto the market. A few of the rooms are set out as museum display rooms to enable visitors to learn something of the Boulton family history and the products and processes in use at the Soho Manufactory. The house, which is Grade II* listed, opened to the public in October 1995.

18 The Soho Manufactory viewed from the latchet works side, the only known image of the building from this angle. (Reproduced by permission of the trustees of the William Salt Library, Stafford)

3: Picturing Soho: Images of Matthew Boulton's Manufactory[1]

Val Loggie

This essay explores the production and use of visual images of Soho Manufactory and the surrounding estate during Matthew Boulton's lifetime. Published views of Soho Manufactory frequently survive removed from their original setting, and as a result, they are now often seen as single images divorced from their accompanying descriptions and intended audience. The essay will argue that these representations formed part of the marketing of Soho's products and the development of what we would now call a brand, as Boulton looked to differentiate himself from other manufacturers.[2] As potential markets expanded geographically, the relationship between buyers and sellers was no longer based on personal interactions and new ways had to be found to influence consumers in their choice of product.[3] As Rita McLean has pointed out in her Introduction, Boulton also faced the problems that Birmingham-made goods had a reputation for poor quality and that he was manufacturing a diverse range of products. There were many strands to his marketing, including salespeople, the showroom, auctions, aristocratic patronage, Boulton's own name and personality, and the development of the Manufactory site as a destination for the industrial tourist.[4] The use of visual images also formed a fundamental part of that branding.

Most published images of the Soho Manufactory focus predominantly on the principal building. This provided a recognisable image of a fashionable building set within a landscaped park, an image that was later to appear in books, directories and magazines, giving the impression of a country mansion.[5] The principal building could be read as symbolising Boulton's social aspirations; mid eighteenth-century manufacturers could seldom afford really grand houses, so grand manufactories came to be a substitute.[6] From the outset the building 'engaged the attention of all ranks of people'[7] and by the summer of 1767 Boulton was writing every day of foreigners or strangers 'who are all much delighted by the extension and regularity of our Manufactory'.[8]

After Boulton acquired the Soho site in 1761, building work on the new Manufactory probably started early in 1762. Development of the site continued for some years, with the principal building reaching completion some time between 1765 and 1767. It was designed by local architect William Wyatt, whom Boulton & Fothergill had already used at the Manufactory site.[9] The use of an architect for an industrial building at this date was unusual; more often engineers or millwrights with little or no knowledge of accepted architectural theory created buildings determined by the size of the machinery, the source of power and local traditional building methods. They were not the kind of men who would have been able to afford either the time or expense of a Grand Tour.[10] Most industrial developments were on a considerably smaller scale than Soho, often due to lack of capital. Boulton, the magnificent self-publicist, described his as 'the largest Hardware manufactory in the World.'[11] In fact Boulton & Fothergill also had limited funds available and the cost of the principal building far overran its estimates of £2,000 to cost £10,000 and cause financial problems which were to plague the firm for years to come.[12]

Why then was Boulton prepared to erect such a spectacular building? It seems likely that it was related to his desire to enhance the poor reputation of Birmingham goods. He wanted to improve the standards of design and workmanship, and it was important that this was represented by the building as well as the goods produced there. He needed to demonstrate his understanding of classical taste and the fashionable market in order to sell the most luxurious of the range of products that were emerging from Soho.[13] This demonstration of taste helped him to build important connections with architects like Robert Adam, William Chambers and James and Samuel Wyatt, who provided designs and influenced potential sales. Robinson suggests that these links with architects gave Boulton a distinct advantage over his rivals.[14] While belonging to the section of society that needed to undertake practical work, he also needed to be able to mix with, understand and impress those who did not have to engage in manual labour, the wealthy aristocrats that he hoped to secure as customers.[15] Additionally, by associating the lower-value 'toys' like buttons and buckles with high-value products like silver and ormolu, and an impressive, fashionable building, he was adding status to those low-value goods. By associating the diverse products and the building, Boulton was beginning to create the Soho brand.[16]

The earliest known images of the Manufactory are a pair showing both the front and the rear of the site (Fig. 19).[17] Demidowicz's work on the buildings has determined that these images show the site between 1766 and 1775.[18] There is archival evidence for a drawing produced by William Jupp, a London

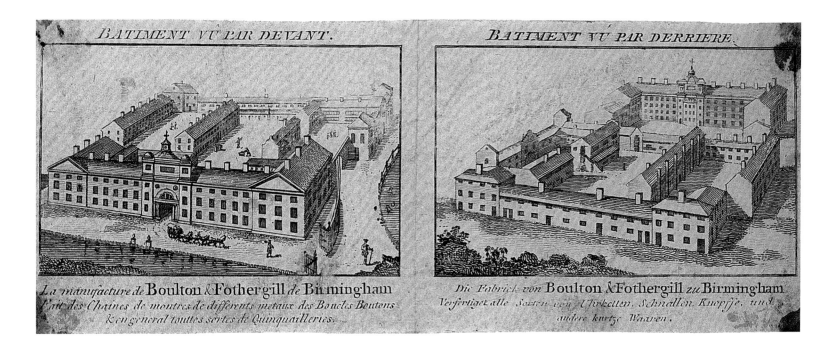

surveyor in 1768, and given to Mr Rooker (possibly Michael Angelo Rooker, an engraver of architectural subjects) to engrave, and of Boulton providing a view of Soho for inclusion in a history of Staffordshire being prepared by the Reverend Thomas Feilde. This work was never completed but the research was later used by Stebbing Shaw in *The History and Antiquities of the County of Stafford* (1798–1801).[19] It is possible that having gone to the time and expense to prepare the view for Feilde, Boulton went on to consider other ways in which it could be used.

From the start of his partnership with John Fothergill in 1762 it is clear that Boulton was keen to exploit the potential market on the Continent, one which he had already begun to explore while working with his father. Indeed, a large part of Fothergill's appeal as a partner was his network and experience in selling in Europe. The French and German captions on these views (see Fig. 19) indicate that they were intended for a Continental audience and are likely to have been connected with sales material which would have consisted mainly of patterns to show designs, samples of the goods themselves, or engraved drawings.[20] The two images show the whole Soho complex, and are the only published images to depict the smaller buildings behind the principal building prior to the 1830s.[21] The *Batiment vu par derriere* shows the workers' housing of Brook Row in the foreground with the rear of the principal building in the background; the *Batiment vu par devant* includes the buildings behind the principal building. It is likely that these views were attempts to emphasise the size of the Manufactory, especially when linked with the caption claiming the manufacture of all sorts of hardware. The front view shows a functioning, busy site with travellers, staff and details of the practical workings of the factory such as the water wheel.

The next known image of Soho began to exploit the principal building more and to make it the main focus of the image (Fig. 20). This aquatint of the Manufactory, now known to be by Francis Eginton, has previously been dated to c. 1781, but recent research has made it clear that this is in fact an early example of an aquatint undertaken in 1773.[22] It was sent to London to be printed amid great secrecy as aquatint was still an innovative technique with the method known to very few, although some examples had been exhibited in 1772.[23] The original version of this print had a French caption, '*Vue des Magasins &c &c appartenants à la manufacture de Boulton & Fothergill Située à Soho près de BIRMINGHAM en Angleterre*'. The production of an aquatint for a French audience is interesting as it had become an established process in France earlier than in Britain.[24] The fact that the earliest known images of the Manufactory are aimed at a continental audience may be a consequence of the increasing need to focus on this market.[25] At a later date the plate was crudely reworked and the caption changed to an English one (see Cat. 361).[26]

Boulton soon turned his attention to images for an English audience as 'The Soho' featured prominently in *The New Birmingham Directory and Gentleman's Compleat Memorandum Book* from 1774 until at least 1777 (Fig. 21). This was printed by and for Myles Swinney, a Birmingham printer and bookseller, and sold in Coventry and London as well as Birmingham.[27] The illustration of Soho is the only one in the Directory and is highlighted on the frontispiece, the volume being 'embellished with a North East view of The Soho neatly engraved on copper'. A description of Soho appears in the preface emphasising the superiority of its products over their French equivalents and 'it is by the Natives hereof, or of the Parts adjacent, (whofe emulation and

19 *Batiment vu par derriere* and *Batiment vu par devant*, c. 1766–75, the earliest known images of the Soho Manufactory. (BCL 82934, vol 1.)

20 Francis Eginton, *Vue des Magasins &c &c appartenants à la manufacture de Boulton & Fothergill Située à Soho près de BIRMINGHAM en Angleterre*, 1773. (Reproduced by permission of the trustees of the William Salt Library, Stafford)

21 *A Northeast View of The Soho*, from Swinney's *Directory*, 1774.

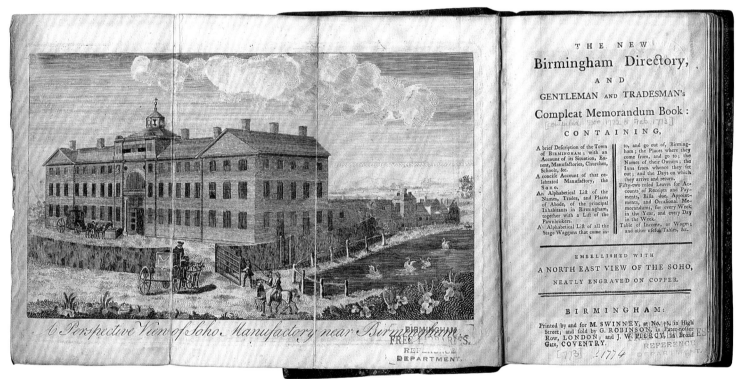

taste the Proprietors have spared no Care or Expence to excite and improve) that it is brought to its present flourishing State.'[28]

Following the production of these views there appears to have been a long gap before an image of Soho was again used for promotional purposes. The next example was an image of the Manufactory associated with the rules of the Soho Insurance Society produced in 1792 (Fig. 22). This is the first dated set of rules but it is now clear that there had been some form of soci-

ety for workers at Soho since before 1782, which allowed staff to make payments into the scheme and to draw benefits from it if they were ill.[29] The plate from which the 1792 version was printed includes only the image, not the rules, which would make it simpler to print the image alone as a single sheet or to alter the rules that went through a number of revisions.[30] In its poster form with the rules listed below the image was accompanied by an explanation of some of its allegorical content:

22 The Soho Manufactory from a poster showing the rules of the Soho Insurance Society (1792). (Birmingham Archives & Heritage, MS 3147/8/47)

A Member of this Society with his Arm in a Sling, is seated on a Cube, which is an Emblem of Stability, as the Dog at his Feet is of Fidelity; he is attended by Art, Prudence, and Industry, the Latter of whom raiseth him with one Hand, and with the other sheweth him Plenty, expressed by the Cornucopia lying at the Feet of Commerce, from whence it flows. Art resteth on a Table of the Mechanic Powers, and looks up to Minerva, Goddess of Arts and Wisdom, who, descending in the Clouds, directs to the SOHO MANUFACTORY, near which are little Boys busy in designing &c., which shew that an early Application to the Study of Arts, is an effectual Means to improve them; the flowers that are strewed over the Bee-Hive, represent the Sweets that Industry is ever crowned with.

Such an explanatory text suggests that the image was aimed at multiple audiences, at the workforce for whom the club was intended (some of whom could read text but were not able to decode the image and had to have it explained), but also at the wealthy customers who were well able to interpret the iconography.[31] The image was used to communicate two messages, firstly to show potential customers that Boulton was a good employer as well as providing quality goods, but also to inspire loyalty among his staff. He was developing an appreciation of the ways in which images could be used to convey particular messages and to assist with his marketing.

Having realised this potential, it is likely that he took it into account when undertaking extensive landscaping work on the estate. From the late 1780s Boulton had been in conflict with George Birch, Lord of the Manor of Handsworth, partly due to Boulton enclosing and improving land he had not leased. Birch had given him notice to quit in 1791 and Boulton considered continuing the business at Soho but moving to the other side of Birmingham to live at Edgbaston Hall (taking over William Withering's lease), or at nearby Aston Hall, or to a house in Handsworth built by one of the Whateleys.[32] He gave the position a great deal of thought; he was reluctant to buy the Soho land unless he could buy enough of it to control the views from his house and prevent others building too close to him. He set out his options in a document headed 'Considerations upon the propriety of buying Soho'.[33] This document and his notebook highlight the importance of those views; he considers views both into and out of Soho, and work was later undertaken to improve the outward views by removing trees or using planting to blot out unsightly buildings.[34] It is likely that in doing this he was again seeking to make a statement about his own social standing. He wanted to create an appropriate setting for his grand Manufactory.

In 1794 and 1795 Boulton made purchases of land from Birch, and embarked on a further programme of improvements in 1795. He planned to 'shut out the sight of the world and make openings to all that is pleasant and agreeable' and to 'make all entrances in to Soho Dark by Plantations and enter through Gothick arches made by Trees'.[35] A separate entrance

23 John Phillp, *View across the roof of the Manufactory*. This view from near Soho House looks towards Thornhill House, the white house at the far right which from 1818 was the home of Miss Anne Boulton. Watercolour on paper, 1796. (Birmingham Museum & Art Gallery, 2003.0031.32)

for the house was created from the main road rather than the old carriage drive which had branched off from the entrance to the Manufactory. New drives and walks were created through the plantations and around the pools, which emphasised the varied scenery and created an illusion of size.[36] John Phillp's illustrations give an impression of the effect (Fig. 23). The Manufactory was not forgotten in this work. Boulton wanted to 'surround my Farm and Works by a Garland of Flowers on one side and by an aquious mirror on the other'.[37] The Works are presumably the Manufactory, which did have a pool to one side, and a 'canal' in front of it that could provide the aqueous mirror.[38] He planned to 'Form the Terras at the front of the Manufactory so as to be always clean and neat.' A new retaining wall was built, the ground levelled and then gravelled.[39] The experience he created for visitors to the Manufactory was still considered important, as he asked himself in 1795, 'How shall I form my Western ground to be handsome in the sight of those going to the Manufactory?'[40] In 1797 he 'put a good pale fence by the side of the road and down to the Manufactory' and the informal five-barred gates were replaced with metal gates and piers.[41]

Boulton took great pride in his estate, sparing no time or expense to achieve the desired effect, calling it 'the Monument I have raised to myself'.[42] Over a period of forty years he transformed what had been a barren heath into an elegant park, a transformation which inspired the poet in more than one visitor:

On yonder gentle slope which shrubs adorn,
Where grew of late rank weeds, gorse, ling and thorn,
Now pendant woods and shady groves are seen,
And Nature there assumes a nobler mien.
Here verdant lawns, cool grots, peaceful bowr's,
Luxurient, now are strew'd with sweetest flowers,
Reflected by the Lake, which spreads below,
And Nature smiles around – there stands Soho![43]

In spite of the large number of visitors to the Manufactory, many of whom also took the opportunity to explore the grounds, little visual representation is known and by far the largest source is the work of John Phillp, mostly undertaken in the 1790s. Phillp was a talented draftsman who was brought to Soho from Cornwall in 1793 at the age of about fourteen. He was sent by a copper merchant, George C. Fox, Boulton writing to Fox that the work he had intended for Phillp when they discussed the matter in Falmouth was now discontinued 'on account of ye unfortunate rupture with France, & I now have no species of Painting done in my Manufacture: however I will find out what sort of employment is best suited for his talents'.[44] The lad received instruction in architectural drawing from William Hollins, who himself had been trained in classical architecture at the London office of George Saunders, and in drawing or painting from Joseph Barber.[45] A number of Phillp's works were collected together in the nineteenth century by a descendant, and it is this body of work which shows the detail of the Soho Estate.[46]

24 The Soho Manufactory, from *The Monthly Magazine* (1796). (Birmingham Museum & Art Gallery)

25 The Soho Manufactory, from Bisset's *Magnificent Directory of Birmingham* (1800 and 1808). (Birmingham Museum & Art Gallery)

A further group of seven views of Soho was produced around 1796 by Robert Andrew Riddell, who wrote to Boulton that he was sending seven views of Soho at five guineas each, some of which suggested possible improvements to the grounds. The views may be those in the list of eight 'pictures of and from Soho' in Boulton's garden notebook.[47] Riddell wrote:

> In the view made to suggest the improvements, you'll observe that simplicity is particularly attended to and can easily be accomplished in nature, the places planted with shrubs are purposely to break the lines of formality, and hide the stalks of the fir trees which give the scene a bare dry look, and to break the edge of the ford when seen from the windows of the house &c. I think the break in the hill when the sand has been dug adds to the beauty of the place [...].[48]

Boulton's estate received recognition when it was selected as one of the gentleman's seats featured in the first edition of *The Tablet* in 1796.[49] This was an illustrated almanac printed by Thomas Pearson, a Birmingham printer, bookseller and stationer who undertook much of Soho's printing. There were twelve gentlemen's seats and a frontispiece engraved from drawings by Joseph Barber, the Birmingham artist who had given lessons to Boulton's daughter, Anne, and John Phillp.[50] Barber's view showed Soho House perched on the top of the hill with the Manufactory buildings just visible within the trees. The work carried out on the estate meant that Boulton was again drawn to the attention of potential customers.

Soon after the resolution of the future of the estate and the new landscaping works, a number of images of the Manufactory were published. The first of these was in *The Monthly Magazine* (Fig. 24), a 'miscellany' first published in 1796 which aimed to lay before the public 'objects of information and discussion', and to propagate liberal principles. It was a radical magazine, a journal of Unitarians and Dissenters.[51] The publisher Richard Phillips wrote to James Watt via James Watt junior 'respectfully to solicit your correspondence [...] especially on Topics of a mechanical nature & relating to those wonderful improvements in the useful arts always making in your manufactory at Birmingham.'[52] The following year Phillips wrote:

> I wish through the medium of the Monthly Magazine to present to the public an engraved view and description of your celebrated Manufactory the Soho. I have no doubt that you are in possession of a drawing such as would answer my purpose, I wish you to have the goodness to send me. How far you have it in your power to assist me in a description I know not; if you can refer me to one already in print, it may serve as the groundwork of that which may be written for my purpose.[53]

The published view is the only image of the Manufactory labelled as belonging to Boulton and Watt, though in fact the buildings were never part of their engine partnership and remained the sole property of Boulton.

The following year Soho was illustrated in the *Copper-Plate Magazine* which was aimed at a very different audience.[54] Each number contained two prints, from 'Original Paintings and

Drawings by the First Masters with Letter-Press descriptions' and cost a shilling an issue. The engravings were by John Walker after a variety of artists. The Soho image was signed by Walker and labelled as engraved from an original drawing but does not name the artist of that drawing. The magazine was designed to be viewed as single plates and would probably have been broken up and integrated into collections. The text for each plate was on a separate page, Soho's entry highlighting 'Mr Boulton at length resolved to render his works a seminary of taste, and at very considerable expense procured the most able and ingenious artists in every branch', thus reinforcing the importance of design and taste to Boulton, and the style of his products. It was rare to include an industrial site in this *Monthly Cabinet of Picturesque Engravings intended to comprise all the most interesting, sublime, and beautiful views of Principal Cities, Royal Palaces, Sea-Ports, Seats of Nobility and Gentry, Forests, Curious Remains of Antiquity, Rivers, Public Edifices, Lakes, Parks, Mountains, Gardens &c in England, Scotland, Ireland and Wales.*[55] Elegant visitors to the Manufactory in a fashionable carriage are displayed prominently in the foreground of this view with some others wandering in the grounds, allowing the viewer to identify with them and so relate more directly to the scene (see Cat. 360). The overall impression is that of a genteel country house rather than a bustling working factory. The extent to which the image is carefully constructed is highlighted by the fact that the latchet works (to the left) was not actually completed as shown until 1825–6. The south bay and wing were built in 1794–5 with the central section in 1797–8 and a single storey north bay thereafter.[56]

Soho was a major focus in James Bisset's *Magnificent Directory of Birmingham* published in 1800 with a second, enlarged edition in 1808 (Fig. 25). This comprised Bisset's 'Poetic Survey round Birmingham' and 'a brief Description of the Different Curiosities and Manufactories of the Place Intended as a Guide for Strangers' with 'names, professions &c. superbly engraved in emblematic plates'. The work was printed for Bisset in Birmingham but also sold by T. Heptinstall, Holborn, London, and 'all other book sellers'. Several different versions were available from the basic one priced at six shillings to one printed in colour, intended for the libraries of gentlemen, priced at two guineas.[57] In the second edition gentlemen, merchants, tradesmen and manufacturers were invited to have their names inserted at 10s. 6d. each, or free of charge if they supplied their own plate.[58] The most prominent of the plates was the one for Soho which is given an entire page; by the 1808 edition it was considered so important that it is mentioned on the title page of the volume. There are eight businesses listed under the image, all but one of which contain Boulton's name. Bisset's *Directory* was primarily aimed at an audience more interested in business and trade, so Boulton took the opportunity to promote the breadth of his businesses including the mercantile trade and the banking business. The inclusion, indeed emphasis, on a business in Staffordshire, in a Birmingham directory highlights the importance of Soho and Boulton to Birmingham and the wider area. Not only is it the only plate specifically named on the later title page, but Soho also appears prominently in the 'Poetic survey round Birmingham'.

A pair of images by Francis Eginton junior and a description of Soho appeared in Part I of the second volume of the Reverend Stebbing Shaw's *History and Antiquities of the County of Stafford* published in 1801. One shows Soho House perched on top of the hill with the surrounding parkland and lake (see p. 14), and a reference to the industrial site visible in the smoking chimney to the left, similar to the view used by Barber for *The Tablet*. The other plate shows the usual oblique view of the principal building (Cat. 68). A series of watercolours exists which may have provided the source for these views, including one which is the only known image of the principal building from the opposite angle (Fig. 18).[59] Like Bisset's *Directory*, Shaw's work was available in a variety of formats: small paper with folded plates, large paper, or a special 'illuminated' version available to subscribers which included 'additional watercolour drawings', the map of the county and some coloured plates. Three volumes were planned but only Volume One and Volume Two, Part I, were completed.

Shaw utilised research undertaken by the Reverend Thomas Feilde for his own proposed history of Staffordshire in 1769, for which Boulton had supplied the image discussed above. Shaw quoted a description of Soho written for Feilde in 1768 by Erasmus Darwin, who had drawn Feilde's attention to Soho as being worthy of inclusion, 'if you admit into your account of Staffordshire the wonders of art as well as those of nature'.[60] Darwin's original description was followed by a 4,500 word description and history of Soho. It highlighted the importance of design and aesthetics:

> Impelled by an ardent attachment to the arts, and by the patriotic ambition of bringing his favourite Soho to the highest degree of perfection, the ingenious proprietor soon established a seminary of artists for drawing and modelling; and men of genius were now sought for and liberally patronised.[61]

Once again emphasis was placed on taste in the text, encouraging readers to view the illustrations of Soho with this in mind, to think of the elegance of the products as well as that of the buildings and their surroundings. The audience was guided towards thinking of the whole Soho brand.

Shaw wrote to Boulton in London in 1798 sending him proofs 'as I wanted as well as promised to publish nothing relating to it without first shewing it to you'.[62] The plates for the illustration were to be supplied from Soho but were considerably delayed, arriving after printing of the other illustrations had begun.[63] Much of the book is taken up with grand houses, there are some industrial sites, but Soho is given priority with a lengthy written description. Boulton's status is reinforced by the inclusion of a plate of his house and park as well as his Manufactory. A very similar image of the Manufactory to that used in Bisset's *Directory* is given a different slant by being dedicated by Shaw to Boulton, and by the inclusion of Boulton's coat-of-arms. This view speaks of Boulton the country gentleman as well as Boulton the successful businessman of Bisset's volume. Shaw died in 1802 and Boulton made efforts to retrieve the plates from his executors, presumably aware of both the promotional use to which he could put the plates and the fact that they could be misused by others.[64] He was seeking to retain control over the tools he had used to create his brand.

It is clear that Boulton did use these images for their marketing potential. He sent his proofs to Joseph Franel, an English merchant in Smyrna saying:

> As I have never extended my dealing to Smyrna or to any part of Turkey, I presume you are little acquainted with my home and my establishments which consist of a merchantile house in Birmingham and the most considerable manufactory of sundry hardware in England situated about two miles from that town, but in order to give you a more particular detail I send you herewith a part of the history of Staffordshire now in a state of publication which the author sent me yesterday a few sheets for my own inspection and remarks. If you should think that any of the branches of my manufactory could find a sale in Smyrna I should be happy to make that through your respectable House in which I have great confidence.[65]

Boulton used a number of methods to market his products but the use of images was an important part of that marketing and a way of linking the diverse aspects of the output, to associate them all with the Soho brand. Images of the Manufactory were used in increasingly sophisticated ways during Boulton's lifetime; one was at the forefront of the development of the aquatint process in Britain; others were used to convey messages about Boulton's social status, benevolence as an employer or the breadth of his enterprises. The examination of the visual images of Soho alongside archival material allows us to shed new light on those images, the intentions behind their production, the intended audience and message. The repeated use of the image of the principal building to reinforce the idea that high-quality, fashionable goods came from Soho was innovatory. The linking of all the metalwares emerging from the site added value to the cheaper goods such as buttons and buckles by giving them an association with the expensive, high-status silver and ormolu. The use of the elegant principal building also ensured that as Soho became increasingly associated with the steam engine there was a reminder that it was a site that still produced a wide range of products, particularly when linked with the list of businesses as in Bisset's *Directory*. Buildings would come to be used frequently on billheads in the nineteenth century when 'the value of a building as a source of status and means of advertising became increasingly apparent',[66] but Boulton appreciated the importance of his principal building's status and used it in his marketing at a much earlier date.

4: The Context of Neo-Classicism

Nicholas Goodison

The architects, artists and designers of the latter part of the eighteenth century did not see themselves as 'neo-classicists'. The word 'neo-classicism' was not coined until the 1880s. They talked rather of antiquity, of the antique style or antique taste, and particularly of the Grecian taste – *le goût grec* – a term which they loosely applied to a wide range of classical and antique models, many of them not Greek at all. The phrase was so catching that purveyors to the fashionable world applied it to all sorts of consumer goods and services.[1]

The new taste was not just a revival of the classical taste of Andrea Palladio and Inigo Jones or of the classical monumentality of the period of Louis XIV. Neo-classical artists saw themselves as partaking in a revolution. Just as thinkers and natural philosophers were striving, with the enthusiasm of revolutionaries, to build a new world of rationality, so artists, architects, and designers strove to purify their designs by returning to classical norms. Their work was a stylistic and moral reaction to the public and domestic hedonism of the rococo. They were standard-bearers of the 'true' style.

The seeds of the change were sown in the circle of young French architects, artists and sculptors who attended the French Academy in Rome in the 1740s. They included the architects J.-L. LeGeay and E-A. Petitot and the sculptor J.-F. Saly. In Rome they feasted their eyes on the classical ruins and sculptures, on the works of Bramante, Michelangelo and other great masters, and on the many more recent late baroque buildings that now adorned Rome. Their published and unpublished projects and designs, not least their ideas for vases, reflected these experiences. The draughtsman C.-N. Cochin particularly singled out LeGeay and said that he was chiefly responsible for the revival of '*un meilleur goust*'.[2] Cochin himself, along with the Abbé Le Blanc and the architect J.-G. Soufflot, went to Rome in 1749–51 with the young Marquis de Marigny, Madame de Pompadour's brother, who was about to become France's influential *Directeur-Général des Bâtiments, Jardins, Arts, Académies et Manufactures Royales*. It was a visit that Cochin later described as a turning point in the return of good taste to France.[3]

Back in Paris there was an increasing volume of criticism, from the late 1730s, of the prevalent rococo taste. J.-F. Blondel, Soufflot, Le Blanc, Cochin, and the archaeologist the Comte de Caylus all voiced their ridiculing (and very readable) opinions of the frivolous excesses of the rococo, with its asymmetric design, elaborate ornament, unconnected motifs and lack of proportion. Le Blanc, writing in 1745 and echoing Blondel, was particularly critical of the 'barbarity' of the decorative arts, the elaboration of which went, in his view, to ridiculous lengths in the exotic use of ornament.[4] These writers left no doubt in their readers' minds that the rococo was simply bad taste.

The combination of these criticisms by leaders of taste and the rediscovery of classical proportion and ornament, stimulated by the sensational discoveries at Herculaneum in 1738 and Pompeii in 1748, led to the revival of classical norms in architecture and the decorative arts. But like most changes of taste, the shift from the rococo to the new style was not as sudden as we, with our backward-looking and compressed view of historical events, might imagine. The most radical among the proponents of the new taste included the designer L.-J. Le Lorrain[5] and the architect J.-F. de Neufforge. Others were more restrained. Many traces of the curvilinear influence of the rococo continued to lurk in contemporary design, especially in furniture and objects of decorative art. Few patrons in any age leap suddenly and wholeheartedly into a new style of decoration. It takes time to reconstruct or redecorate rooms, to build new houses, to commission new furniture; and some no doubt liked the comfort of their curvilinear seat furniture. But by the late 1760s the new taste, with its accent on rectilinear forms, was pervasive.

The French architect Pierre Patte described this pervasiveness of the new taste in 1777:

> … at last the return of the Antique taste having extended its influence over all our decorative arts, especially during the last fifteen years or so, it can be said that the interior decoration of suites of rooms and the style of their furnishings have become in some sense a new art.

Patte understood that the new taste was really a return to the classical idiom that had been the mainstay of architects and decorative artists before the rococo enjoyed its exuberant interlude, but with a difference. He welcomed the return to 'sager, less eccentric forms', following the excesses of the rococo, which he described as 'tormenting interior decorations in every possible manner' and as a 'torrent of fashion':

...to the correct style of the last century's decorations has been added less severity, more delicacy, more variety in the forms ... the ornaments that are most admired in the best works of antiquity, like acanthus and laurel leaves, festoons, ovate ornaments, ogees, shells, flutes, guilloche work, Vitruvian scrolls, medallions etc., have been applied to interior decorations ...[6]

English architects, artists and designers were quick to follow where their French counterparts had led. English eighteenth-century fashion tended to follow France, although in somewhat tamer mood. French designs reached English designers, craftsmen and their patrons through many channels. Among these were the works of art imported from France both during the Seven Years' War and, in increasing quantities, after its ending in 1763. Engravings and folios of designs published in France, such as de Neufforge's *Recueil Élémentaire d'Architecture* (1757–72) and J-C. Delafosse's *Nouvelle Iconologie Historique* (1768), which contained a large collection of designs for frames, medallions, trophies, vases, friezes, pedestals and brackets, were influential. But home-grown architects too, who were themselves influenced by their contacts and travels in Italy and France, were actively promoting designs in the new taste from an early date: and there were influential source books published in England. The antique taste had a firm foothold in the work of English designers from the mid-1750s. James Stuart's patently classical design for a wall of the proposed hall at Kedleston in Derbyshire,[7] which contains a classically rectilinear table, was contemporary with some of the earliest furniture designed in the '*style antique*' in France.[8]

Exponents of the new taste were not just copyists. The essence of neo-classicism was its eclecticism. The Grecian taste was not enough. Nor was the slavish copying of antique models. Imitation was the rule. The young Venetian Giambattista Piranesi, who arrived in Rome in 1740, became one of the most ardent and influential promoters of classical models to artists and collectors during their visits to Rome, publishing copious engravings of bold designs. He wrote in his *Diverse Maniere* in 1769:

... an artist, who would do himself honour, and acquire a name, must not content himself with copying faithfully the ancients, but studying their works he ought to shew himself of an inventive, and, I had almost said, of a creating genius; and by prudently combining the Grecian, the Tuscan, and the Egyptian together, he ought to open himself a road to the finding out of new ornaments and new manners.[9]

Although Piranesi wrote these words in the context of his attack on the assumed supremacy of Greek architecture as the sublime model for aspiring architects and designers, they convey the creative eclecticism of the new taste. Imitation was not despised, as it might be today, but admired. It was a form of invention. Sir Joshua Reynolds echoed this neo-classical doctrine of the creative use of imitation when he said in a lecture at the Royal Academy in 1774 that not only variety but even 'originality of invention' was produced by imitation:

Even genius, at least what generally is so called, is the child of imitation ... A mind enriched by an assemblage of all the treasures of ancient and modern art, will be more elevated and fruitful in resources, in proportion to the number of ideas which have been carefully collected and thoroughly digested. There can be no doubt that he who has the most materials has the greatest means of invention ...[10]

The expansion of Boulton & Fothergill's manufacturing business in the 1760s coincided with this fashion for the 'antique taste' and for the mining of antiquity for decorative motifs. Their workshops had made rococo objects – in what Boulton called the 'French style'.[11] Examples survive from both the 'toy' business (Cats 81–4) and the silver and plate business (see p. 52, Fig. 42 and Cat. 145), although the full three-dimensional extravagance of the rococo was ill-suited to some of Soho's manufacturing methods, particularly die-stamping.[12] The factory continued to make pieces of silver and plate decorated with rococo motifs well into the 1770s:[13] but they were ghosts of the rococo past and were now a small minority of the firm's production of silver and ormolu.[14] By the time the Manufactory was launching into the ormolu and silver-plate businesses on a serious scale in the early 1770s, the new classical taste was dominant.

The repertory of classical ornament appealed to Boulton because it appealed to his patrons. He was particularly expressive to his friend and patroness Mrs Montagu. In a letter to her in 1772 he said:

Fashion hath much to do in these things, and as that of the present age distinguishes itself by adopting the most elegant ornaments of the most refined Grecian artists, I am satisfyed in conforming thereto, and humbly copying their style, and makeing new combinations of old ornaments without presuming to invent new ones.[15]

Boulton was thus an inventor in the eighteenth-century sense, like many of his contemporaries. His work in ormolu, silver and plate must be seen in this context.

He was far from immune to the influence of his French competitors. Apart from visiting his patron's houses and the showrooms of shopkeepers who imported works of art from Paris, he saw their work in the new taste when he visited Paris in 1765. He could hardly have avoided it. As Baron Friedrich von Grimm had said two years earlier: '... *la décoration extérieure et intérieure des bâtiments, les meubles, les étoffes, les bijoux de toute espèce, tout est à Paris à la grecque ...*'[16]

Boulton was not a slavish imitator of French designs. Little of his ormolu or silver work is wholly French in feeling. The highly ornamented work of his contemporaries in France was too extravagant for his English taste. He hankered after emulating the quality of the French work, particularly the quality of the chasing and gilding, and the fine contrast between matt and polished surfaces. But he wanted to simplify the ornament. Whatever the approved taste of his patrons, he said on one occasion: 'whether it be French, Roman, Athenian, Egyptian, Arabesk, Etruscan or any other ... I would have elegant simplicity the leading principal, whereas in my opinion such of the *orfèvre* of the French as I have generally seen is *trop chargé*.'[17]

Boulton's quest to satisfy his patrons' love affair with the new taste, with a typically English restraint, was centred on ormolu, silver and plate. He must have been pleased both by the encouragement of his English patrons and by the courtly flattery of the Empress Catherine of Russia, who in 1772 was reported to have said that his ormolu vases were 'superiour in every respect to the French'.[18]

The French influence on Boulton's designs was considerable, but usually at second hand. In its most direct form, it is clear in his silver and ormolu 'lion-faced' candlesticks (see p. 42, Fig. 33 and Cat. 146), the ormolu versions of which were fitted with branches. The candlestick is a design usually attributed to the well-known *doreur-ciseleur* Pierre Gouthière. Ormolu versions without candle branches attributed to him are thought to date from the mid-1760s or earlier.[19] Boulton's branches are also French in feeling and clearly owe more to the style of his French rivals than to the work of his predecessors in England. Their exact source is uncertain. Several *bronziers* in Paris used similar branches with a squared bend and acanthus decoration, not least the *fondeur* and furniture designer Jean-Louis Prieur.[20] The lions' masks and feet reappear on the vases illustrated in 'Ormolu ornaments' (p. 57, Fig. 48), illustrating the way in which, from the first, Boulton's modellers made 'new combinations of old ornaments.'

The inspiration of French models is apparent in several other ornaments and decorative motifs. The very idea of using vases as the central theme of Boulton's decorative ormolu and silver came from France. The inspiration of the rather poor figure of Venus ('Ormolu ornaments', p. 56, Fig. 46) may well have been François Vion's figure of *La Douleur*, or a clock case based on it,[21] although it seems to me more likely that it was derived from a print of a French painting published in England.[22] But whatever the actual source, the inspiration was French. The fitting of a horizontal movement to a vase clock, as Boulton did with his Venus clocks, also came from France.[23] The use of classical figures to decorate clock cases, and the use of vases and obelisks, were common themes of French clockmakers. And the parallels with individual mounts are legion – lions' masks, satyr masks, rams' and goats' heads, snake handles, coiled snakes to symbolise healing, allegorical figures, urns and medallions, were all familiar motifs to French designers, as was the whole repertoire of classically inspired friezes and borders.

Some of the books from which Boulton culled his ideas for classical designs also came from France. Like other makers and designers of his day, he would have been aware of the much-circulated classical miscellanies, such as de Neufforge's *Recueil Élémentaire d'Architecture* and Delafosse's *Nouvelle Iconologie Historique*, both of which I have mentioned above. Evidence of his interest in French source books lies in his library. This included furniture designs by Delafosse and designs of vases by Delafosse, Saly and others.[24] In 1764 he bought a copy of another great classical miscellany, Bernard de Montfaucon's *L'Antiquité Expliquée et Représentée en Figures*, first published in Paris in 1719.[25] There are several images in Montfaucon's celebrated work which could have inspired Boulton or been the origins of his designs.[26] He also appears to have owned the seven volumes of the Comte de Caylus's *Recueil d'Antiquités Égyptiennes, Étrusques, Grecques et Romains*, which was published in Paris in 1761–7 and was another of the miscellanies of classical objects and ornaments, including everyday objects, which acted as a source book for artists and designers.[27] As with Montfaucon, de Caylus was probably the source of at least one set of images used by Boulton's modellers.[28]

Boulton probably owned Antoine Desgodetz's influential account of the ancient monuments of Rome, *Les Édifices Antiques de Rome*, which was first published in Paris in 1682 and was reproduced in an English edition in 1771.[29] He also had a collection of thirteen works by Gabriel Huquier (1695–1772) on architectural detail, vases, trophies, fountains and other ornaments by a number of artists, many of them in the rococo tradition but including le Geay.[30]

But classical source books published in France did not have a monopoly of Boulton's attention. Thus he ordered the first three volumes of Antonio Francesco Gori's *Museum Florentinum* (1731–66) in 1771.[31] This was another source book well known in England. It was an illustrated and annotated survey of gems, statues, coins and paintings in the most important collections in Florence. It was the source of several of Boulton's ornaments, including his figure of Titus (Cat. 176).[32]

He subscribed to George Richardson's *Iconology*, published in London in 1778, a collection of the 'most approved emblematical representations of the ancient Egyptians, Greeks and Romans and from the compositions of Cavaliere Cesare Ripa'. Ripa's *Iconologia* had first been published in Rome in 1593, and was reprinted in many editions in the next two centuries. Richardson's colour illustrations were a useful source of design to the many artists and craftsmen who, like Boulton, subscribed to the volumes. This volume was published too late to have any effect on designs for ormolu, but Boulton may have been aware from its predecessors of some of the iconography associated with classical figures, for example the winged head of Science who is seen teaching the laws of Nature on the sidereal clock (Cat. 170).

Boulton is also likely to have seen at least some of Piranesi's engravings, the first series of which appeared in 1756. Some of the mounts on Boulton's ornaments echo drawings in Piranesi's works.³³ He bought and borrowed unidentified books of vases³⁴ and acquired a set of the coloured prints of the ancient vases in the collection of the British ambassador at Naples, Sir William Hamilton, produced by 'Baron d'Hancarville', or at least the first volume which was published in 1767. One of the compiler's aims in publishing the prints was to encourage manufacturers to use the correct classical models.³⁵ Boulton's chief practical interest in d'Hancarville's prints probably centred on the decorative borders to the plates, some of which are echoed in the friezes of his silverware and ormolu, and on the use of the book as a means of instructing the young draughtsmen whom he employed in the factory.

Boulton actively sought publications that depicted classical sites, and particularly the discoveries at Herculaneum and Pompeii. He had to be up with the times if he was to capitalise on the new taste. By 1771 he owned the first five volumes of *Le Antichità di Ercolano*, which was published by the Accademia Ercolense from 1757 onwards.³⁶ He was a subscriber to Thomas Martyn and John Lettice's *Antiquities of Herculaneum* (1773), a one-volume summary of *Le Antichità*, as were Josiah Wedgwood and William Duesbury. He was also familiar with the two important books on classical sites by the 'Greek' architects in Britain, namely the first volume of James Stuart and Nicholas Revett's *The Antiquities of Athens* (1762), and Robert Adam's *Ruins of the Palace of the Emperor Diocletian at Spalatro in Dalmatia* (1764). And although there is no firm evidence, he must surely have seen other contemporary sources of the classical idiom, such as the celebrated surveys of classical sites in Robert Wood's *Ruins of Palmyra* (1753) and *Ruins of Balbec* (1757), and Thomas Major's *The Ruins of Paestum* (1768).

Boulton's eagerness to have these works was symptomatic of his intellectual and commercial interest in the antique. He was not unusual in this. Such books were used practically as pattern books by many craftsmen and designers, all over Europe. The books were a prolific source of 'invention'. They contained many details that may have led Boulton and his modellers to create particular mounts, borders or friezes – especially the guilloche, Vitruvian scroll and other friezes that adorn his silver and plate. What is certain is that these books generally influenced Boulton's repertoire of classical ornament and put a stamp of authority on his work.

Some of Boulton's designs came from contemporary architects in England, especially the royal architect William Chambers, James Stuart and James Wyatt. Chambers was one of the leaders of the revival of the classical idiom. He was the principal carrier of the Palladian inheritance. His strong preference for Franco-Italian sources of classical design was formed during his years of academic grounding in Paris and Rome in the early 1750s (Fig. 26). While attending Blondel's *École des Arts* in Paris in 1749–50 he met, as fellow students, such revolutionary French architects as Charles de Wailly and Marie-Joseph Peyre, both of whom remained friends for the rest of his life. His studies in Paris

26 William Chambers, *Franco-Italian Album*, p. 42, 1750s. (Victoria and Albert Museum)

27 William Chambers, *Treatise on Civil Architecture* (1759), facing p. 36 ('Persians and Caryatides').

and Rome, and especially his contact with the artists and architects who had studied at the French Academy in Rome, gave him a solid grounding in the early forms of the '*style antique*'.

Chambers's preoccupation with Roman, or rather Italian, models set him apart from architects such as James Stuart and Robert Adam, whose styles were based more on Greek models. He defended his preference for Rome over Greece with vigour in his renowned *Treatise on the Decorative Part of Civil Architecture* (Fig. 27), in which he said that the Greeks were not a suitable model:

> Our knowledge ought not to be collected from them; but from some purer, more abundant source; which, in whatever relates to the ornamental part of the art, can be no other than the Roman antiquities yet remaining, in Italy, France or elsewhere: vestiges of buildings erected in the politest ages; by the wealthiest, most splendid, and powerful people in the world.³⁷

The *Treatise* was a panegyric of Roman imperial building, although Chambers did add an obeisance to more recent architects:

> The labours of the celebrated masters of the fifteenth, sixteenth and seventeenth centuries may, perhaps, be added to enrich the stock; and we may avail ourselves of their labours, to facilitate, or shorten our own; but, it should always be remembered, that though the stream may swell in its course, by the intervention of other supplies, yet it is purest at the fountain's head.³⁸

In fact, Chambers was a more eclectic architect than these passages might suggest. But his preoccupation with Roman models and his studies in Italy and France, where he became familiar both with the trends in French ornamental design and with the published works of French designers such as Saly, Beauvais, de Wailly, Vien, Petitot, and le Geay, did mean that his designs for buildings and objects centred on the somewhat heavier neo-classicism associated with French artists. This influence is apparent in some of Boulton's early ormolu ornaments, and coincided with Boulton's own fascination with French models.

Chambers became architect and tutor to the future George III, and remained the king's preferred architect after his accession to the throne in 1760. He was effectively the royal architect during the later years of the eighteenth century, concentrating on public buildings while Adam built a substantial private practice for the aristocracy. Boulton met Chambers in 1770. The royal connection must have greatly attracted Boulton in his quest to sell his new products to the fashionable world.

Chambers drew many of the architectural decorative details, statues, busts, friezes, vases, furniture and other ornaments that interested him during his travels in Italy and France in 1749–55. Over five hundred of these drawings are pasted into his so-called 'Franco-Italian Album'. This album is one of the most important documents for the study of the revival of the 'antique taste' in Europe. His later work shows that he often referred back to these drawings, and the inspiration for some of the mounts used on Boulton & Fothergill's ornaments can be traced to them. The rams' heads and swags, and the spirally grooved foot on the king's clock, are two examples (Cat. 163). Grooved feet appear on other Chambers-inspired ornaments.[39] It is likely that this vase foot was given to Boulton by the architect over breakfast at their first meeting in 1770.[40] It was followed by Chambers's design for the king's clock case, which shows clearly Chambers's familiarity with French models[41] and by his designs for the vases made for the king (Cats 160, 164), which Chambers exhibited at the Royal Academy in 1770.[42] The fish (or dolphin) handles of the ewers on top of the clock, with their lips biting the rims of the ewers, appear in a drawing in Chambers's *Treatise*, in his Franco-Italian drawings and in a drawing by Chambers's assistant John Yenn of a ewer probably designed by Chambers.[43]

Chambers's influence on Boulton went further than the objects commissioned for the king and queen. The figures supporting Boulton's 'Persian' vases (Fig. 28) derive from Chambers's drawings of 'Persians and Caryatides' in his *Treatise on Civil Architecture* in 1759 (Fig. 27). He may also have given Boulton models for some of the other decorative motifs that appear on Boulton's vases, including the winged figures on the sidereal clock (Cat. 170). He supplied no designs for silverware.

James 'Athenian' Stuart was also a source of one-off designs. Stuart too studied in Rome, but his chief claim to fame was his part in the publication of the first volume of *The Antiquities of Athens* in 1762.[44] This was the first accurate survey of Greek classical buildings and it became an important source book for architects and designers, as Stuart intended. In 1769 Stuart recommended to his patron Thomas Anson that 'my friend Mr Boulton' should cast the great tripod that was to stand on top of the 'Lanthorn of Demosthenes' which Stuart had designed for Shugborough.[45] The making of this tripod, the legs of which weighed five hundred-

28 Boulton & Fothergill, Persian candelabra, ormolu, blue john and white marble, height 32 inches (81.25 cm). (Victoria and Albert Museum)

29 Boulton & Fothergill, Tripod perfume burner with candelabra, the tripod and stand after a design by James Stuart, ormolu and white marble, height 25.2 inches (64 cm). (Private Collection)

30 Boulton & Fothergill, Pattern Book I, p. 111, tureens, 1762–90, Admiralty tureen, upper centre. (Birmingham Archives & Heritage, MS 3782/21/2)

31 Boulton & Fothergill, Door escutcheon and knob, the design by Robert Adam, ormolu, height 6.3 inches (16 cm), 1776–8. (National Trust, Kedleston Hall)

weight, brought Boulton into closer contact with Stuart at a time when he was actively seeking ideas and designs.

It is a measure of Boulton's eclecticism that he did not overtly favour Chambers's Franco-Italian taste, even though it was the chief influence on his major early ormolu ornaments, over Stuart's '*gusto greco*' despite Chambers's contempt for it. He put four tripods 'after a design of Mr Stuart's' into the sale at Christie's in April 1771.[46] They were probably similar to the tripod perfume burners which have survived at Kedleston Hall, Althorp and elsewhere, the original design of which should be attributed to James Stuart. I have argued elsewhere that the tripods designed by Stuart were made by Diederich Nicolaus Anderson, who exhibited 'a tripod, from an original design of Mr Stuart' in 1761.[47] Boulton's version of this tripod is illustrated in Fig. 29.

Other designs that can be attributed with some confidence to Stuart are the 'tripodic' tea urn made by Boulton & Fothergill in 1770, examples of which survive in the Royal Collection (Cat. 167) and at Syon, and the candle branches on these tea urns.[48] These designs were an important contribution to the style of Boulton's early ornaments. The candle branches reappear on other vases from Soho.

The vases illustrated in Cat. 158 were probably also designed by Stuart and I have tentatively attributed their manufacture to Anderson.[49] The case for Stuart rests not only on stylistic grounds, but also on the survival of pairs of them from Spencer House and Hagley, two houses at which Stuart designed interiors. If the vases were designed by Stuart, they represent a further influence on Boulton's designs, since the spiral stem with its floral collar appears on early vases made by Boulton & Fothergill (see 'Ormolu ornaments', p. 58, Fig. 49).

Stuart's direct influence on silver designs seems to have been less. His only known involvement was in the commissioning of the silver soup tureen for the Admiralty in 1771 (Fig. 30). It seems highly likely that he was also responsible for the design.[50]

Neither Chambers nor Stuart took interior design as far as Robert Adam. To Adam, the design of a building's interior and its contents was an integral part of the architectural scheme. For many of his commissions Adam included chosen craftsmen in his plans. His success with many of the leading patrons of the time has led to a traditional view that his influence on makers was pervasive, a view that Sir John Soane advanced in his well-known lecture in 1812:

> It is to the activity of the Messrs Adam that we are more particularly indebted for breaking the talismanic charm, which the fashion of the day had imposed, and for the introduction from ancient works of a light and fanciful style of decoration … This taste soon became general; everything was Adamitic: buildings and furniture of every description … To Mr Adam's taste in the ornaments of his buildings, and furniture, we stand indebted, in-as-much as manufacturers of every kind felt, as it were, the electric power of this revolution in art.[51]

Curiously, although the Adam brothers admired Boulton as a metalworker – so much so that James Adam suggested that they should cooperate in the production of ormolu and silver plate – and although they corresponded about the possibility of Boulton taking a property in London from them for a showroom,[52] there are no objects from Soho in either ormolu or silver that resulted from cooperation between them. Even on those occasions when Boulton worked to their designs they do not appear to have given him the orders or the designs. It was Samuel Wyatt, then clerk of the works at Kedleston, who sent Boulton the commission for the door furniture in 1765 (Fig. 31).[53] At the very least the Adams' work must have had some effect in training Boulton's eye in the classical repertory, and it is clear that as the 1770s wore on Boulton & Fothergill's ornamental metalwork, especially silver, took on a lighter and more delicate feeling more in keeping with the repertoire of the Adams than with the earlier French-influenced and more monumental classical taste. This is especially true of many of Boulton's works in silver, but some of the ormolu of the later 1770s followed suit.

One of the main reasons for this was the architect James Wyatt. Boulton remained close to the Wyatt family throughout his career. Their building firm constructed the Soho Manufactory, Samuel Wyatt was the architect of the Albion Mill in London in the 1780s, and his brother James provided designs for silverware in the 1770s.

James Wyatt too spent time studying in Italy during his formative years, in Venice and Rome. He returned to England in about 1768 when he was twenty-two, and learned much from

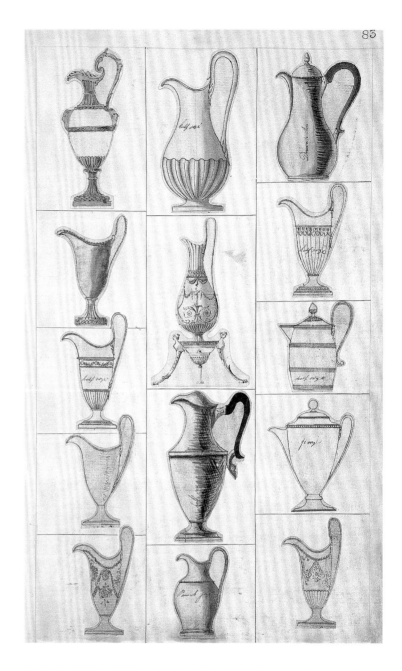

32 Boulton & Fothergill, Pattern Book I, p. 83, showing in the centre a jug on stand to a design by James Wyatt, 1762–90. (Birmingham Archives & Heritage, MS 3782/21/2)

the work of Robert Adam at houses such as Kedleston where his brother Samuel was working. By the mid-1770s James was an established architect, following his precocious success with the Pantheon in Oxford Street in 1772. He went on to design a large number of public and domestic buildings and interiors. Like other architects of his day, including Chambers, Stuart and Adam, his work was eclectic and, despite his criticisms of the work of Adam, he drew to a large extent on Adams' repertoire of ornament, but he tended to simplify the decoration.

Some of James Wyatt's drawings appear in Boulton and Fothergill's pattern books.[54] The design of the silver coffee pot (Cat. 207), which is hallmarked 1769–70, can be attributed to

Wyatt, reflecting as it does his drawing of a tureen and stand in an album of designs in the collection of the Vicomte de Noailles.[55] The supporting sphinxes reappear on an ormolu perfume burner and in the Boulton & Fothergill pattern book.[56] Another Wyatt triangular design, for a jug on a stand, also appears in the pattern book (Fig. 32), copying another Wyatt drawing in the Noailles album:[57] its legs clearly inspired perfume burners that Boulton & Fothergill made in both silver and ormolu.[58] It is clear from the archives that in the middle-1770s Wyatt designed several pieces of silver, including coffee pots, an epergne,[59] a tea urn, a teapot and possibly a tureen.[60] One of Boulton & Fothergill's candlesticks was called a 'Wyatt's pattern silver candlestick'[61] (Cat. 148) and closely mirrors a drawing by the architect.[62]

The architects thus had some influence on Boulton & Fothergill's designs. But none of them had a dominant role. None of them was a systematic supplier of designs. None of them commissioned Boulton to supply metalwork as part of an overall decorative treatment of an interior.

In summary, Boulton did not have a single or even a dominant source of design for his metalwork. He drew from any and many sources. He and his modellers were entirely eclectic. He swam with the times and with the eclecticism of the new taste, accommodating his patrons' desire for fashionable objects. He searched for ornamental designs all over the place. He visited the houses of the aristocracy in order to assimilate their tastes, scoured London and its shops for models and ideas, asked his travelling agent in Italy to observe what taste and fashion prevailed in the various parts of Europe,[63] bought books of engravings, bought from other manufacturers, including Wedgwood, and culled the designs of architects. The possibility of finding a fresh idea, a fresh motif, or a new material was constantly in his mind. He would take a design from any classical source if he thought that it would promote the business, and he encouraged his modellers to combine the same ornamental motifs in new designs.

The new taste remained the fashion for the rest of Boulton's life. The ormolu business did not long outlive the 1770s, but the plate business continued into the nineteenth century, while progressing (or regressing) from the fashion for Adam's and Wyatt's delicate designs to the heavier interpretations of the Regency. The designs for silver and plate in Boulton's pattern books show how his modellers mastered the application of the taste, from the simplest early Corinthian column candlesticks to complex candelabra, from workaday tea urns to elaborate epergnes, from plain dishes and cutlery to delicate dish rings and bread baskets.

Boulton was not a leader of taste. He took a lively entrepreneurial interest in creating the markets for his products, and one of his more impressive talents as a businessman was his determination to relate design to marketing prospects. He understood the importance of keeping abreast with fashion. He said in 1776 that he only wished 'to excell in the execution of that taste which my employers most approve'.[64] That was the key to his enthusiasm for the new taste.

5: Matthew Boulton's Silver and Sheffield Plate

Kenneth Quickenden

Matthew Boulton was a major producer of silver and Sheffield plate, a substitute first developed in Sheffield in 1742 with silver fused and rolled onto one or both sides of copper, which Boulton called 'plated' wares.[1] Although he maintained the use of traditional silversmithing techniques, he was also a pioneer in the application of industrial methods to plate manufacture, and he was personally involved in these developments in the 1760s and 1770s. This led directly to him being mainly responsible for founding the Birmingham Assay Office in 1773. There is no evidence for the production of large pieces of either silver[2] or Sheffield plate in Birmingham before Boulton.

Boulton inherited his father's business in Birmingham in 1759. Production gradually transferred to the Soho Manufactory, about two miles to the north of the town, after he bought, in 1761, the lease on the land and buildings, which he extended considerably over the following years.[3] By the 1750s,[4] silver as well as other metals were being used extensively at Boulton senior's factory,[5] and elsewhere in Birmingham, to make small 'toys' like snuffboxes, filigree items such as thimbles,[6] and chains, hooks, buttons and buckles. Sheffield plate was also in use in Birmingham by this time for small items such as snuffboxes, and this usage increased.[7] A large part of Soho's later activity was in making 'toys' such as boxes in silver or tortoiseshell,[8] chains of silver-plated or gilt metal,[9] and buttons in a wide range of materials such as enamel,[10] as well as silver and silver-plated metal patterns.[11]

The production of such items, though considerable, was insufficient to meet Boulton's ambitions to transform Birmingham's low reputation as a centre of manufacture or to establish a large eminent manufactory.[12] Reinforced financially – and socially – by his marriage in 1760 to his second wife, Ann Robinson (whose fortune was then £6,000 and which increased to £28,000 after her brother's death in 1764) and also by forming a partnership with John Fothergill in 1762, Boulton was not only able to make Soho an unusually impressive Manufactory, which became a major attraction,[13] but he was also able to make more ambitious items. Demand for silver plate had increased after an Act in 1758 withdrew a duty of 6d. per oz.;[14] by the late 1750s Sheffield makers were producing large tablewares in Sheffield plate[15] and Boulton followed these leads.

Boulton claimed in 1763 that he had recently set up a workshop to make silver candlesticks and other tablewares; however, the first recorded order was in 1766, while the earliest surviving silver is from the assay year 1768–9, among which is a pair of swirling, rococo candlesticks with flowers and leaves (see p. 52).[16] There is evidence that Boulton went to Sheffield to learn about the technique in 1764[17] and he made Sheffield plate candlesticks by 1765.[18] Early manufacture seems to have been confined to candlesticks, and designs were taken from the trade: a Boulton 'gadroon' Sheffield plate candlestick (which, somewhat misleadingly, shows the same makers' marks – B&F and two crowns – that the partners used on their silver) follows a pattern used earlier in Sheffield (Fig. 33).[19] Boulton was using a technique, developed in Sheffield, of making candlesticks cheaply by joining thin die-stamped sections, and strengthening each candlestick with an iron rod and fillings of substances such as resin; this took advantage of eighteenth-century improvements in steel for dies and the use of rolling mills. The method was used both for Sheffield Plate and silver candlesticks, and, for the latter, thin rolled sheets greatly reduced the amount of silver used in comparison with the traditional technique of casting.[20]

By the early 1770s Sheffield Plate became one of Soho's 'principle [sic] branches', and included a wide range of tablewares including plates and tureens.[21] But Boulton felt that his ambitions to become 'a great silversmith'[22] were frustrated by the lack of an assay office in Birmingham. While many small lightweight 'toys' such as thimbles, clasps and filigree were not required to be assayed, larger items weighing ten pennyweights or more had to be sent to an assay office to check the quality of silver (which had to be 925 parts per thousand for the sterling standard, to which Boulton worked).[23] In the absence of an assay office in Birmingham, Boulton was obliged to send silver plate elsewhere, and he chose the Chester Assay Office, which added its own mark, the date letter and the lion passant denoting sterling silver to his wares. Nevertheless the inconvenience remained, and perhaps because of this, from the late 1760s Boulton's ambitions to make fine metalwork were largely diverted to the production of ormolu (usually vases, often with stone bodies and with rich, gilded mounts), but when that failed from c. 1771, he diverted his energies to silversmithing and to obtaining an assay office.[24]

Since an Act of Parliament was necessary for the latter, Boulton had to persuade both Houses and also combat the

33 Boulton & Fothergill, Pair of Candlesticks, Sheffield plate, c. 1770. Height 27.9 cm (Birmingham Assay Office 374) (Cat. 179).

34 Boulton & Fothergill, Pair of 'Lyon' Candlesticks, silver, 1768–9. Height 32.7 cm (Collection of The Speed Art Museum, Louisville, Kentucky)

opposition of London's makers, who feared the development of silversmithing elsewhere, and highlighted what they regarded as the limited quantity and quality (they had in mind stamped candlesticks) of silver in Birmingham and Sheffield (which also wanted an assay office); the Londoners doubted the claims of these provincial makers to combine quality with low prices. Boulton countered in a petition that London's prices were 'unsaleable to all but a few rich people' and restricted exports.[25] These were strong arguments to those politicians concerned about the economy and the development of the implied new markets, and the further employment that would ensue, but the arguments were also calculated to allay London makers' fears of losing work by pretending to avoid their traditional market for the rich.

Boulton's low-pricing claims were several times confirmed in practice in the early 1770s, but for influential clients. In 1771 Boulton had secured a prestigious commission from the Admiralty for a tureen requiring hand-raising and elaborate chasing, which was designed by the leading neo-classical architect, James Stuart. Boulton had, however, secured the commission by making it at a loss;[26] later, he stressed a low pricing policy to the Duke of Richmond,[27] and on one occasion offered to make a coffee pot at 3s. 0d. per oz., which was half what he reckoned any silversmith would have charged elsewhere.[28]

Boulton's campaign triumphed in Birmingham, as did Sheffield's, when the new Assay Offices were opened in both towns in 1773 but questions remained. Could low prices be sustained and were they linked to extending the market to the non-rich and exports (as he argued), or, as his courting of the rich implied, were his prices aimed at stealing orders which would otherwise have gone to London's makers? Was Sheffield plate the material Boulton really had in mind for the wider market? Why, given the large demand for cheaper silver items like spoons[29] and forks, did Boulton quietly either reject[30] or factor[31] such orders in the early 1770s, when they were amongst the more reasonably priced items that the non-rich might be able to afford?

Boulton stood a good chance of making profitably at a reasonable price items made with the aid of mechanisation, such as candlesticks. As if to confound Londoners' contempt for stamped candlesticks, he had introduced to his range by 1768–9

35 Boulton & Fothergill, soup tureen with cover, silver, 1776–7, height 25.4 cm. One of a pair made for Elizabeth Montagu. (Collection of The Speed Art Museum, Louisville, Kentucky.)

the 'Lyon' candlestick (Fig. 34). It derived from Court circles in France, the most important design centre in Europe, exhibiting a rich early neo-classicism, using classical details such as the term stem and the Greek key on the nozzle on a high (12¾ in., 32.4 cm) piece. This was an expensive pattern; nevertheless, because Boulton used thinly rolled silver, his candlesticks only required thirty-eight ounces of metal compared to the 108 ounces needed for cast candlesticks as made in London. This meant that his price was £17 2s. 0d. as against the price for cast ones of £44 11s. 0d. per pair (silver then costing 5s. 9d. per ounce). Only occasionally did Boulton produce the heavier cast versions. Even so, only the aristocracy and gentry are known to have bought the lighter-weight pattern in silver, but it was available in Sheffield Plate at £7 17s. 6d., though even here Lady Hertford and Lord Ravensworth were among customers.[32]

However, cheaper candlestick patterns were available: one cost £6 0s. 0d. in silver[33] and £1 7s. 0d. in Sheffield Plate (both per pair),[34] but significantly in 1774 out of a range of six patterns used for Sheffield plate only two had been made in silver.[35] Demand for candlesticks was larger than for any other plate items[36] and was sufficient to employ a specialist group of makers,[37] so that maximum advantage was gained from the investment in dies. Apart from frequent sales of Sheffield plate candlesticks to the middle classes,[38] some were also sold to them in silver.[39]

Services of silver plate provided opportunities for important work, and Boulton made five full services in the 1770s, mainly of silver, all for the aristocracy or, as with Mrs Elizabeth Montagu, the gentry.[40] She was a celebrated bluestocking, with an annual income of £7,000, who bought a service for £1,039 14s. 4d., though she was also credited with £74 6s. 7d. for old silver. She visited Soho and Boulton, and corresponded with him over details before delivery in 1777. For her elegant dinner parties, for guests numbering rarely more than a dozen or so, at first in Hill Street and later Portman Square,[41] in London's increasingly fashionable West End, she purchased a large quantity of plate including a monteith (to fill with ice to cool wine glasses placed in it) and an ice pail, both in unspecified metals, plus two further ice pails (for cooling wine bottles) in Sheffield plate. These were refinements introduced in France in the late seventeenth century, as was Mrs Montagu's use, along with wealthy contemporaries in England, of *service à la française*, whereby individual place settings accompanied the symmetrical arrangements of dishes in the centre of the table so guests could help themselves. Mrs Montagu ordered two soup tureens to provide a choice of soups (Fig. 35); these had linings which were filled in the kitchen and carried to the tureens on the dining table, to avoid damage to the tureens, along with ladles and twelve soup plates. The first course also included fish, meat and salads. For the first and second courses eight silver salts, seven dozen silver plates and more than nineteen dishes, at least twelve of which were in silver, together with at least two covers in silver and four more in Sheffield plate, were provided (Cat. 143). The use of Sheffield plate covers appealed since they cost one eighth of those made in silver, and, since they

36 Design for an Aimé Argand lamp from the Boulton pattern books. These were made in both sterling silver and Sheffield plate. (Birmingham Archives & Heritage)

chased band has been applied to the top edge of the bowl.[44] The second course normally included cooked puddings, which would also have required the use of plates and dishes. A dessert course was also usual, for which glass and ceramic plates and dishes were the norm. The meal concluded with coffee or tea in a drawing room, where Mrs Montagu may well have used the Sheffield plate tea urn she had bought from Boulton for £9 9s. 0d. in 1773.[45]

For mainly hand-made items, often involving considerable discussion about design and prices, and using the same traditional methods as in London, profitability was in doubt, especially as Boulton recruited some silversmiths from the capital who, at least in one case, received the same wage at Soho. Having secured the Birmingham Assay Office, Boulton pushed up charges, but still maintained that his prices were lower than those of London makers. In Mrs Montagu's case he charged 1s. 2d. per ounce for making plates as opposed to the 1s. 6d. he reckoned a London firm would have charged, and he made the plates lighter at sixteen ounces rather than twenty to twenty-two ounces.[46] The principal basis on which Boulton thought he could compete was that London prices were inflated by retailers' mark-ups,[47] whereas he generally supplied the public direct. In 1776 he priced a tea urn, similar in weight and simplicity to one supplied by the London makers Ansill & Gilbert to the London retailer Parker & Wakelin; Boulton charged £17 or 4s. 0d. per ounce for fashioning,[48] whereas Ansill & Gilbert charged Parker & Wakelin £14 2s. 0d., or 3s. 0d. per ounce,[49] but the retailer added a mark-up of £8 3s. 10d.[50] There was thus a slim basis for Boulton to make the limited profits he was content with on each article (because Soho made so much).[51]

But profit was also dependent on prompt payment. In 1773 losses on all of Soho's trading led to a deficit of £10,000. To try to avoid additional interest charges through buying bullion, Boulton negotiated a three-month period of credit with his dealer in 1772 and resolved to make customers pay promptly. However, too many failed to do so. A cup and cover, sent to the Irish MP Cornelius O'Callaghan with some other silver in 1777 was not paid for, despite reminders, until early in 1779, and then only £100 was paid out of the original charge of £123 4s. 5d. This was not unusual and an enquiry into the silver-plate business concluded that it was unprofitable because the interest on bullion payments was greater than the fashioning charge.[52]

Boulton's most positive attempt to sell silver plate to the London trade came before 1773, when he once invited eight London silversmiths to see a show of silver and Sheffield plate in 1772,[53] but sales came primarily from the latter.[54] This was largely due to Boulton's own actions. He refused trade discounts on silver, while allowing at different times fifteen or twenty per cent on Sheffield plate, with a further discount for prompt payment. Moreover, after the passing of the Assay Office Act, he was admitting that his prices for common silver items were no cheaper than London's and insisting that the firm wanted to concentrate on ornamental plate, so silver orders from trade

were quickly removed by footmen, there was little chance of a silver cover being appreciated.

The second course was lighter and more subtle, with game birds, vegetables and ragouts (French-inspired stews combining many ingredients). The silver asparagus tongs (a recent innovation for this most favoured of vegetables), six silver sauce ladles and eight silver sauce boats (see p. 54), an integral part of *service à la française*, were probably mainly used for this course.[42] The design of the sauceboats (Cat. 143) was new and probably based on designs by James Wyatt, whose light and elegant neo-classicism[43] was the dominant influence on Francis Eginton, Soho's in-house designer (and partner in plate manufacture) in the mid 1770s. Although the ribbon and reed border was probably die stamped (the pattern was much used at Soho), the piece was predominantly made with traditional techniques: the base, bowl and cover were all hand-raised, while the repoussé festoons of husks in the fluting were shaped with specially shaped punches, and a pierced and customers for plain spoons, tankards and mugs were deflected, yet it was often items of these kinds that trade customers and the wider public wished to buy.[55]

Very little Soho silver was exported. To support his apparent determination in 1773 to export silver plate, Boulton notified at least one merchant[56] to that effect, but thereafter his enthusiasm waned. An agent in Switzerland, in contact with Boulton since 1775, expressed surprise in 1781 to have discovered from travellers that Soho had been making silver plate for many years.[57] Boulton's apparent reluctance may have been based on his fears of not receiving payment,[58] especially from abroad on expensive items, but demand was also limited since the intrinsic standard of silver plate on the Continent was lower than in England, though Boulton was not clear about that until the mid-1780s.[59] Sheffield plate, but not silver plate, was regularly sent to countries like Holland[60] and France.[61] Most silver plate was exported through direct contact with the nobility: the Duke of Holstein-Gottorp visited Soho and bought a service.[62]

In all, despite Boulton's claims in 1773, of the known sales of silver plate during the Fothergill partnership only ten per cent (by value) was distributed by the trade; who then would have bought it is unclear, as is the case with about four per cent of sales by the partners direct to the public. Of the rest, most was bought by the rich (£6,612 or forty-five per cent to the aristocracy, £4,299 or twenty-nine per cent to the gentry), with only twelve per cent going to the middle classes and less than £20 worth going to the lower classes. (These figures exclude sales where the value of the sale is unclear.) While these figures probably give an exaggerated impression of the proportion of silver bought by the rich (because more documentation survives for them), they undermine Boulton's implication in 1773 that he would be substantially interested in the non-elite market. Could Boulton ever really have supposed that even at his prices, the middle classes, earning no more than £600 per annum, could have afforded much silver plate?[63]

Boulton's commitment to silversmithing declined with the start of the steam-engine partnership with James Watt from 1775, and with the need to overcome Soho's desperate financial position (the deficit on the Bill Account was £25,000 by 1777), so from 1777 the unprofitable silver business was run down and Fothergill was adamant that the business must concentrate on Sheffield Plate. The production of assay silver, which reached a high of 11,831 ounces in the assay year (July to July) 1776–7, slumped to 263 ounces in 1782–3.[64] The proportion of Sheffield plate to silver sales in the mid-1770s is not known but in 1780 (if plated buttons that totalled 548,888 and silver buttons totalling 5,652 are excluded) 11,234 items in silver and Sheffield plate were sold (including 'toys' and buckles), ninety-one per cent of which were in Sheffield Plate.[65]

Following Fothergill's death in 1782, the Matthew Boulton Plate Company was formed, with a separate company for buttons, but Boulton was personally little involved with either.

37 Matthew Boulton Plate Co. Bread basket, silver, 1788–9. Height 29.4 cm, width 32.5 cm (Birmingham Assay Office 569) (Cat. 154).

He concentrated on the steam-engine business, and from the mid-1780s devoted much attention to the Mint. The key figures in the Plate Company became the manager, John Hodges from 1783, and a new London agent, Richard Chippindall, from 1784.[66] Sometimes Boulton was obliged to settle disputes between them; in 1799, Boulton, against Hodges's advice, approved Chippindall's request for a big stock increase. Generally Boulton was slow to invest in the Plate Company: a new rolling mill to improve efficiency in making sheets of plated metal, thought essential in 1782, was not completed until 1786, and while Chippindall wanted more fully-trained staff to speed up production, Boulton preferred the cheaper alternative of training up apprentices.[67]

While determined to push silver filigree items like tea measures[68] and small articles like latchets (which pinched the two sides of shoes together, for which Boulton formed another partnership in 1796)[69] that were made in both silver and plated metal,[70] Hodges told customers they were declining larger silver plate in favour of Sheffield plate.[71] However, there was flexibility in implementing this policy. Capacity was found to make for Catherine the Great[72] an improved oil-lamp (designed by Aimé Argand and patented in 1784)[73] in silver; this was also available in Sheffield plate (Fig. 36; Cats 193–4). Further silver plate was also made for the Duke of Holstein-Gottorp, more 'for honour than profit',[74] while orders for silver spoons for trade customers were factored.[75] Chippindall was told to search for Sheffield plate orders and that silver plate orders would only be acceptable if there was a lull in making Sheffield plate.[76] Silver plate prices were pushed up at Soho.[77] Demand was reduced nationally as economic depression followed war with America from 1775,[78] and this was accentuated by the reintroduction of a duty of 6d. per

38 Matthew Boulton Plate Company, Cup and cover, silver gilt, 1803–4, height 42 cm, width including handles 31 cm. (Birmingham Assay Office 559.)

ounce in 1784.[79] At the same time makers of Sheffield plate, who from 1773 had not been allowed to put marks on their wares (to avoid confusion with hallmarks on silver), were permitted to register marks at the Sheffield Assay Office, as long as they were clearly different from marks on silver: thus the date letter, the Birmingham Assay office mark (an anchor), the sterling silver mark (a lion), and Boulton's initials used on his silver, were replaced on his Sheffield plate by two suns or 'BOULTON' and one sun.[80]

In parallel with more ambitious firms in Sheffield, whose prices Hodges kept in step with, the emphasis was on raising the standard of Sheffield plate, to make it look more like silver. There were frequent references to the use of 'silver mouldings' and 'extra plated metal' to overcome the effects of surface wear which revealed the underlying copper. In 1798 the Plate Company was happy to quote prices for ecclesiastical wares in Sheffield plate, whereas in the 1770s Boulton had felt only silver was appropriate for such work. Chippindall obtained very large orders from London firms: in 1798 Rundell & Bridge purchased £247 1s. 0d. of Sheffield Plate, but only £8 0s. 9d. of silver.[81] A large proportion of silver orders were for the firm's reasonably priced candlesticks.[82] There were many Sheffield plate orders for a wide range of wares for middle-class customers: in 1800 a Dr H. Edgar bought three pairs of candlesticks and one pair of branches, a waiter, an epergne (an elaborate dining table centerpiece for sweetmeats), a tray, a teapot, a toast rack, two muffineers (sugar shakers), three pairs of salts, a cream jug, and three pairs of bottle stands, which with the glass dishes for the epergne and salts came to £37 11s. 6d.[83] There was also a market for Sheffield plate 'toys' such as wine labels,[84] and exports were sent as far as the West Indies.[85]

Sales were achieved despite what Chippindall regarded as a lack of commitment to design at Soho. This view was partly due to the slowness in introducing new designs and because the firm was content to follow[86] trends elsewhere towards greater lightness (and competitiveness) where wire work became prominent,[87] as at Soho (Fig. 37). There was less ambition, no use of important designers and less artistic skill at Soho than there had been in the 1770s. Chippindall, to prove his criticisms, took a cup and cover, made at Soho in 1803 (Fig. 38), to prominent firms in London for their opinion. Though the design shows an awareness of the fashionable Regency tendency to contrast areas of simplicity with confined areas of heavy decoration, the verdict was not entirely encouraging: the base needed to be larger, the leaf under the body should have come up higher and the grapes were too small for the strong, beautifully chased stems.[88]

In the few years before Boulton's death in 1809, despite the Napoleonic Wars (1796–1815), demand for Sheffield plate and silver was strong.[89] Boulton's prosperity, primarily from his other enterprises including buttons,[90] meant that silver-plate manufacture no longer suffered from any financial weakness elsewhere at Soho. Production of assay silver recovered to 10,016 troy ounces in 1805–6 largely because of Chippindall's success with trade customers in London. The profits for the Plate Company increased from £1,197 in 1787 to £3,564 in 1805.[91]

In the 1770s silver plate and Boulton's successful efforts to establish the Birmingham Assay Office, had gained him prestige and valuable connections, which was what he intended. 'Toys', buckles and buttons, in silver, plated metal and other metals, continued to provide a steady income and reached a wide market, as did Sheffield plate, the appeal of which spread across the board from the middle to the upper classes. Together with Roberts, Cadman & Co. of Sheffield, Boulton became the most respected maker of Sheffield plate.[92] Though there were larger makers of silver plate in London[93] Boulton was the largest in Birmingham[94] and when staff left Soho to go elsewhere they contributed to the development of the industry in Birmingham.[95] Boulton's son, Matthew Robinson, sold the Plate Company in 1833.[96]

6: 'I Am Very Desirous of Becoming a Great Silversmith': Matthew Boulton and The Birmingham Assay Office[1]

Sally Baggott

Of all the enterprises in which Matthew Boulton was involved, the Birmingham Assay Office is remarkable as the only one that is still in existence. A permanent reminder of Boulton's legacy in Birmingham and beyond, the Office still carries out the statutory duty of assaying and hallmarking precious metals with which it was charged over 230 years ago. The Office was established in 1773 by an Act of Parliament entitled, 'An Act for appointing Wardens and Assay Masters for assaying Wrought Plate in the Towns of Sheffield and Birmingham,' and the success of the bill was in the most part due to Boulton's efforts.[2] A logical progression from the manufacture of Sheffield Plate, silver production had begun at the Soho Manufactory in 1766 under the aegis of Boulton's partnership with John Fothergill, yet from the outset they faced one major obstacle to the silver business.[3] Since 1300, for precious metal articles to be saleable, the law has required that they are assayed or tested to ascertain their precious metal content, and hallmarked in order to guarantee that the metal is of a legally required standard. Along with every other manufacturer working in silver in Birmingham, Boulton & Fothergill were obliged to send their goods for assay and hallmarking to an Assay Office, the nearest being Chester, some seventy-two miles from Birmingham. Not prepared to tolerate the increase in costs, the risk of damage or robbery, the problems caused by delays or the danger of designs being copied, as early as 1766 Boulton had begun to consider the possibility of obtaining an Assay Office in Birmingham.

The story of how Boulton overcame this disadvantage and was successful in establishing an Assay Office in Birmingham has already been told from a range of perspectives. Written during his term as Assay Master at Birmingham, Arthur Westwood's account from 1936 remains the most comprehensive to date; although each account draws on material in the Matthew Boulton Papers, and each gives a clear sense of Boulton as an ambitious, intelligent and articulate man, skilful at argument, and competent in making use of his friends in high places.[4] However, the documents shed light on the wider cultural context of the period; the story of Boulton and the Birmingham Assay Office is also one of tension between the metropolis and the provinces, between craftsmen and manufacturers, between the old and the new.

Without Boulton, it is likely that the Birmingham Assay Office would not have been established; his campaign was certainly more successful as a result of the support he solicited from the large network of influential people with whom he was acquainted. In 1766, the Earl of Shelburne and his wife were entertained on a visit to Birmingham by the industrialist and ironmaster Samuel Garbett. This was their second visit to Soho, and afterwards the Earl of Shelburne contributed an account of Birmingham to his wife's diary. Celebrating the innovative developments that manufacturers in metal were making in Birmingham, he continued:

> Another thing they are in great way of is an assay-master, which is allowed at Chester and York; but it is very hard on a manufacturer to be obliged to send every piece of plate to Chester to be marked, without which no one will purchase it, … It would be of infinite public advantage if silver plate came to be manufactured here, as watches lately are, and that it should be taken out of the imposing monopoly of London.[5]

Kenneth Quickenden has noted that these remarks 'were almost certainly prompted by Boulton'. Moreover, in the light of Garbett, Boulton's ally in the 1773 campaign, having acted as host to the earl, and the correspondence between Boulton and the earl during this time, these are surely Boulton's words.[6] Furthermore, the implication of disadvantage at Birmingham contrasted with the 'imposing monopoly of London' is a point on which Boulton would later rest his argument for an Assay Office at Birmingham.

In a letter some five years later in 1771, Boulton apologised to the Earl of Shelburne at length for the delay in sending an order for two pairs of candlesticks which the Earl had commissioned. According to Boulton, the candlesticks had been returned to Soho damaged due to 'the wilful or careless bad packing of them at Chester,' and he wrote:

> I am so exceedingly vex'd about the disappointment and loss which have attended the two pairs of Candlesticks that altho' I am very desirous of becoming a great Silversmith, yet I am now determined never to take up that branch in the Large Way I intended, unless powers can be obtained to have a Marking Hall at Birmingham.[7]

Quickenden has argued that Boulton would often exaggerate the delays caused by the necessity of sending work to Chester for assay and hallmarking in order to conceal the time involved in the production of silver at Soho. The fact that there is not one letter in the Matthew Boulton Papers from Boulton to his agent in Chester, James Folliot, in which any instance of damage is mentioned, indicates that this was Boulton both effectively explaining away any delays resulting from his own operation, whilst also stressing to his influential acquaintance the need for an Assay Office at Birmingham.[8] It was surely not a coincidence that the candlesticks supposedly damaged at Chester were a commission for the Earl of Shelburne whom Boulton had pressed some five years before in 1766 on the matter of an Assay Office.

Boulton's original plan had been to obtain an Act of Parliament to allow for an Assay Office in any city or town where the quantity of silver manufactured warranted it, and to close any existing offices which were unjustified by the same principle. However, by January 1773, lobbying for the lesser regional cause was taking up most of his time in London.[9] Following exposure of his campaign in the press, the Clerk to the Sheffield Cutlers' Company, Gilbert Dixon, had contacted Boulton, and on 1 February the Sheffield Cutlers presented their petition for an Assay Office in Sheffield; the Birmingham petition was presented the following day. The petition from Birmingham stressed the inconvenience, expense, hazard and delay of sending silver to Chester, along with the constriction of the Birmingham trade in terms of expansion and improvement, and the lack of access to domestic and foreign markets caused by the absence of an Assay Office in Birmingham.[10] Boulton also circulated a document, prepared and signed by him, among Members of Parliament.

He gave it the title of *Memorial relative to Assaying and marking wrought plate at Birmingham*, and he stated economic, historical and legal reasons, confidently citing the legislation, to make a case for an Assay Office in Birmingham.[11] He reiterated the disadvantages suffered by the Birmingham trade, adding that 'their fresh Designs, which have often cost them considerable Sums of Money, and always Pains and Time, are communicated to Rivals before the Inventors have reaped Benefit from them.'[12] Further, he stated that an Assay Office in Birmingham and the resulting expansion and improvement in the trade would not only benefit the ever-increasing number of precious-metal workers in Birmingham financially, it would contribute to the health of the national economy. He presented the historical lack of 'incorporations' or craft guilds in Birmingham as a positive factor; 'the Inhabitants are too sensible of the Disadvantages of such Societies to wish for any,' he said, and he praised the quality of Birmingham production.[13] Taking a swipe at the craft tradition, here he was implicitly celebrating the emerging industrial character of Birmingham. The law made Assay Offices necessary, he said, and it had provided for towns with precious-metal trades to have the power to assay, but, he pointed out, it did not seek to dictate where gold and silversmiths might live and work.[14] Through the metropolitan press, the London goldsmiths had given Boulton more than a hint of their objections to the Bill.[15] In his biography of Boulton from 1937, H. W. Dickinson describes the *Memorial* as 'masterly…logical…thoroughly Boultonian', and when Boulton reaches his final point, he skilfully undermines the objections London goldsmiths would make in their petitions.[16] He ends:

> Objections may possibly be made by the Corporations of goldsmiths in London and in other Marking Towns to such a Grant; which, it may be said, may prove injurious to them, because it may enable others to work in Plate with as much Convenience as they now do. But as Birmingham is not near to any Market for Plate, it can deprive the other Towns of no Part of their Trade, except by working better than they do and cheaper; and against Losses of Business by these Means the proper Securities are not Privileges, but Excellence in Design and Workmanship, and moderate Prices.[17]

On 17 February, the Goldsmiths' Company in London presented their petition to the House of Commons. Exploiting their historical standing as the Guardians of the standard of wrought plate and the standard of the national coinage, and the fundamental relationship between the metropolis and the nation, they asserted that the duty they had carried out for 'time out of mind' was '…a very great and important trust, and ought to be committed to such persons, and in such Places only, where the same is likely to be executed and discharged with the greatest of care and fidelity.'[18] They attested that as Birmingham and Sheffield had few goldsmiths or silversmiths, no incorporations, nor any companies of goldsmiths, the establishment of Assay Offices there would result in 'various frauds and deceipts in the manufacturing of gold and silver wares…in the said towns.'[19] More seriously still, the potential for the debasement of the standard, they stated, would lead to '…the injury of his Majesty's subjects, the Detriment of the Fair Trader and the Diminution of the Wealth, the Credit and the Commerce of this Kingdom.'[20] A further petition was presented the following day by 'the Goldsmiths, Silversmiths and Plateworkers, of the city of London, and places adjacent.'[21] Here it was argued that the establishment of Assay Offices in Birmingham and Sheffield would bring the trade into 'an irretrievable state of Discredit and Suspicion', and it would '…impair the trade of the metropolis, diminish his Majesty's Revenue, and sensibly affect the honour and riches of the kingdom in general.'[22] Birmingham and Sheffield plate workers, they argued, were few in number, confined to the 'least important branches of the said manufactory,' and insufficiently conversant with or skilled in the trade.[23] In London, though, the trade was carried on '…in all its variety of useful and curious branches, in much greater perfection than in any other place in these Kingdoms, and to an extent sufficient to supply the greatest demands.'[24]

On 7 February, before the Parliamentary Committee, with Thomas Skipwith, M.P. for Warwickshire, in the chair, Samuel Garbett, the only Birmingham representative to be questioned by the Committee, made much of the disadvantages under which the Birmingham trade laboured.[25] The day went well for Boulton and for Birmingham, with Garbett making many sound suggestions – he would have undoubtedly discussed these with Boulton beforehand – as to how the business of the Assay Offices could be improved, such as, for example, the introduction of systems of locked boxes to prevent designs being copied.[26] However, accusations regarding irregularities in the manufacture and assay of silver were made during these sittings of the Committee, and a Special Parliamentary Committee was appointed on 26 February whose remit was 'to enquire into the Manner of conducting the several Assay Offices in London, York, Exeter, Bristol, Chester, Norwich and Newcastle-upon Tyne.'[27]

In a further petition presented to Parliament by the London Goldsmiths, Silversmiths and Plateworkers, on 8 March, the accusation was levelled at Birmingham silversmiths of making silver wares which contained more iron than was legally allowed, and of impressing marks upon their plated wares to make their work 'appear like real plate marked at an Assay Office.'[28] The charge that Birmingham silver contained too high a content of iron was found to relate to some candle-snuffers manufactured by Benjamin May and sold to London retailers, and this was the sole instance of this charge that was proven.[29] John Scasebrick, the Assay Master at Chester, when asked 'Did Messrs Boulton and Fothergill whose name [*sic*] are entered in your book ever send you any Plate to be assayed that was above or under standard?', answered 'It has generally been 2 or 3 dwts [pennyweights] above standard.'[30] It was also during the investigation by the Special Committee that the distinction between the silver produced in London and Boulton's silver, and the fundamental reason for London's concern with the competition from Birmingham and Sheffield, came to light. There was a great deal of questioning as to whether the silversmiths in the two towns were manufacturing 'heavy' or 'slight' candlesticks, to which most London witnesses, some of whom had worked for Boulton at Soho, replied 'slight'.[31] The difference referred to is that between candlesticks manufactured by casting, an older but much more expensive technique on account of the amount of metal used, or by die-stamping, a newer method developed and used by Boulton at Soho that was significantly cheaper when used to produce articles in quantity. Inigo Wakelin, a London silversmith of Haymarket told the Committee that 'Those of 30 ounces cast would cost more than those stamped on account Workmanship,' whilst stamped candlesticks 'would be less useful on account of their thinness.'[32]

The Goldsmiths' Company of London presented a further petition to the House of Commons on 6 May, and the Goldsmiths,

39 Boulton & Fothergill's entry in the Birmingham Assay Office first sponsor's mark register, 1773. (Birmingham Assay Office)

Silversmiths and Plateworkers of London presented another the day after, each one repeating their attacks on the Birmingham trade.[33] Having set himself up as the defender and mouthpiece of the trade in his home town, Boulton was never going to take these insults lightly, and, in answer, he circulated his *Reply of the Petitioners from Birmingham and Sheffield to the Case of the Goldsmiths, Silversmiths, and Plateworkers of London and Places Adjacent*, and another document entitled *Observations Relative to The Standard of Wrought Plate*.[34] In the first, he made it plain that the London goldsmiths and silversmiths had enjoyed unfair advantages for too long, and that

> …were they *really* conscious, as they pretend to be, that their productions were more elegant, more useful, and of more intrinsic value than those of all others … this Superiority of their Wares, must infallibly secure to them, better than *all the Laws on Earth, the whole Trade in Plate*.[35]

40 Boulton & Fothergill's first entry in the Birmingham Assay Office first Plate Register (Cat. 206), showing a large consignment of wares brought for assay on 31 August 1773, the first day of business at the Assay Office. (Birmingham Assay Office)

The emphasis is Boulton's, and he stated that the facts proved otherwise:

> The Plate made in and near London is not of *uncommonly intrinsic* Value, the Workmanship is not so *masterly*, the Designs are not so *elegant* or *convenient*, nor is the Price *so reasonable* as to admit of its rivalling *foreign Plate*, and becoming a *considerable* Article of Commerce with the *neighbouring* Nations.[36]

Shrewdly, he argued that the Birmingham and Sheffield petitioners were mindful of the responsibility of overseeing the standard of wrought plate, and that they would seek to avoid even the slightest of suspicion by appointing officers that were above reproach.[37] Competition not monopoly, he said, would guard the trade against fraud and bring prices down, thus driving demand up and increasing employment in the trade, and with Assay Offices in Birmingham and Sheffield more plate could be made which would, in turn, increase revenue.[38] Furthermore, if the manufacturers of plate in Birmingham and Sheffield were only few, whether their work was good or bad, they could not possibly present a threat to London manufacturers.[39]

Boulton saved his masterstroke, though, for the *Observations Relative to the Standard of Wrought Plate*. Here, drawing on evidence given before the Special Committee, he takes the London Goldsmiths' Company to task for passing sub-standard silver and giving it their hallmark: 'The Company however, daily authenticate by their Standard Mark Silver Plate worse by Two Pennyweight in the Pound than the legal Standards.'[40] He had calculated that in doing so, the London Goldsmiths' Company were defrauding those who bought plate marked at London by five thousand pounds a year, and that the 'Londoners' consequently had an advantage over silversmiths in the rest of the country of one per cent.[41] He concluded that the lack of checks on those involved with assaying at London allowed their fraudulent practices to go undetected.[42] Boulton appears to have wholly deserved the compliment paid to him in a letter from Joseph Wilkinson, written earlier in the campaign. On 11 March, Wilkinson had written to Boulton '... [I] am happy to find your Manoeuvres in Turning their own Batterys upon themselves – May that ever be the case where such mean and dishonourable Arts are adopted!'[43] It was left to Garbett to communicate the imminent victory over London manufacturers through the organ of the local press. In a passage repeated from a letter he wrote to Boulton only two days earlier, published in *Aris's Birmingham Gazette* on 10 May, Garbett's tone is unmistakably triumphant when he writes of:

…the Company which it is hoped will be established in the Town, to Authorise a Mark to be stamp'd on our Silver Wares, by which we may obtain the Honour thro' Europe of making Wrought Silver of better Standard (as well as of better Workmanship) than is generally marked at Goldsmiths' Hall in London.[44]

Indeed, the Special Committee spent days questioning witnesses, who included many senior officials, such as the prime warden and the deputy warden to the Goldsmiths' Company, on how assaying was actually carried out at the Assay Offices.[45] The result of their investigations surely had a negative impact on the London case. Shortly before they reported their findings on 13 May, the Committee had sent representatives incognito to purchase silver that bore the London hallmark; these items were subsequently independently assayed, and of twenty-two articles, twenty-one were found to be substantially below standard.[46] If that was not enough to damage the reputation of the London contingent, during the investigations it was implied that certain ex-employees of Boulton's at Soho, now employed in London, had been offered bribes to blacken Boulton's character and prevent the Bill being passed. On 8 March, Thomas Cliff, an ex-apprentice of Boulton's, gave evidence that he had been offered five guineas if he would state before the Committee that Boulton had paid his workers at Soho in counterfeit money.[47] It was alleged that Robert Boyd and William Bickley had offered a similar bribe to Joseph Kettle, another ex-employee of Boulton's.[48] Boyd and Bickley appeared before the Committee on 10 March, and denied the allegations which were never proved, but the very suspicion that they had sought to fill 'Counsel's mouths to be a blot on Mr. Boulton's character' cannot have furthered the case of the London petitioners.[49]

The Bill successfully passed through the Commons on 18 May 1773, and received Royal Assent in the Lords on 28 May. Not only had Boulton been successful in gaining an Assay Office for Birmingham, but the Committee had made amendments to include increased checks and regulations in order to reduce abuses of the system and fraud.[50] Having been the first to register their sponsor or maker's mark at Birmingham on 31 August 1773, the day the Assay Office opened, Boulton & Fothergill were its first customers. On that same day, they submitted some 104 articles for assay and hallmarking, in addition to tea-vase furniture and thirty pairs of buckle rims (Figs. 39 and 40).[51] The Birmingham Assay Office began life in two rooms rented from the landlord above the King's Head Inn on New Street. The inn sign still exists, and as a reminder of its humble beginnings, a replica hangs on the staircase of the Birmingham Assay Office's current premises (Fig. 41). Boulton & Fothergill paid towards the cost of obtaining the Act and of establishing the Office, for which they were later reimbursed.[52] During Boulton's campaign in London, business had been conducted in the Crown and Anchor Tavern in the Strand. When

41 The sign of the King's Head, the inn in New Street above which the first Birmingham Assay Office opened in 1773. (Cat. 205) (Birmingham Assay Office)

the Act was obtained, it seems reasonable to conjecture that the Birmingham and Sheffield contingents chose to adopt the sign of the tavern for their respective marks. How the two marks were decided upon is unknown; the suggestion that the matter was decided by the toss of a coin seems to have originated with Arthur Westwood, but it was written into the Act that Sheffield's mark would be the crown and Birmingham's, the anchor. (After 1903, when the Sheffield Assay Office was first allowed to assay and hallmark gold, it adopted the Tudor rose as its mark on gold wares to avoid confusion with the crown symbol for gold, and since 1975 has used the Tudor rose as its silver mark, too.) Birmingham still uses the anchor, a mark that for over 230 years has signified integrity, independence and a commitment to consumer protection. Boulton had achieved his objective, and the number of articles submitted in his first consignment suggests the partners had waited to get that all-important anchor mark, for which Boulton had worked so hard, on their silver.

42 Boulton & Fothergill, pair of candlesticks in the rococo style, silver with Sheffield Plate nozzles, hallmarked at Chester, 1768–9, height 25.5 cm. (Cat. 145) [BAO 1140]

43 Boulton & Fothergill, finely pierced mazerine or strainer dish, silver, hallmarked at Chester Assay Office, 1769–70, diameter 30.5 cm [BAO 23], and below: detail of the centre.

In June 1782 the Boulton & Fothergill partnership was formally dissolved after Fothergill's sudden death, but Boulton continued manufacturing silver under his own name.[53] Instead of the mark 'MB IF' which the former partners had used, Boulton used the mark of 'MB' alone.[54] Following Boulton's death in 1809, there was no immediate change to the mark put on silver, made now by his son Matthew Robinson Boulton at Soho, but in 1820, on re-registration of the mark, the company's name is given as 'Boulton & Plate Co., Soho'.[55]

The Birmingham Assay Office Silver Collection holds in total sixty objects marked for Boulton & Fothergill, a further sixty marked for Matthew Boulton, and twenty for the Matthew Boulton & Plate Co. Arthur Westwood began the Silver Collection and the Library while he was junior Assay Master and his father, Henry Westwood, was Assay Master at Birmingham. The Birmingham Assay Office Act of 1902 officially sanctioned the Collections, and the Office continues to acquire objects relating to Boulton.[56] Significant regionally, nationally and internationally, Boulton silver features in the collections at Birmingham Museums and Art Gallery, the Victoria and Albert Museum, Windsor Castle and at the Speed Art Museum in Louisville, Kentucky, U.S.A. The Birmingham Assay Office Silver Collection holds 140 items in metal produced at Soho, the majority of which are in silver, with a few in other metals, such as steel and Sheffield

Plate. In addition, the original registers where the mark of Boulton & Fothergill was registered on 31 August 1773, and the first consignment of wares received from them for assay and hall-marking recorded on the same day, are held in the Archive.[57]

Examples of Boulton & Fothergill silver with the mark of Chester Assay Office, predating the establishment of the Birmingham Assay Office, are particularly rare, but the Birmingham Assay Office has two pieces that are worthy of note. The pair of candlesticks (Fig. 42) date from 1768, and are silver with Sheffield Plate nozzles. The design of the candlesticks is unreservedly rococo, and they stand in stark contrast to Boulton's later neo-classical style. These candlesticks bear a remarkably close resemblance to a pair in the Victoria and Albert Museum collection that were made by Francis Butty and Nicholas Dumée.[58] Shirley Bury, in an article from 1968, considers two possibilities as to how the similarity arose; the first being that Boulton and the Butty and Dumée partnership used the same source books or variants thereof, or that Boulton had copied another manufacturer's work![59] The second is that Dumée was employed by Boulton during this period, and Bury cites evidence of a letter in the Matthew Boulton Papers from Dumée to Boulton, dated 15 March 1773. Addressed to Boulton in London whilst he was involved in the Birmingham Assay Office campaign, the letter expressed Dumée's intention to resign from Soho as he had been prevented from chasing some coffee pots in Boulton's absence.[60] Another Chester-marked item is the mazarine, 1769–70 (Fig. 43). A type of straining dish, the mazarine is a rare circular example, and is finely decorated with machine-piercing. It is interesting for the set of marks that it bears. In addition to Boulton & Fothergill's mark that was registered at Chester, 'B&F' with a crown either side, it has a third crown mark also struck above the maker's mark.[61] The arms of Smith of Theddlethorpe, Lincolnshire, are hand-engraved in the centre with the motto VIDEO MELIORA PROBOQUE, with which Boulton must surely have concurred: 'I see and approve better things.'

A set of four candlesticks (Cat. 148) have a strong documentary provenance. Marked for Birmingham, 1774–5, they are almost entirely based on a design by James Wyatt, the architect whose own work was heavily influenced by Robert Adam.[62] Candlesticks in this style became a popular line for Soho, but the partnership of Wyatt and Soho was not an altogether happy one.[63] Boulton & Fothergill often had to wait for Wyatt's designs, and the delay sometimes angered customers.[64] Boulton used Wyatt patterns more than those of any other designer, but when Sir Harbord Harbord complained at the late delivery of a coffee pot he had commissioned from Soho, Boulton showed no hesitation in blaming Wyatt.[65] In typically Boultonian style, the partners wrote back that if they had been responsible for the design themselves, the commission would have been completed earlier![66] The jug which dates from 1776–7 (Fig. 44), also from a design by Wyatt, is typical of the simple, elegant neo-classicism

44 Boulton & Fothergill wine jug, silver, hallmarked at Birmingham Assay Office, 1776–7. Height 30.5 cm [BAO 915]

which Boulton became so skilled at producing in silver.[67] Quickenden has noted that although the jug bears some resemblance to ancient examples of 'ewers', the partners were more than likely relying on Wyatt's own interpretation of antiquity.[68] At Fig. 45, a pair of sauce tureens were made by Boulton for Mrs Elizabeth Montagu.[69] They are marked for 1773–4, the year the Birmingham Assay Office opened, and they are a striking example of the older craft techniques of silversmithing that existed at Soho.[70] Boulton's association with Mrs Montagu, a highly connected bluestocking, is well documented in their correspondence. Boulton also made a dinner service for Mrs Montagu a few years later, plates from which are now owned

45 Boulton & Fothergill, pair of sauce tureens made for Elizabeth Montagu, silver bearing the mark of Birmingham Assay Office, 1773–4. Height 13.7 cm, width 12.3 cm (Birmingham Assay Office 335) (Cat. 143b)

jointly by Birmingham Museums and Art Gallery and the Birmingham Assay Office.[71]

Boulton's most significant legacy, however, would seem not to be the silver that was produced at Soho but his more far-reaching contribution to the Birmingham trade in establishing the Assay Office. From the toymakers who were Boulton's contemporaries, through the great manufacturing companies of the nineteenth century, such as Elkingtons, to the arts and crafts firms of the early twentieth century, such as A. Edward Jones, it is arguable that Birmingham silver manufacturers would not have become so successful, so prolific or so well-known on the international stage had it not been for Boulton. In June 2008, the Birmingham Assay Office hosted an event, 'A Celebration of Birmingham Silver' as part of British Silver Week. Here, the work of contemporary silversmiths, still working in Birmingham's Jewellery Quarter, was on display alongside some of the best pieces ever to come out of the Soho Manufactory. It was a testimony to 235 years of silver made, assayed and hallmarked in Birmingham. The Birmingham Assay Office is now the busiest Assay Office in the world, and its expertise is exported around the world, to assist in the establishment of Assay Offices from Egypt to Uzbekhistan. Although it is thus primarily a working organisation, the Birmingham Assay Office welcomes on average some seventy visitors a week who come (by appointment) to see the silver collection and the library.

The story of Boulton and the Birmingham Assay Office, then, displays the conflict that existed in the late eighteenth century between London and the provinces. Despite the London contingent drawing heavily on their national standing as the capital of the country, and on their historical status as protectors of their trade and guardians of the standard of wrought plate, Birmingham's emerging industrialists had succeeded in gaining an Assay Office – a mechanism by which the quality of their work and their integrity as manufacturers would be indisputably guaranteed. The new had challenged the old and won. Boulton's own silver business was not entirely successful; he did not become the 'great silversmith' in either the scale or scope he had intended, but in skill, undoubtedly he did. However, the manufacture of silver, under the name of Boulton did continue at Soho until Matthew Robinson Boulton sold the Plate Company in 1833.[73] By 1775, Boulton had entered into the partnership with James Watt that was to create the most widely-held perception of him.[74] From the story of the Birmingham Assay Office, however, Boulton emerges as a modern figure, with the majority of the skills and qualities in place that would serve him so well throughout his career; a master at public relations or what is more commonly now known as 'spin', and adept at gathering around him influential and prominent people who might prove useful to him, or rather what we now call 'networking'. It is these modern attributes that render Boulton such an attractive and charismatic figure to us in the present.

7: Ormolu Ornaments

Nicholas Goodison

Matthew Boulton's gilt ornaments, along with his silver and plate, are the products for which he is especially well known today to collectors, art historians, dealers and salerooms. He established a name for them in the 1770s among the aristocracy and throughout the known world.[1] He sold them to the leaders of fashion in England, including George III and Queen Charlotte, and exported them to many parts of Europe, including to the Court of the Empress Catherine at St Petersburg. He was minutely involved himself in the processes of their design and production, and took the lead in marketing them.

The French term '*or moulu*' means literally 'ground gold', that is to say, gold reduced to a powder so that it can be easily amalgamated with mercury for use in the process of fire gilding. The contributor to Diderot's *Encyclopédie* defined it as '*l'or qui a été amalgamé avec du mercure, pour appliquer sur des pièces d'argent ou de cuivre que l'on veut dorer solidement*'. Boulton was probably aware of the definition in the *Encyclopédie*, parts of which he owned, and he would have come across the phrase '*d'or moulu*' when he was looking into the work of French *bronziers*. The phrase had marketing attractions for him, the French makers being his chief competitors. Boulton seems to have been one of the first English makers to use it.

Boulton claimed that he was the first to introduce the manufacture of ormolu ornaments into England.[2] If we define ormolu, in its widest sense, as gilt brass or bronze ornament for almost any application, this claim cannot be justified. There was a long tradition of ornamental mount making in England before Boulton arrived on the scene. There were, for example, the makers of fire-gilt bronze mounts for clock cases and furniture, all of which could have been described as 'made in *or moulu*'.

But there were reasons for Boulton's boast. He was the first English manufacturer to attempt the speculative production of ormolu ornaments on a large scale. Making the models was expensive. So were the materials. The mercury gilding process was tricky and liable to costly mistakes. Despite these risks Boulton and his partner John Fothergill were the first manufacturers to offer such a large and rich choice of ornaments. Second, Boulton was the first English manufacturer to mount an assault on the French *bronziers*' domination of the market. Third, Boulton was one of the first manufacturers to exploit the new fashion for the 'antique taste', which today we call neo-classicism.

The new taste brought with it a craze for vases. Wedgwood said in 1769 that 'an epidemical madness reigns for vases, which must be gratified'.[3] Later in the same year he was told by an eyewitness that at their showroom in Newport Street in London 'vases … vases was all the cry',[4] and in 1770 he said that 'vase madness' was like a disorder which should be 'cherished in some way or other'.[5] Boulton too saw the disorder as a commercial opportunity.

In France the opulent taste of the period of Louis XV had brought with it a fashion for decorating vases with rococo pedestals, rims and handles. Many vases were heightened both in stature and in elegance, and some were transformed from mediocrity by the addition of ormolu mounts. The fashion persisted when classical designs began to come into favour in the 1760s. Apart from being used as ornaments, vases were often turned into objects of utility. A dull vase, or a fine one, became an elegant ewer or a perfume burner. Vases were also converted into candelabra and timepieces. The combination of ornament and utility was calculated to appeal.

Boulton seized on this idea and produced candle vases of many designs, some to hold a single candle, usually with a reversible nozzle (Cat. 171) so that when out of use the vase could become an ornament, and others with ormolu branches for two, four or six candles. He also produced timepieces with vases in their designs (Cats 163, 176), many of them using a watch as the timepiece and some of them fitted with horizontal watch movements in imitation of the French. Many of these vase candelabra and timepieces could serve also as perfume burners (Fig. 46; Cat. 177). But although the candle vases, perfume burners and vase timepieces produced by Boulton & Fothergill had their uses, their designs were emphatically biased towards ornament. Boulton's letters and advertisements dwelt on the ornamental aspects of his vases as their chief attraction.

So it was Boulton's production of ornamental vases, on a large scale and in competition with the French, which justified his boast that he had been the first to introduce the manufacture of ormolu ornaments into England. Vases were by far the most numerous of his products. His other ormolu products, such as candlesticks, ice pails, girandoles, knife urns, tripods, timepieces with vases or obelisks, and the more elaborate clock cases, which also aped his French rivals, were not made at Soho in anything like the same numbers.

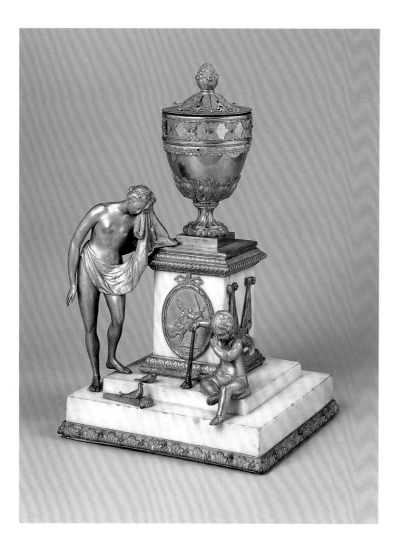

46 Venus vase perfume burner, Ormolu and white marble, c. 1770, height 11.5 inches (29.2 cm) (Private collection).

47 Candle vases, Ormolu and green enamel on copper, c. 1769, height 7.7 inches (19.5 cm) (Private collection).

48 Lion-faced candle vases, Ormolu and blue john, c. 1770–1, height 17.5 inches (44.5 cm) (Private collection).

Boulton's interest in vases began to crystallise in 1767. After the end of the Seven Years War in 1763, the English aristocracy again began to purchase French furniture in some quantity, and during the second half of the decade an increasing number of objects, including ormolu-mounted vases, flowed into England. Boulton would have seen many of them during his visit to Paris in 1765 and his journeys to London. He was incensed that the French metalworkers seemed to have the market to themselves. According to Wedgwood, Boulton told him early in 1768 that he would be surprised to learn the extent of the trade that had

> lately been made out of vases at Paris. The artists have even come over to London, picked up all the old whimsical ugly things they could meet with, carried them to Paris, where they have mounted and ornamented them with metal, and sold them to the virtuosi of every nation, and particularly to Millords d'Anglise, for the greatest rarities.[6]

By the end of 1767 Boulton had decided that the mounting of vases was a practical proposition. He thought that ceramic bodies would be suitable, and was attracted by Wedgwood's 'Etruscan' ware. He sent an order to Wedgwood for 'some bodys of vases for mounting.'[7] He began to make enamelled vase bodies, using green, white and other enamels on rolled copper (Fig. 47). In the late autumn of 1768 he suggested a joint venture with Wedgwood. As usual it brought out Wedgwood's worries about his friend's competitive zeal. He feared that if he refused Boulton might be affronted and start making pottery vases himself.[8] Wedgwood received some orders from Boulton before the end of the year for 'one, two or three dozen pair of the vessell part of some good formed vases for the small candlestick.' Boulton wanted the vases to be strong for fear of breakage when fitting the metal mounts, and to be some of 'black Etruscan clay, some green, some blew or any other simple coulor you think proper'.[9] Early in 1769 he sent another order to Wedgwood which made it clear that he was producing mounts in quantity from standard patterns. The vases had to fit.[10]

But although Boulton flirted in 1768–9 with the idea of using ceramic bodies for his vases, it didn't last. He produced very few mounted vases with china bodies. Wedgwood still feared in 1769 that Boulton might source ceramic vases from one of the other potters, or worse still that he would start making ceramic vases himself, but he needn't have worried.[11] Boulton seems to have

decided at an early stage to concentrate on materials other than china, which was not well suited to bearing the weight of candle branches or to being pierced so that handles or decorative swags could be pinned in position. It was too fragile.

Enamelled bodies were one answer. They could be white or coloured. But Boulton had found much more attractive and stronger materials for his vase bodies – the fluorspars and marbles of Derbyshire, and especially the colourful and decorative fluorspar known as blue john, which was mined near Castleton. It was probably known from outcrops as early as 1700 and the local lapidaries began to make ornaments from it, such as obelisks and vases, when mining started in about 1760. Boulton may have seen vases or ornaments made out of this stone in the houses of local patrons, or in shops or workshops.

In December 1768 he wrote to his friend John Whitehurst, who was well known for his interest in local geology, that he had 'found a use for Blew John which will consume some quantity of it. I mean that sort which is proper for turning into vases.' He wanted to buy a lease on the mine, or at least to get hold of 'any of the best and largest sort of the produce of it'.[12]

Boulton did not acquire a lease of the blue john mines, although he later boasted to a gullible shopkeeper in Bath that he had an exclusive right to their production.[13] He went to Derbyshire in the spring of 1769. He noted the names of several suppliers in his diary and recorded a few of the places where the stones were mined. He bought blue john and other stones, and some vase bodies. His biggest purchase was more than fourteen tons of blue john.[14] This was a great quantity of stone. There is no other order for materials suitable for vase bodies of anything like this size in the surviving archives. It was followed, further-

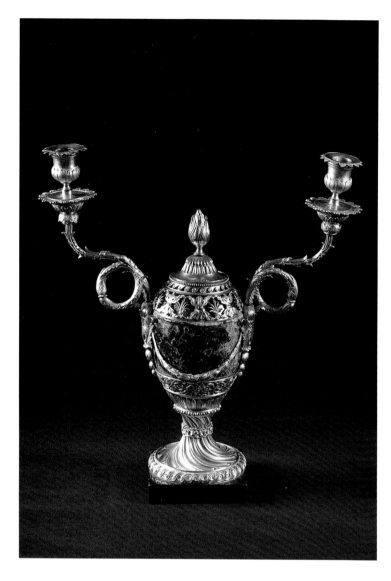

49 Candle vase, ormolu and blue john, height 16.5 inches (42 cm) (Cat. 159) (Birmingham Museum and Art Gallery).

more, by the purchase of two casks nine months later from a stone merchant who promised to do his best to purchase any 'very large, very good, or very beautifull' pieces which might be found.[15] Boulton meant business.

There is no evidence that vases were produced at Soho in any quantity in 1768. But by the beginning of 1769 Boulton was planning to apply some of the principles of mass production to the manufacture of vases. He was pleased with the quality of his mounts, reckoning, somewhat optimistically, that it already exceeded that achieved by his French rivals, and he considered the mounting of vases to be 'a large field for fancy'.[16] He was thinking hard about the forms in which the vases should be made, for example standing on steps, plinths, pedestals or tripod feet.[17]

From the start his approach to the design of his ormolu ornaments was entirely eclectic. He borrowed ideas from French designs, and some of his early vases show the decided influence of the early form of neo-classicism in France (Fig. 48). He borrowed from books and engravings, from architects (especially the Royal architect William Chambers who also reflected the Franco-Italian forms of neo-classical design), from other manufacturers (Fig. 49), from sculptors and plaster shops, from anywhere and anybody. The vases and other ornaments in the Boulton Bicentenary Exhibition show how, like Wedgwood and other manufacturers, he absorbed the new taste and drew widely from the repertoire of classical ornament – lions' masks, rams' and goats' heads, winged figures, sphinxes, satyrs, allegorical figures, ox skulls, lyres, Greek fret, guilloche and Vitruvian scroll friezes and bands, paterae, bows and husks, laurel and drapery swags, acanthus, gadrooning, ribbon and laurel borders, medallions, and so on. The vases and other ornaments also show the variety of sources that he culled for figures and medallions and other motifs. His eclectic approach to design was in accord with eighteenth-century ideas of invention.[18]

At the same time Boulton was actively learning about the gilding processes, and especially about the methods of imparting colour to gilt work. He wanted to learn about gilding for his trade in buttons, snuffboxes and other toys, but there is little doubt that in 1768-9 his researches into gilding were directed towards the production of mounts which could rival the best work of the French. He thought of visiting metalworkers in London in order to pick their brains, and he tried his hand at some industrial espionage in Paris through his agent there, whom he asked to obtain some recipes for him even if they cost 'a few guineas'.[19] The results of his researches and experiments can be seen in the characteristically rich gilding of his best ornaments.

In 1769 Boulton was beginning to broadcast his new line, and to claim that it was already worthy of the attention of the most sophisticated patrons, even in Paris. There are records of orders in 1769 in the archives and the number of designs was increasing. Among the early orders were goat's head vases (Cat. 160) and Cleopatra vases (Cat. 161). The orders show that, besides extending the number of vase designs, Boulton was considering what other objects besides vases might be fashioned in ormolu. One of his correspondents, who had taken soundings among those who 'have nothing to do but to copy or invent new modes of luxury and magnificence and who have lived amongst the French', recommended 'ice potts' and 'clock-cases, of which there are a thousand fancies, for chimneypieces etc., and just now these fancied clock-cases of the ormullie, inlaied and enamelled etc. are all the *mode-Francaise* which they sett on tables, bracketts, toiletts, chimneypieces'.[20]

It is the first occasion in the surviving manuscripts in which ice pails are mentioned. The first reference to an order for ice pails does not occur in the archives until 1772. Clocks entered the repertoire in 1770, when Boulton received the order from the king (Cat. 163), although none were offered for sale publicly until 1771.

It was not easy to sell these expensive ornaments through the normal agents who handled sales of buttons, buckles and 'toys'. The best prospect was to sell direct to patrons, both in London and through the showrooms at Soho and in Birmingham. The remarkable flow of visitors to the Soho Manufactory, which was one of the wonders of the industrial world, helped to stimulate purchases and orders. In 1769–70 Boulton was beginning to have some success with aristocratic patrons. He cultivated them assiduously.

Then, in March 1770, Boulton achieved the summit of his ambition when he visited the Royal Family and received an order for several vases from Queen Charlotte.[21] The royal architect William Chambers was the key figure in the design of these objects. The vases for the Queen included the resplendent King's candle vases, made of large pieces of veined blue john (Cat. 164), and almost certainly the pair of sphinx vase perfume burners (Cat. 165). Boulton was also asked to make a clock case for the king to the design of William Chambers (Cat. 163); blue john was used for the panels and the pedestal for the classical urn on the top, to very pleasing effect. The patronage of the Royal Family gave a great stimulus to Boulton's marketing plans. The objects with which he supplied them were sophisticated designs with cast figures and other motifs, rich gilding and attractively coloured crystalline stone. They were something completely new in the repertoire of English metalworkers.

Encouraged thus by the patronage of the fashionable world and of the first family in the land, Boulton & Fothergill built up the production of ormolu, or at any rate of vases, at Soho. During her visit to Soho in May 1770, Dorothy Richardson reported seeing 'the greatest variety' of blue john 'embellished with gilt, and silvered, ornaments, many of the tops take off and form candlesticks'.[22] According to Wedgwood, John Fothergill said in 1769 that 'the vase trade would be inexhaustable, it would be impossible to supply the demand for good things in that way',[23] and in 1770 Boulton reckoned that they would need twenty more good chasers to cope with the expected demand.[24]

One of the centres of the fashionable world was James Christie's saleroom in Pall Mall in London. Boulton decided to stage a sale of his new ornaments there as part of his campaign to attract the nobility and gentry to his new products. No catalogue for this sale, which was held on 6–7 April 1770, has yet been found and there are few references to it in the archives. The advertisements for it, however, list 'vause candlesticks, branches, arms, tea and other vauses, perfume and essence pots', all 'of exquisite workmanship, and finished in the antique taste'.[25] The sale seems to have achieved its purpose. Like similar exhibitions mounted in London by Wedgwood and others, it attracted the attention of people of fashion. Horace Walpole reported that a 'tea-kettle' sold for an extravagant price.[26]

The sale illustrates Boulton's policy of selling ornaments direct to his customers rather than relying on merchants and retailers. It also indicates that vases were now being made at Soho in some quantity. Lord Shelburne thought that the production of ormolu at Soho was progressing so well that he made a point of mentioning it to James Adam and said that it 'only wanted a variety of elegant designs to make it one of the most magnificent manufactures in Europe'.[27] For a while, as a result of Lord Shelburne's suggestion, the Adam brothers and Boulton flirted with the idea of a joint venture for a showroom in London, but nothing came of the negotiations. The Adams took the plan seriously, recognising that Boulton had become an important manufacturer of ormolu ornaments and that he was the one manufacturer in England capable of producing the 'useful and ornamental' metal furniture for which they foresaw a large market. The suggestion must have pleased Boulton. But he chose not to fall in with their building plans.[28]

The new business was not without its growing pains. In the early part of 1770 Boulton was not happy with the quality either of the designs or of the workmanship. He wrote to his partner from London complaining that the gilt work was tarnishing for want of better drying after burnishing and that the vases were 'so carelessly got up in every particular, that I fear I shall have great difficulty to sell those I have here [i.e. in London], … For all of them are now tarnished and spotted to such a degree as to render them unsaleable'. He was very critical of the manager in charge.[29] Such attention to the details of production and the quality of his products was typical of Boulton. He could not hope to fulfill his boast of competing effectively with the French makers of decorative ormolu unless he could offer objects that approached the quality of French chasing and gilding. New designs, even if they appealed to the more sober English taste, were not enough. It is a tribute to his determination that some of the products of 1770 were as fine as they were.

In his desire to produce the best, at this time Boulton paid little regard to costs. It must have been an exciting prospect to become known for producing artistic ornaments of superior quality when his accustomed trade had consisted largely of practical hardware, and to provide objects that decorated famous men's saloons and dining rooms rather than merely fastened their shoes, uniforms and waistcoats. It is easy to see how the encouragement of Lord Shelburne and James Adam might have gone to his head. And so he rushed headlong in 1770 into improving the production of ormolu ornaments, pricing his vases by rule of thumb and assuming that somehow some of the money received from sales would be profit.

By the end of 1770 the number of vase designs had increased, some of them devised by the simple expedient of arranging the same components in different combinations. The quality of the chasing and gilding was improved. Boulton was able to plan another exhibition and sale of ormolu ornaments at James Christie's saleroom in London in the spring of 1771 with some confidence. He told the Earl of Warwick that he hoped to show 'specimens of many new things in our or moulu'[30] and convinced Wedgwood that he would have 'a superb shew of vases for the spring'.[31] He wrote to his noble patrons that 'the variety and taste of the models, the neatness of the workmanship, the richness and durability of the guilding, exceeds my first essay very much',[32] a claim repeated and expanded in the preface to the catalogue:

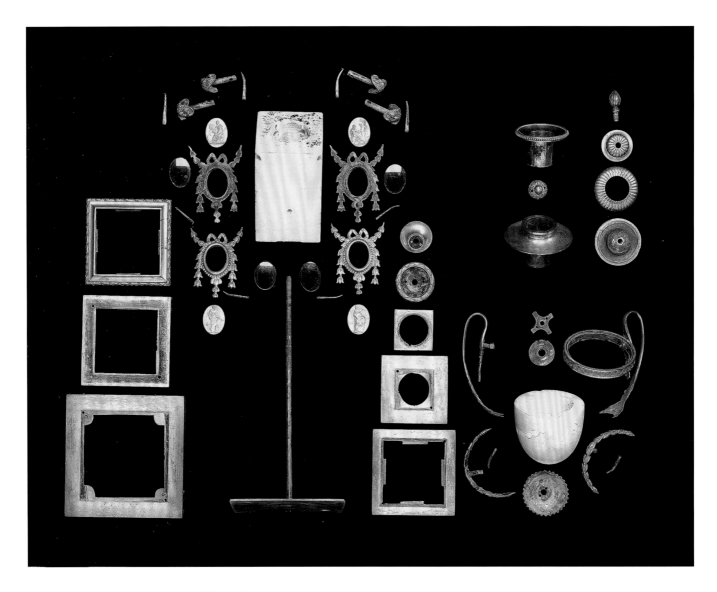

50 Candle vase dismantled (similar to Cat. 175 but without branches), Ormolu and white marble, height 11.9 inches (30.25 cm) (Private Collection).

… The color of the gilding is considerably improved … It is now intermediate betwixt the color of the French gilding, which was observed to be too near that of brass, and the color of our own former gilding, which was very justly censured for being too red; the gilding of the present assortment is at least equal to any that we have ever seen of the manufacture of any other country, either in richness of appearance, or in durability.

The articles now exhibited, will be found greatly to exceed those of the last year, in design, sculpture, and execution in general.

The preface went on to express the hope that with the continued 'liberal encouragement' and assistance of 'persons of taste', Boulton & Fothergill would be enabled

to make further advances in elegance, correctness and execution: and to establish and extend this new manufactory, by which not only large sums may be prevented from being sent abroad for the purchase of a foreign commodity, but also a considerable branch of commerce may (as we have great occasion to believe from our last year's exports) be established.[33] (Cat. 166)

The preparations for the sale, which took place on 11–13 April, were thorough. Boulton advertised it widely, arranging for announcements in the London newspapers and writing flattering letters to the nobility and gentry inviting them to a private view. The advertisements mentioned 'clock cases, candle branches, perfume and essence pots, and many other ornaments, replete with elegance and true taste, and most elaborately furnished'.[34] As in 1770 the advertisement talked of the produce of 'Mr Boulton's Or Moulu Manufactory at Soho in Staffordshire', omitting his partner's name and demonstrating his personal handling of the ormolu business.

The exhibition was a display of which Boulton could justly be proud. It contained nearly four hundred objects. Only eight were not ormolu. It is not surprising that the workshops in which the ormolu ornaments were made at Soho had little time for anything else in the early months of 1771.[35] The display ranged from large quantities of smaller candle vases and perfume burners (forty-seven pairs of one model and twenty-eight of another) to ultra-expensive showpieces. Among these were blue john sphinx vases (Cat. 165), perfume burner vases supported by griffins, large candle vases with branches supported by caryatic figures, vases with branches standing on altars, grand candle vases like the king's with its 'demi satyrs' (Cat. 164) and, last and grandest of all, 'a magnificent Persian candelabra for 7 lights' which was reserved at £200 and was the last lot in the sale (see Chapter 4, Fig. 28, p. 36). There were also five allegorical pieces representing Venus weeping at the tomb of Adonis (Fig. 46), the forerunner of the many allegorical watch stands made in later years. Other ornaments included a clock case representing Minerva unveiling a votive vase, which was reserved at £150, several table candlesticks, most of them made entirely of ormolu, four tripod perfume burners with candle branches 'after a design of Mr Stuart's' (see Chapter 4, Fig. 29, p. 37) and a triton candlestick.[36]

Most of the vase bodies were made of blue john. Others were made of gilt or enamelled copper, china or marble. The vases comprised over thirty different designs. Some of them represented a substantial investment in models, materials and working time. Others were practically mass-produced – a term which, in the context of Soho, meant specialist workshops performing each function (modelling, casting, rolling, chasing, gilding and colouring) with advanced techniques of handcraft.[37] The characteristic assembly of a vase from its component parts can be seen in Fig. 50.

It is clear from the vases that can be identified today that the firm's technical competence had advanced considerably by 1771. The exhibition thus enhanced Boulton's reputation for quality among the fashionable classes, even though some did not like his prices. Mrs Delaney said that she had 'seen a fine show at Christie's and am much pleased with the neatness and elegance of the work, but it bears a price only for those who have superfluous money…'[38]

She would have been in the same two minds about the sale in 1772, for which Boulton asked Christie for more space.[39] No catalogue for this sale has yet been found, but we know from correspondence that it included a number of identified ornaments such as goat's head vases, a griffin vase, a sphinx vase and a Persian vase, all perhaps left over from the previous year. It also included a large number of the cheaper, or perhaps I should say less expensive, vases that formed the major part of the firm's production. There were several grander ornaments, including glass-bodied wing-figured candle vases (Cat. 172), a Titus clock (Cat. 176), and a clock modelled on the one made for the king; it was typical of Boulton to copy objects designed and made for one patron for sale to others. They must have been a glittering sight, and Boulton was looking forward to the nobility's appreciation of the 'quite new and elegant things'.[40] He was especially pleased with his two 'philosophical clocks', the geographical (Cat. 169) and the sidereal (Cat. 170), the movements of which he had commissioned from his friend John Whitehurst. They were the grandest things he had made. Both were sumptuous examples of his eclectic approach to design. They were the climax of the sale.

With the two Christie's sales in 1771 and 1772 Boulton intended to impress noble patrons and make money. They certainly achieved the first objective. They increased Boulton's reputation among 'fine folks', several of whom made purchases. In 1771 the buyers included the Earls of Bessborough, Exeter, Fitzwilliam and Kerry; Lords Arundell, Melbourne and Orwell; Lady Elliott, Lady Godolphin; and Sir Watkin Williams Wynn. In 1772 purchases were made by the Prince of Wales, the Dukes of Manchester and Northumberland, the Earl of Sefton and the banker Robert Child (See Cat. 166).

But financially the sales were a failure. A large number of the lots in 1771 remained unsold, and the sale raised only about half what Boulton hoped for. The 1772 sale was an even greater failure. Several of the larger lots were unsold, and Boulton was obliged, he said, 'to knock down many things much under their value' owing to lack of support, in order to ensure the employment of the artists he had trained at great expense.[41] The failure to sell the two 'philosophical' clocks was the bitterest blow. He had devised them to exploit the growing interest in scientific enquiry and the concurrent fashion for the antique taste. 'I am determined,' he had written to Whitehurst, 'to make such like sciences fashionable among fine folks.'[42] After the sale, disillusioned by the frivolity of the rich, he wrote to his wife, 'I find philosophy at a very low ebb in London…If I had made the clocks play jigs upon bells and a dancing bear keeping time, or if I had made a horse race upon the faces I believe they would have had better bidders…'[43] It was not just a bitter comment on aristocratic taste. It was a turning point.

The ormolu business went into decline after 1772–3 because the costs of production, and of the raw materials, were too high. The ornaments were expensive luxuries. There wasn't a large enough market for them. John Fothergill, always the gloomier of the two partners and usually more aware of the financial risks, reckoned even before the 1772 sale that the vase business would be 'very detrimental to the profits of this year and consume all our labours.'[44] Worse, the gentry who bought ornaments at the 1772 sale proved slow at paying.[45] Business was also depressed later in the year by the repercussions of the bankruptcy in June of the bankers Neale, James, Fordyce and Down. Optimism faded, and never again was Soho to produce ormolu ornaments in such quantities as in 1770–2.

Boulton did not admit publicly that the ormolu trade after 1772 was disappointing. That would not have been in character.

He even said to patrons that he would not be holding a further sale in 1773 because the factory was too busy fulfilling orders, both for plate and ormolu.[46] But the truth was that another sale was not worth risking, and orders for ormolu were lagging. John Fothergill wrote to Boulton in the spring of 1773 that everything at Soho had 'the most gloomy aspect, except being full of button orders.' He wanted to sell some of the vases in London and abroad, especially in Russia.[47]

The economy duly picked up again, and the archives show a reasonably steady stream of orders for ormolu in 1774–7. There were sales to several members of the nobility. Perhaps the most remarkable commission during these years was for the mounts for the cabinet made by Mayhew and Ince for the Duchess of Manchester (Cat. 174). The firm claimed to have 'several large orders' in hand in 1775[48] and was still ordering blue john bodies in some quantity.[49] The archives also reveal positive efforts to sell more abroad. Further samples were sent to the Empress Catherine of Russia in 1774, which appear to have led to a 'very large order',[50] and several pieces were delivered to agents on the continent in 1776.[51] Wedgwood reported that a surprisingly large proportion of the ormolu ornaments were sold abroad, and that Boulton and Fothergill hoped to supplant their French rivals in the Russian market.[52] But there were nothing like enough sales to reduce the stocks to reasonable levels. Meanwhile Boulton himself was increasingly preoccupied with other trades, particularly silver and plate after his successful efforts to establish the Assay Office in Birmingham in 1773, and the engine trade with James Watt from 1775.

Ormolu ornaments were among the list of products that Boulton planned to sell in the showroom that he thought of establishing in London in 1776, but they ranked below silver and plate. By 1778 there was a problem of excess stock. At the beginning of the year there were finished vases valued at £1300 and several more waiting to be gilt.[53] Fothergill called it, somewhat alarmingly, a 'prodigious stock' and suggested selling some of it at Christie's.[54] The sale, when it was held in May, which was not a good month for the gentry who customarily departed to the country for the summer, was a failure despite being well advertised. The total number of ormolu pieces was 166. Most of them were unsold.

The descriptions in the catalogue were more sparse than in 1771, but several designs can be identified.[55] Many of the lots were based on designs that the firm had been making in 1771–2, including Titus clock cases, a Minerva clock, a Persian candle vase, five Venus vases and a king's clock. But they also included some later designs, including allegorical timepieces depicting severally Cleopatra at the tomb of Mark Antony, an offering to Diana, Narcissus admiring himself in a fountain (Cat. 178), Penelope petitioning Minerva, and Urania the Muse of Astronomy. These allegorical conceits were another example of Boulton's conscious emulation of the French *bronziers*. There were also Apollo and Diana candelabra, and a Bacchanalian vase (Cat. 177), perhaps the most elegant vase design produced at Soho.[56] Many of the lots showed that Boulton & Fothergill were making greater use of white marble, and of obelisks instead of vases, confirming impressions gained from the archives.

The failure of the 1778 sale undermined any remaining hopes that the ormolu business would pay. There are records of orders after 1778, but they grew less and less frequent. Ormolu was eclipsed by the continuing button, plate and toy businesses. Boulton himself was engrossed in the engine business and was flirting with a new fancy, the production of mechanical pictures, which had at least some connection with the ormolu trade because besides being used in door and shutter panels the pictures were sold in ormolu frames. Ormolu pieces still figured in the showroom when visitors came to Soho, but the word was dropped from the description of the business, which from 1779 was called 'Silver, plated etc. goods' in the firm's ledgers.

The shrinking of the business is illustrated by the sparse references to ormolu objects in the firm's ledgers and journals, and by the inventory of stock occasioned by the death of John Fothergill in 1782. There were a mere fourteen ornaments in the toy room, and very few in the main workshop. Several pieces in the warehouses were damaged and some were not completed.[57] Some of the objects were those that were not sold in the 1778 sale. It is not a picture of a thriving trade. After 1782 there are only a few references to ormolu in the archives. The turnover of the Manufactory consisted mostly of buttons, toys and silver plate. Ormolu orders were still sought, but they were few and far between.

Boulton never revived the ormolu business, even when his London agent Richard Chippindall tried to persuade him in 1790 that there was a market in London and that the clockmaker Benjamin Vulliamy had more orders than he could satisfy.[58] In the words of Boulton's friend James Keir, the ormolu trade was 'too expensive for general demand, and therefore not a proper object of wholesale manufacture.'[59] The ormolu business was a modest part of the total business at the Soho Manufactory. It lasted for not much more than a decade, but despite its brevity it remains a fascinating study. It consumed resources and stretched the firm's finances, but it enhanced Boulton's reputation among the leaders of society and was thus helpful to him when he wanted to establish the Assay Office and extend Watt's patent on the steam engine, both of which required Acts of Parliament. As Keir said, it may not have been profitable, but it 'greatly tended to his celebrity and admiration of his various talents, taste and enterprise.'[60]

The ormolu trade also shows how works of art were produced in a factory environment at the height of the 'neo-classical' period. It shows how a Birmingham metal manufacturer could, with determination, attention to detail and fashionable designs, capture the imagination of 'fine folks'. And Boulton's later successes in the production and sale of engines and coinage owed not a little to the lessons learned in the earlier business.

8: The Soho Steam-Engine Business

Jim Andrew

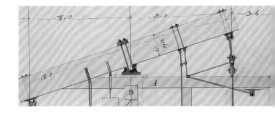

Matthew Boulton's prime purpose in moving his metalwares business from Birmingham town centre to Soho in 1761 was to apply water power in the larger manufacturing operation he planned to establish. It was the presence of a water mill that decided him on the site. But the new water-powered Soho Manufactory was already running short of power when he heard that a Glasgow instrument maker, James Watt, had ideas that promised to increase greatly the power and efficiency of the existing design of steam engine. From the mid-1760s Boulton and Watt began corresponding, though they did not meet until 1768 when Watt was journeying between London and Glasgow on business. Boulton had already been experimenting with steam power with his friend and collaborator, Dr William Small; and Watt and Small also became regular correspondents, greatly to Watt's encouragement. Boulton was interested in Watt's new design, both to power the Soho Manufactory and as a business venture to supply engines to other users. Watt's development of his ideas was being funded by John Roebuck, who had many business interests in Scotland. Roebuck's businesses were developing financial troubles and, following much deliberation, Boulton was able to take over Roebuck's interest in Watt's inventions, move Watt to Birmingham in 1774 and establish a partnership with him in 1775.

Steam engines were being built from about 1710 to the design of Thomas Newcomen in which steam was condensed in the steam cylinder to produce a low pressure under a piston so that atmospheric pressure forced the piston down. One wonders how soon Matthew Boulton and James Watt realised that Watt's design of engine with its separate condenser was such an advance on the earlier Newcomen design that it called for a different kind of business. In the eighteenth century the greatest cost in building a steam engine was the cost of the steam cylinder, while the greatest cost in operating the engine was the amount of fuel burnt to supply the engine with steam. Even Watt's experimental engine gave twice the power from its cylinder, the work it could do, and three times the efficiency, the work done for the coal burnt, compared to a normal Newcomen engine with the same size cylinder. How could Watt and Boulton satisfy the expected demand for the engine, stop others using these improvements and obtain reasonable reward for their endeavours?

Watt already had a patent on the separate condenser, which they were able to extend to 1800 by an Act of Parliament, but there had been two reasons for the slow progress in perfecting his engine before Boulton brought him from Scotland to Birmingham in 1774. Firstly he was a rather diffident inventor, always trying further improvements, and secondly he needed a cylinder made to far greater accuracy than those used in Newcomen engines. Boulton was a great grasper of opportunities and a great networker of people. He got John Wilkinson, the ironmaster, to use his cannon-boring machine to produce an accurate eighteen-inch diameter cylinder for the experimental engine, then encouraged him to develop an accurate machine for boring larger cylinders. Boulton also constantly pressed Watt to be decisive and to finalise the design for each engine, going back to improve it later if necessary.

At this time between twenty and thirty Newcomen engines were being built each year, with perhaps as many as ten firms active in engine manufacture and a number of engineers such as John Smeaton acting as consultants.[1] These engines were of considerable size, often 'house-built', where the building supported the major components, and were mostly mine-drainage or water-supply pumping engines. Thus the engine-building firms would supply components, buying in some as necessary, to their customer's site, where the engine was erected into the building during its construction. While Boulton and Watt spoke of setting up an engine manufactory they could, in the meantime, make use of the existing network of engine builders, and this is how they proceeded.

Acting under the patent, Boulton & Watt, as consulting engineers, would design each engine and would not normally allow anyone else to design an engine with a separate condenser. They would then specify both the supplier and the quality of components which their customers had to use, with the customer making payments direct to these suppliers. Boulton & Watt also supplied, from the Soho Manufactory, specialist components such as the engine valves, for which they made a separate charge. Boulton & Watt's fee for their design work, and to operate the patented design, was based on the saving in fuel from using their design compared to an earlier Newcomen-type engine doing the same work.

In the early years, Boulton & Watt would occasionally undertake to obtain all the parts for an engine which they would then

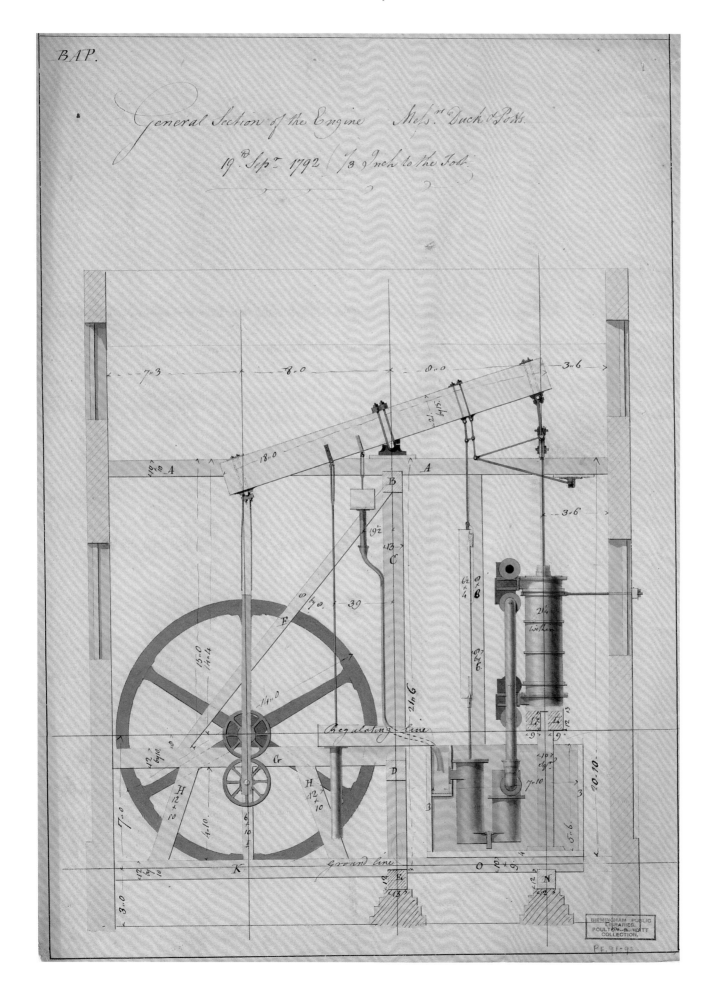

set up for the customer, but this was an exception to their usual consultancy role. Watt and his assistants would produce drawings of the engine installation and components with various recommendations to the customer. Quite often, an engine erector would be sent from Soho to supervise the engines' construction, with the customer paying the erector and Boulton & Watt separately, but in other cases customers could either use their own technical staff or a freelance engine erector, so long as Boulton & Watt were happy with the person selected.

As time went by several engines were erected at the Soho Manufactory, where they returned water to water wheels or, using later designs, drove directly onto machinery. But all were in some degree experimental and Boulton & Watt used customers' engines for some of their development work and development testing, as with the Birmingham Canal Company's engine at Smethwick.[2] With the introduction of each new design of engine, there was rapid progress in improving the designs supplied to successive customers and Boulton & Watt, either as after-sales service or to protect their reputation and that of their engine, ensured that earlier engines could be updated with some of the later improvements, although it is not clear whether the cost of this work fell entirely on the customer or the partners.

As Rita McLean has demonstrated in her Introduction, Boulton was already taking a theoretical interest in steam power as early as 1765, though he had not yet brought it to practical application. His interest stemmed, to a considerable extent, from his need for power at the Soho Manufactory where water power was used but supply could be inadequate in the dry season. Using a pumping engine to raise water from below a water wheel back into the pool above the wheel was one way of overcoming this problem; indeed the early experimental engines at Soho were used in this way, but this was not very efficient. By 1780, Watt's pumping engines had become established and Boulton returned to the need for efficient factory power, where he felt there was a large market among factory owners. Watt's early idea for a direct rotary engine where rotation was produced straight from the action of the steam, did not seem to offer a successful solution to this need, and Boulton pressed him to develop a rotative engine based on the beam-engine design of pumping engine, converting its linear motion to the smooth rotary motion needed to power machinery.

Others had attempted to drive machinery by Newcomen engines, using a crank, but it was soon seen that these crude devices might be satisfactory for winding coal but were too irregular to work machinery. Watt realised that designs were needed that would not only turn reciprocating motion into circular motion, but do it with the smooth controlled operation needed to drive machinery, as was already available from waterwheels. Watt, Boulton and others in the business contributed ideas and during the 1780s the rotative Boulton & Watt engine emerged with double-acting cylinder, parallel motion, sun and planet gear and centrifugal governor (Fig. 51).

As the factory power unit was perfected, Watt had to indicate the engines' capabilities to customers and needed to compare steam engines with some well-known source of rotative power. Water wheels were not that well understood by customers but many factories had been using horses to drive their machinery and Watt developed the standard 'horse power', both as a means of showing what various sizes and designs of engines could achieve and also as part of the assessment of Boulton & Watt's fee for using their designs, which had been covered by a number of patents taken out by the partners. It was not long before the rotative engines were giving smooth enough motion to drive textile machines, probably the most delicate large-scale manufacturing process at that time and one which was to become a major market for steam engines.

Many of the rotative engines designed by Boulton & Watt were considerably smaller than their pumping engines and their smaller components were more within the capability of the steam engine section at the Soho Manufactory, which supplied a steadily greater proportion of engine components. Being able to produce them in-house both brought in more income and allowed better control over the quality of the small engines, again safeguarding their reputation.

The partners had always intended to set up a steam-engine works once their engines were established, and certainly they needed to do this before all their patents expired because they would then lose the fees for working engines under those patents. This move had been put off while they were so busy with the business and with defending the patents against infringements, but they had to act when their main source of accurate cast and machined cylinders and other cast iron components was threatened by a legal dispute between John Wilkinson and his brother. In the 1790s Boulton and Watt took their sons into the partnership and developed a site near the Soho Manufactory, on the banks of the Birmingham Canal. The Soho Foundry came into production in 1795 although it took some years to raise the quality of its work to that of the Wilkinson business at Bersham, North Wales.

Once Boulton & Watt had realised that their design of steam engine represented such a leap in performance, they naturally wondered how to charge for the privilege of using it. Royalties for using a patented idea are often a single fee on each product, or sometimes licensees pay an annual fee during the life of the patent, but so confident were Boulton & Watt of the savings that would result from using their design that they offered 'no saving, no fee' terms – in exchange for a portion of that saving. Initially it would appear quite easy to identify the saving since the engines could be compared to Newcomen pumping engines which were

51 Drawing for a Boulton & Watt 16-horse sun & planet rotative engine to provide power for the cotton mill of Duck & Potts, Manchester, 1792. (Birmingham Archives & Heritage, MS3147/5/92)

already in use. Indeed in some cases the Watt engines would directly replace a Newcomen pumping engine which could be tested before it was decommissioned. In reality, establishing the performance of different designs of engine, to the satisfaction of both the customer and Boulton & Watt, was to be by no means as straightforward.

Having settled the saving from using the new design, Watt's initial suggestion was that this saving would offset the extra cost of building the Watt engine, because of its more complicated design, and then the customer would pay Boulton & Watt a third of the rest of the saving during the life of the engine or until the patent expired.[3] It seems that the suggestion to offset the extra cost was soon dropped, and an analysis shows that the Watt engine probably did not cost more than a Newcomen engine built to similar engineering standards.[4] There was probably enough saving from having a smaller cylinder and boiler on a Watt engine to cover the extra cost of the condenser and the more complicated valve gear, compared to a Newcomen engine. This left the third of the saving in fuel to be paid to Boulton & Watt.

It is unfortunate that this rather equitable scheme led to a great deal of argument because the customers often did not trust the calculations made by the partners, or indeed understand the methods used to test the engines, no easy matter with the limited instrumentation available at that time. Nor did they see why they should pay very much for a patented idea. Meanwhile the special situation at collieries led Boulton & Watt to 'create a saving' in their calculations by setting a minimum coal cost of five shillings a ton, a typical cost of coal to other customers.[5] The difficulty for Boulton & Watt was that collieries fed their steam-engine boilers with coal that was of poor quality with little or no sale value. It was said that often the only cost to the colliery was raising the coal to ground level rather than leaving it underground, but this would leave no premium for Boulton & Watt so they imposed this notional fuel value of five shillings a ton. An early settlement of such a dispute was in 1779 when the owner of Bedworth or Hawkesbury Colliery near Coventry, Richard Parrott, thought £30 a year rather than Watt's calculation of about £270 would be a reasonable fee. Parrott's Newcomen engine was refurbished to work reasonably and tested against his new Watt engine with arbitration by Samuel Garbett, who supervised the tests and checked the calculations to establish that the fee should be £217 a year.[6]

Comprehensive tests, such as those on engines at Bedworth Colliery and Poldice in Cornwall, allowed Watt to establish typical performances for Newcomen pumping engines of various sizes and powers, which he then compared to the actual test results from working customers' engines to calculate the fee, or premium. The calculations for Bedworth did, however, incorporate the imposed five shillings a ton fuel cost and this did mean that collieries were a poor market for Boulton & Watt engines. Only in situations where a colliery could install one powerful Watt engine, rather than two Newcomen engines, was there a sound economic case for using a Watt engine at a colliery before 1800. This resulted in most Watt colliery engines being of over fifty inch (1.25 metre) cylinder diameter and so Richard Parrott had made the right choice because he had installed a fifty-eight-inch (1.5 metre) Watt engine.

Where there was an established pattern of use for steam power, the calculated premium could be an annual fee, as at Bedworth Colliery, but many customers would have fluctuating demands, and various arrangements were made for the reporting of the actual use made of particular engines. Eventually sealed counters were developed to fit to the beams of engines so that a calculation could be made from the number of engine strokes in the accounting period, and hence the premium to be paid could be established. If the customer was certain that their engine would have a long working life at the stated performance, they could make a single payment which was initially set at ten years' worth of premium,[7] and this was slowly reduced, for new engines, to match the remaining years of patent protection up to 1800. Finally, certain customers with close links to the partnership were offered discounted single payments. The Birmingham Canal Company paid a much-reduced single premium on the Smethwick and Spon Lane engines, which partly reflected Boulton & Watt's use of the engines for tests and partly the partners' personal investment in the canal company.[8] It is clear that many fees were eventually settled by negotiation and the oft-reported third of the saving was only the starting point for settling Boulton & Watt's income from these engines.

Engines installed at Cornish mines presented a particular problem for the partnership because of the high cost of mine pumping in the county where the nearest coal was in south Wales. Many of the Cornish mineral deposits were deep below ground and in areas where flooding was a serious problem. The Cornish market for Boulton & Watt's engines was large and complex, with the added difficulty of the distance from Birmingham. The partners made extensive and long visits to Cornish mining areas in establishing a number of powerful pumping engines. In late 1779 the partners sent William Murdock, one of their established engine erectors, to live in Cornwall (Fig. 52). Murdock (1754–1839) was naturally inventive. In his spare time he began experimenting with gas lighting at his house in Redruth, which became the first house to be lit by gas in 1792. He spent nearly twenty years in Cornwall for Boulton & Watt, supervising the erection of engines and advising on their maintenance. The Cornish mine owners were suspicious of Watt's calculations of the premium and resented paying for the privilege of running an engine that they had already bought. Many owners took every opportunity to put off paying the premium, and there were many arguments about how long the engines worked and how the saving should be assessed.

Unfortunately this was also the time when the Cornish copper mines found difficulty in competing with copper produced from deposits in Anglesey, which were far easier to extract. Boulton &

Watt found it necessary to accept reductions in fees in situations where their full fee might seriously affect the customers' trading position, for of course if the customer went out of business the partners would not receive any fees at all. In some cases the partners invested in Cornish mines as a means of supporting their market for engines in that area. Recent research has concluded that the partners probably received only about forty per cent of fees, at a third of the saving, which they might have hoped for from the Cornish engines between 1778 and 1798.[9] In 1799 the partners won a long-drawn-out legal case over engines which infringed their patent, and were able to collect their fees both from users of these 'pirate' engines and those who had been putting off payments while the validity of Watt's patents was proved.

Boulton & Watt's rotative engines were, of course, quite unlike any earlier design of engine, and it seems that the earliest engines were operated under a simply negotiated fee but, as Watt introduced the horse power, it became possible to set an equitable level of fee for these engines. The chosen rate was £5 per horse power per year for engines outside the London area, and six guineas (six pounds six shillings) per horse power for engines in the London district.[10] Again there were negotiations about the fees for many engines, particularly where an engine was the first in an industry or area that the partners felt would bring further business, or where the customer was uncertain of how much use they would make of an engine, for example where it was to assist at a water-powered factory. With the establishment of the Soho Foundry to supply all the components for complete engines, the partners ceased to specify a separate licence fee on new engines, but built this into the cost of the engine components much as any other supplier would accommodate their overheads, including a licence to operate under any patent protection.[11]

It took only a few years for news of the benefits of Watt's improved design of engine to spread abroad and for enquiries to be received by the partners. Watt's patent only gave protection in Great Britain, so the partners set about obtaining patent protection in foreign markets, but this proved to be a protracted process even in countries where it was theoretically available to them. They wanted to have a similar situation to the one at home, where the design was protected and they could obtain licence fees for engines built to Watt's design. Eventually they did make suitable arrangements in some countries, but there is every likelihood that pirated engines were also built in those countries by simply copying legitimately-built engines or by using plans obtained from England. Between 1775 and 1800 just seventeen engines were built abroad: five in Spain, four in France, three in the Netherlands, two in Russia, two in Germany and one in Italy.[12]

The patent for the separate condenser expired in 1800. After that the partners could no longer extract licence fees for the use of the separate condenser and any engine builder could construct engines using it and Watt's other patented designs, while the opening of the Soho Foundry allowed Boulton & Watt to supply complete sets of engine components to customers both

52 John Graham-Gilbert, *William Murdock*, oil on canvas, 1823–7. 137.9 cm × 107.9 cm. (Birmingham Museum & Art Gallery, 1885 P 2543)

in Britain and abroad. In the five years to 1805 they supplied twenty-one engines to overseas customers[13] and this export trade grew as time went on. The West Indies soon became a significant new market with engines being sent to power sugar mills and other machinery. In percentage terms, foreign markets peaked around 1816 with over sixty per cent of the firm's engines going abroad, and foreign sales remained important throughout the nineteenth century, taking over twenty-five per cent of total sales.[14] Foreign markets introduced two rather different requirements from the home market for steam engine manufacturers. Environmental conditions, termites and other problems affected the life of many engine components, while the distance from Britain complicated the supply of replacement parts. As a result, the components for many Boulton & Watt engines for abroad were substantially more robust and of more durable materials than those supplied at home, thus protecting the reputation of the firm but at some extra cost to the customer.[15]

It cannot have been long after going into partnership that the sheer volume and complexity of correspondence between the partners when they were apart (as they often were) and with their

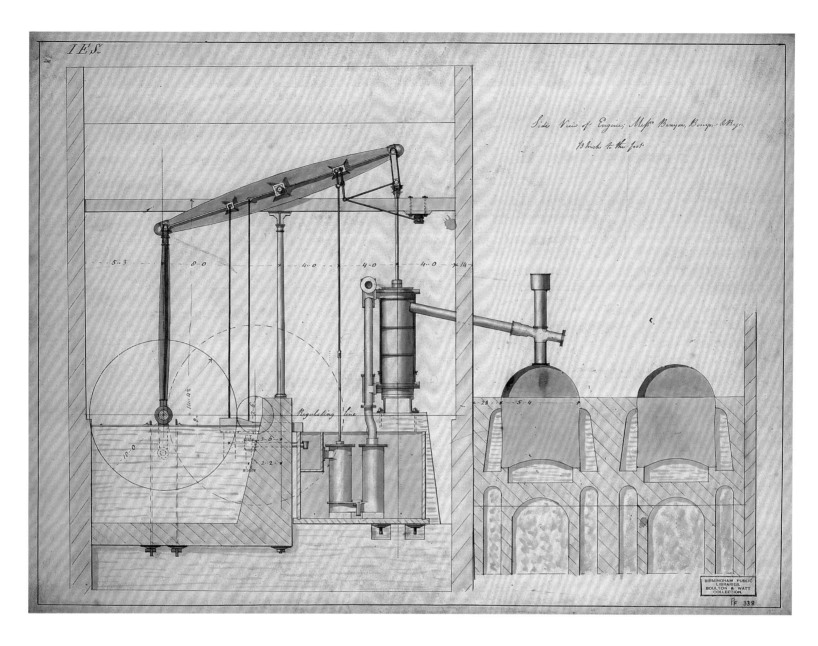

53 Drawing of a Boulton & Watt engine built for Benyon, Benyon & Bage's Shrewsbury flax mill. Ink and colour wash on paper, 1803–4. (Birmingham Archives & Heritage, MS 3147/5/338)

customers, began to cause them difficulties. The business was complicated so there was a need to keep copies of what had been written, which in those days meant the longhand copying of letters; drawings also had to be manually copied. Apart from the time involved, there was the potential for inaccuracies to creep in. James Watt's fertile inventiveness saw a need for some better way of taking copies and Boulton supported him through various experiments in the later 1770s. Watt's endeavours led to a patent in 1780 for a process of writing using special ink and then lifting a copy onto damp tissue paper, which could be read through the paper (see Cat. 355).[16]

Line drawings were soon also being copied but onto damp cartridge paper, giving a reverse image which was no problem to their engineering staff. All the notes and dimensions were then added to both the original and the copy drawings. This copying process not only served the business well through most of the next century but also presented a business opportunity. A new firm of James Watt & Co. was set up to sell copying machines and the special ink and paper to work the process. It was reasonably successful and aspects of Watt's process were still being used in the middle of the twentieth century. The copy process produced a true facsimile of the original document so, while legibility could sometimes be a problem, there could not be any errors of transcription, an important difference from copy clerks' work.

The results of performance trials for Watt engines, and for the Newcomen engines, water wheels and other sources of power which they were to replace, needed to be preserved because they were the basis of claims for payments to Boulton & Watt. At first the results of tests and the calculations of performance and fees were recorded in notebooks, but such was the need to preserve this information that it was then copied onto separate sheets, which were filed; some time later it was transcribed onto standard printed forms. These were also adopted for setting out the specification of each engine as it was ordered, with further forms for the costs of components supplied and other details of the engines.

By 1779 the partners were finding that the instruction of engine erectors and those who would operate the engines for their customers was becoming a major task in itself; because of the distance from Birmingham, this was particularly true for the Cornish engines. Watt wrote *Directions for Erecting and Working the Newly-Invented Steam Engines* as a guide to those working on the engines. It had engraved illustrations and a hundred copies were printed in what is thought to be the first published text on steam engine construction. Today we are used to companies producing manuals, catalogues and standard specifications, as well as using a host of printed forms to record production details, but in the eighteenth century this was quite innovative and places Boulton & Watt in the forefront of developments in business systems.

After 1800 there was less need to keep quite so much information, but it appears that the procedures had become so well established that they tended to continue, and in any case the growth in international orders for engines and the growing complexity of the designs ensured that quantities of records continued to build up. The result of this accumulation, together with the actions of those who preserved the archive after the firm ceased trading in 1895, is the reason that the Archives of Soho hold a uniquely complete record of the activities of an eighteenth- and nineteenth-century manufacturer. The recording of customers' earlier, or competing, sources of power also means that the archive has relevance to studies of other power sources used at this period.

Looking back two hundred years to the steam-engine business of Boulton & Watt, it is not easy to establish all their achievements with great precision. Despite the breadth of the archive of the partnership, there continues to be discussion of the actual number of engines built to their design because engines were moved about, leading to double counting; engines were also given major rebuilds which can look like new orders, while other engines were ordered but not built. The impact of Watt's designs must include pirated engines built by a variety of manufacturers both in the UK and overseas. At the moment the best guess for total engine production within Boulton & Watt's business between 1775 and 1800 is about 450, to which can be added eighty-three pirate engines known to Boulton & Watt and about twenty Newcomen engines improved with separate condensers without licence from the partners.[17] Thus during this period the total of known Watt engines comes to something over 550 against around 1,000 Newcomen engines built in the same period.[18]

There is an aspect of the impact of Boulton & Watt that continues to cause controversy and that is the extent to which they blocked developments in steam engines by other inventors and manufacturers until the separate condenser patent expired in 1800. Boulton & Watt's policy was to design the engines made under this patent and to control the manufacture of those engines, partly to ensure adequate quality to safeguard their reputation and partly to keep the market and some of its income to themselves. While this may have been an extreme example of how patentees behaved, it was well within the spirit of the patent system, where a monopoly is granted for a period of years in exchange for making the innovation public and in this way inventors can gain a reward for their inventiveness.

Many other patentees also kept a tight rein on the use of their patents and were similarly accused of stifling development, but the patent system was intended to reward inventiveness and avoid secrecy in exploiting innovations. Rather than considering how Boulton & Watt stopped actions that infringed their patent, one might consider how soon other inventors would have made similar advances with steam engines, if James Watt had never thought of the separate condenser. From the 1730s there had been a realisation that the Newcomen engine had shortcomings and that improvements should be possible.[19] About the time that Watt was perfecting his invention, John Smeaton, much the most significant improver of Newcomen engines, managed to double the engine's efficiency by paying great attention to detail, but it remained a Newcomen style of engine with no prospect of significant further improvement.[20]

There is little evidence of others contemplating the separate condenser before Watt's ideas became known, and the majority of the successful steam engineers who were to challenge the patent had experience of Watt engines before designing their own competing or improved engines. During the life of Watt's patent, a good deal of effort was put into building engines which improved on the Newcomen engine's performance without, it was claimed, infringing the separate condenser patent, but few could deliver really significant performance, either because the design was flawed or because they were in advance of technical developments needed to work with higher pressure steam. Many engine makers were to benefit during the life of Watt's patent because his engines were actually built by these engine makers, who thus derived income from this engine business while they also continued to supply Newcomen-type engines.

In presenting the case for other innovative steam engineers, many commentators have failed to understand either the workings of the patent system at that time or the characteristics of early steam engines.[21] Various Acts of Parliament during the nineteenth century tightened up on the working of the patent system with better administration, clearer rules over the level of disclosure needed and, eventually, technical assessment for actual innovation. Eighteenth-century steam engines, particularly the large low-pressure pumping engines, were quite different from most later engines with their fixed stroke, greater speed, better boilers and more complex valve gear. It is all too easy to misinterpret the effect of what appear to be interesting innovations when they would have serious shortcomings as applied to engines of Watt's day. Boulton & Watt had plenty of problems with owners of engines and their staff who failed to operate the engines correctly or naively altered the engine installation, resulting in loss of performance or damage to components. Such action reinforced the partners' commitment to retaining control of the use of the patent to safeguard their reputation.

Boulton & Watt's steam-engine partnership can be seen as one of the great businesses of the industrial age. The exploitation of a great advance in steam power, coupled with pioneering business systems and an attempt at a fair way of charging for the use of their innovations, appears to raise them above most firms of their age. By the time the extended patent for the separate condenser expired in 1800, several of their other patents had also expired, so that they became engine builders rather than consulting engineers and patentees. The business now came increasingly under the control of the two sons, Matthew Robinson Boulton and James Watt junior, and even after their deaths it continued to prosper during most of the nineteenth century. The firm developed a fine reputation for steam engines, including innovative designs of stationary and marine engines as well as other engineering products such as the gas lighting systems which developed from the work of William Murdock, one of their longest-serving employees. A total of around 1,750 engines were sold in all markets between 1800 and 1895 (Fig. 53).[22]

Towards the end of the nineteenth century the firm contracted under the control of aging directors who had taken over from the sons of Boulton and Watt. The Soho Foundry was sold up in 1895 and was bought by Avery's, the weighing machine manufacturers. Avery's also bought the firm's name of James Watt & Co., continuing to build a few steam engines for a further ten years. Thus the steam-engine business of Boulton & Watt lasted some 130 years, with a great influence on the exploitation of steam power.

The only pity is that with such a fine record of the partners' activities in the Archives of Soho, we have so little information about the business activities of their competitors. In the absence of comparable archives of both business records and technical data for their competitors, one really is left wondering just how we would assess Boulton & Watt if we had a clearer picture of the other firms, which competed with the Soho steam-engine business.

9: 'I had L[or]ds and Ladys to wait on yesterday …':[1] Visitors to the Soho Manufactory

Peter M. Jones

When Matthew Boulton opened the gates of his new Manufactory to visitors in 1765, Birmingham was already a recognised stopping-place on the industrial tourist map of provincial England. The first recorded foreign visitor to the Manufactory seems to have been Jean-Baptiste Carburi, a chemistry professor from Turin. He sent Boulton a note of thanks for the hospitality he had received at Soho in July of that year, and to judge from the tenor of the letter the two men used the encounter to discuss the startling advances in the field of electricity during the preceding decade. Matthew Boulton passed for the local expert in this new science. Five years earlier, whilst still residing in Snow Hill, he had demonstrated the results of a home-grown experiment into the non-conductivity of glass to Benjamin Franklin, and it was to Franklin that he referred his new Italian acquaintance.

Whether Carburi found much to see in the Manufactory may be questioned, for the main premises were not yet fully roofed. Not until the very end of 1765 would Boulton's partner, John Fothergill, report that 'the buildings now begin to look so very sumptuous as to engage the attention of all ranks of people.'[2] It seems more likely that Carburi was entertained at nearby Soho House, where Fothergill was currently living. But by the following summer the Manufactory was certainly up and running, for Lord and Lady Shelburne paid a visit to Birmingham in May 1766 and reported to this effect. They embarked upon an inspection of the larger industrial premises of the town in what would soon become an established tourist circuit: Taylor's workshops producing gilt buttons and enamelled snuffboxes, Gimlett's watch warehouse, Farmer and Galton's gun shops in and around Steelhouse Lane, and Mrs Baskerville's japanned goods works in Moor Street. It is clear that John Taylor, whose business acumen in catering to the fashion for decorative boxes would make him a fortune, was regarded as the biggest and most admired manufacturer in town at this time. But Matthew Boulton was about to outstrip him inasmuch as his new Manufactory was a purpose-built edifice, rather than a series of workshops located in the crooked back alleys of the town. Moreover, it was designed with an obvious eye to the social pretensions of the affluent consumer. 'The Front of this house is like the stately Palace of some Duke,' another visitor would report, adding that 'within it is divided into hundreds of little apartments, all of which like Bee hives are crowded with the Sons of Industry.'[3] After an escorted tour of Taylor's premises, the Shelburnes took a chaise to Soho, a couple of miles outside town, therefore. Here they purchased watch chains and other trinkets 'at an amazing cheap price'[4] before repairing to Soho House and afternoon tea with Mrs Boulton.

Birmingham was fast becoming a boom town in the late 1760s. During the third quarter of the century its population doubled (from around twenty-four thousand to forty-eight thousand inhabitants), and the thrust of its economy moved away from utilitarian ironmongery and in the direction of domestic consumer goods. In the process the town's manufacturers found themselves increasingly reliant on extra-regional, indeed international markets. At the peak of the European trade cycle – in 1791 – a quarter and maybe as much as a third of the town's stock-in-trade buttons, buckles and other 'toys' were selling overseas. It made good economic sense to cater for the shifting tastes of this fashion-sensitive clientele, then; the more so if the marketing exercise could be coordinated with the developing vogue for internal travel among Britain's gentry and expanding middle classes. Matthew Boulton was not alone in spotting this opportunity. Along with fellow manufacturers, he worked tirelessly to make his native town attractive to visitors, reasoning that the more there was to see and do, the longer travellers would stay and the larger the sums of money they would spend. Indeed, by the 1780s, there was so much to see and do that the Marquis de Bombelles advised, on the strength of his tour of the West Midlands, that visitors should set aside eight days if they wished to take in the full range of the manufactories in Birmingham and its neighbourhood. By this date both Taylor's and Baskerville's premises were *passé*, however. Instead, the recommended itinerary included Samuel Ryland's pin manufactory, a business concern that would be taken over by his nephew Thomas Phipson in 1785; George Simcox's white-metal button and finger-ring workshops in Livery Street; Bridgman's whip manufactory; Eginton's painted-glass establishment situated on the outskirts of town only a short distance from Boulton's own works; and Henry Clay's unforgettable *papier mâché* decorative goods manufactory in Newhall Street. Clay was a close contemporary of Boulton and rivalled him for entrepreneurial flair. An American Quaker visitor to Birmingham described Clay's warehouse as 'like a

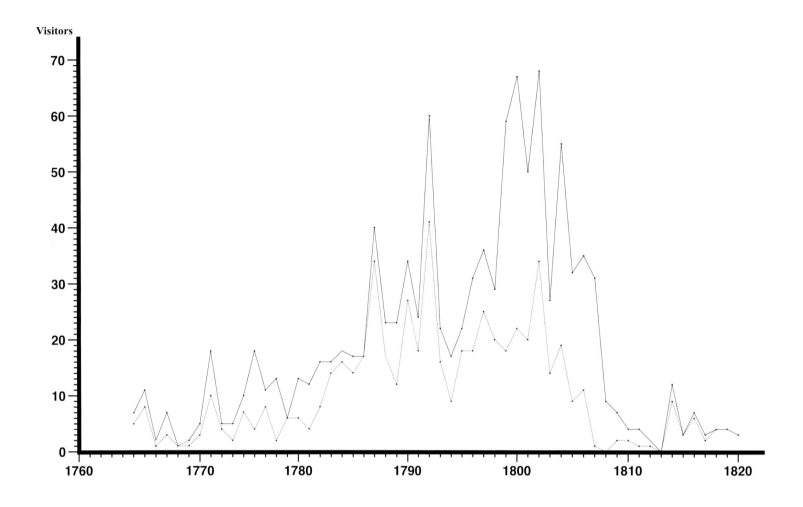

54 Total recorded visitor flows to the Soho Manufactory, 1765–1820 (dotted line indicates foreign visitors.) Source: MS 3147; MS 3219; MS 3782 (Birmingham Archives & Heritage).

Nobleman's Room hung around with paintings: Flowers, Landscapes, Fruit Pieces, Beasts, Portraits and History are exhibited on the ware with nice and beautiful Taste and in great Perfection.'[5]

But Boulton's great Manufactory on Handsworth Heath would become the jewel in Birmingham's crown. For three decades and more it attracted visitors, many of whom came to the Midlands town with no other object than to learn about the industrial and labour innovations being pioneered at Soho and to meet its creator. Sightseers, Grand Tourists, *savants* and accredited technologists, whether native or foreign, intermingled and, until the mid-1780s, were allowed more or less unfettered access. Like Clay, or for that matter Josiah Wedgwood at Etruria, Boulton actively catered for his well-heeled visitors. Early in 1771 he added a showroom to the Soho complex, mainly as a marketing tool to underpin the move into ormolu manufacture and silversmithing. Wedgwood reported it was a 'superb Gallery' featuring not only ormolu vases, but silver and silver-plate 'of the best forms I have seen.'[6] Visitor curiosity was also whetted by another of Boulton's side-lines – his venture into reproduction fine art known evocatively, if inaccurately, as mechanical painting. The Marquis de Bombelles, who was not hugely impressed by Boulton's efforts in the realm of ormolu, was allowed sight of the printing process in 1784. Two years later his compatriot, the French *savant* Grossart de Virly, whose reason for coming to Soho had more to do with science than with shopping, took the opportunity nonetheless to order several 'mechanical' pictures. It is probable that he, too, was given access to the technology involved in copying paintings, for by this time Boulton had concluded that there was little money to be made out of reproductions, whether of Old Masters or of the more cheerful and elegant themes of fashionable artists such as Angelika Kauffmann.

Some idea of the flow of visitors to Soho can be gleaned from Figure 1, whereas Figure 2 records the geographical origins of overseas tourists to Birmingham. The information has been extracted from the Archives of Soho which contain hundreds of letters of introduction. Letters of introduction or recommendation circulated like a form of moral currency in the eighteenth

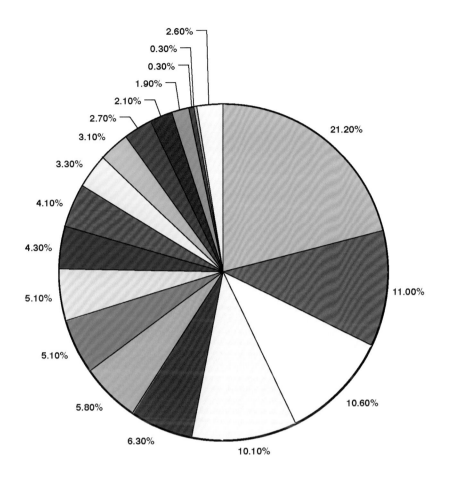

55 Foreign visitors to the Soho Manufactory by origin, 1765–1820. Source: MS 3147; MS 3219; MS 3782 (Birmingham Archives & Heritage).

century. They signalled genteel status, parity of social esteem and allowed bearers to exploit the presumption of civility on which the discourse of the Enlightenment was predicated. Even if he had so desired, Matthew Boulton could scarcely have withheld his hospitality from a traveller whose identity and impending visit had been properly announced. But there is little sign that the proprietor of Soho wished to curtail the flow of visitors – not during the first twenty years or so at any rate. Titled visitors were all part of the marketing strategy, whilst visiting *savants* gave substance to the claim that Birmingham was an important stopping-point in a circuit of scientific sociability linking the West Midlands to London, Edinburgh, Paris and other European seats of learning. Matthew Boulton's elaborate courtesy rituals, therefore, extended beyond the requirements of the aristocratic honour code. They had the transactional features of a commercial society in the making. In devoting considerable amounts of time to showing visitors around his premises, he anticipated receiving something in return: orders for his luxury goods; contacts that could be turned to political advantage; good conversation over the dinner table and, above all, recognition as England's greatest manufacturer of metalwares. This craving for praise and fame, which in the final analysis seems to have outweighed all other considerations (power, wealth, an abiding legacy), provides the linking thread of his adult life. 'Last week', he once informed a German business associate:

> we had prince ponitowski ye Nephew of the King of poland, the French, the Danish, the Sardinian & ye Duch [*sic*] Ambassadors we have had this week Count Orlof one of the 5 Celebrated Brothers who are favourites with ye Empress of Rusia & who have conducted her War against ye Turks & yesterday dined with me the Vice Roy of Ireland not a day passes but we have some Nobleman or other.[7]

Interruptions on this scale and the social responsibilities they entailed could be tiresome. But this was in 1772 when Boulton still had high hopes of his ormolu product-line, and tea and showrooms were actively pandering to the susceptibilities of the higher ranks of tourists. Soho was booming, the workforce having increased from around five hundred in 1765 to one thousand in little more than seven years. As such, it was probably the biggest single-site manufactory in the western world until the structural use of iron, gas lighting and rotary steam power made possible far larger, multi-storeyed industrial premises in the cotton towns of the northwest from the turn of the century. As Figure 1 makes clear, recorded visitor numbers were apt to ebb and flow in tandem with the health of the town's economy. Since Birmingham in general and Soho in particular were highly dependent on export markets, international strife tended to have a depressing effect. Judging from the correspondence files of the Archives of Soho, visitor flows contracted sharply at the nadir of the American War of Independence. Indeed the flow of tourists from overseas all but dried up in 1778–81. Employment at Soho would be cut by half during this period, too. But Birmingham's highly tuned industrial economy was transformed by the news of the signing of a peace treaty in Paris in 1783. Visitor numbers, and particularly foreign visitors now that the sea lanes were open once more, recovered swiftly. We know, too, that the workforce at Soho expanded in proportion – notwithstanding a flurry of international trade embargoes. By the end of 1787 Boulton would inform his Cornish agent that he was employing 'a thousand under my own roofs'[8] once more.

However, international strife did not automatically curtail the practice of industrial tourism, or place industrial premises out of bounds. Until the chronic struggle against revolutionary France was engaged in 1792–3, wars tended to be waged by governments, and tourists often went about their cultural activities unhindered provided land and sea travel remained relatively secure. The same was true of businessmen. Matthew Boulton continued to trade with France, using both official and unofficial channels, throughout the conflict in the American colonies, even though France and England were formally at war. Every visitor to Soho was a potential customer, even if a number were also potential spies as we shall see. But this relaxed cosmopolitanism drained away after 1793. When Britain joined the Europe-wide coalition against revolutionary France, the knock-on effects were felt in Birmingham immediately. The local economy, which was rooted in the extraction of raw materials from the Black Country and further afield, and their conversion into 'toys' and decorative goods, contracted sharply and remained in the doldrums for the remainder of the decade. With the streets of Birmingham filled with idle and increasingly unbridled workmen and women, visitor numbers plummeted. Unsurprisingly, the flow of foreigners presenting letters of recommendation at the gates of Soho declined the most sharply, for all governments involved in the conflict with revolutionary France were now seeking actively to impede the movements of their citizens across international borders.

Matthew Boulton weathered this downturn better than most. Ever since the 1780s the visitor flow to Soho had been driven principally by the desire to inspect his steam-power technology, and by the 1790s orders for Watt's improved steam engine were coming in thick and fast. The new Soho Mint (completely rebuilt and fitted out with pneumatic presses in 1797–98) also served to swell the volume of travellers anxious to catch a glimpse of the miraculous working of semi-automatic machinery. But tourist confidence as a pan-European phenomenon did not really recover until the autumn of 1801, by which time it had become apparent that the war with France had reached a stalemate. The Peace of Amiens (signed on 25 March 1802) was widely, if rather wishfully, interpreted as evidence that the Enlightenment was not over after all, and as a signal to resume the cultural practices of the pre-revolutionary age. Matthew Boulton was by now nearly seventy-four years old, and ailing. Since March of 1802 he had been often bed-bound and quite incapable of providing the conducted tour which each new arrival at Soho had come to expect. In a letter to his partner James Watt senior he raged impotently at the unending flow of visitors, yet still with a characteristic touch of pride and humour:

> I presume you know that I had a visit from Sir Joseph, Lady & Miss Banks' [president of the Royal Society and family], who came with 2 Friends. At the same time 4 Dumergues visited (& servants). Then there followed the facetious Mr Lee for 6 days, also a Russian nobleman, also the Emperor's favourite General Hitroff [Khitrov] & Co – I was likewise particularly requested by John Motteux to entertain & show every civility to the prince Carrini & Mon.^{sr} le Commander d'Aceto Kn.^{t} of Malta. I have also bestowed a day on my old friend Mr Tough of Palermo & his friend, to say nothing of English & Scotch friends & military visitors, also Mr Werberg & Mr Brendel. I long for deep snow that Travellers may be confined at home & I to my bed.[9]

In the event the rapid negation of the peace brought its own solution. Within eighteen months, hostilities between Britain and France had recommenced. Any lingering hopes that cultural intercourse might withstand the pressures of 'total' war were dashed when Bonaparte – as Consul for Life – gave orders that all British travellers in French-dominated Europe be rounded up and detained for the duration of the conflict. Soho's foreign visitor flow dwindled away to nothing, whilst domestic travellers also came to acknowledge that the great age of uninhibited industrial tourism had come to an end.

*

The code of Enlightenment civility demanded transparency. It could not apply to visitors travelling incognito, under assumed names and with ulterior motives. Matthew Boulton and his principal business partner, James Watt senior, first confronted the possibility that they might be extending hospitality to industrial plagiarists and spies in the 1780s. Yet they were reluctant to face up to the issue. At one level competitive spying was routine among Birmingham's 'toy' manufacturers where 'improvements' could scarcely be protected by patents. Steam engine technology was another matter, however, if only for the reason that huge sums of money had been expended on research and development. As men of science, Boulton and Watt were reluctant to impose limits on knowledge gathering and exchange. To have done so would have undermined the seamless character of the Republic of Letters in which even knowledge with remunerative potential was supposed to be free for the asking and the taking. But as businessmen, matters were not quite so clear-cut. Under the cloak of civility, visitors to Soho often took advantage of Boulton and Watt. Johann Georg Büsch of the Hamburg Chamber of Commerce would boast in 1786 of how he and his brother had managed to collect intelligence about Watt's modifications to the steam engine by passing themselves off as mathematics professors when receiving hospitality in Soho nearly a decade earlier.

Yet the partners usually knew when they were being exploited by visitors. As the flow picked up speed in anticipation of peace in the autumn of 1801, Boulton's son would remark in fairly light-hearted tones upon the arrival of the 'usual quantum of travelling cognoscienti [sic] with one or two adventurers, but who performed their part so awkwardly as not to hide the cloven hoof.'[10] His jocularity was rooted, perhaps, in the belief that the firm of Boulton & Watt possessed an unassailable technological lead. However, the remark also hints at the dilemma faced by these men who were both *savants* and leading technologists: the cultural reflexes of the Enlightenment scarcely permitted fine distinctions to be drawn between knowledge and 'sensitive' know-how. Every visitor, whatever his credentials or provenance moreover, was also considered to be a potential customer. Nevertheless, there are grounds for supposing that the partners became more cautious in their dealings with visitors in the 1780s. Were they anxious that their hard-earned knowledge might be 'kidnapped', or were they simply reacting to several flagrant abuses of their confidence and hospitality? Lev Fedorovich Sabakin, a talented Russian craftsman whom Empress Catherine had sent to the West, believed it to be the former. On visiting Soho in 1786 equipped with a letter of introduction from the Russian ambassador, he noted that his guide seemed keener to take him through the gardens than the industrial premises, adding 'I expect that they gave warning in advance what was secret and where I should not be taken.'[11]

The case that caused the most enduring sense of affront was that of Carl Friedrich Bückling. This official of the Prussian state mining administration first visited Soho in 1779 with the ostensible motive of purchasing four of the newly improved steam engines. In the absence of Matthew Boulton, James Watt showed him around and entertained him in his own house. No doubt Bückling's assumed baronial title and apparent lack of knowledge of steam-power technology helped to nurture the burgeoning relationship. However, the partners soon learned that their Prussian guest had bribed Soho workmen in order to obtain access to parts of the engines that Watt had neglected to show him. They also learned that on returning to Prussia, Bückling had caused a copy of the Watt engine to be built at a pit near Hettstedt in the Mansfeld district. However, the pirated technology failed to perform adequately, and in 1786 Bückling returned to England in order to resume his snooping activities among Boulton & Watt's customers. At this point the barely contained resentment of the partners boiled over. Watt described his erstwhile guest as a man 'who violates the duties of hospitality,'[12] recalling for good measure how a nugget of tin had disappeared from his mantelpiece shortly after a request from Bückling that it be given him as a gift. When reports reached Birmingham that summer that Bückling was trying to entice one of the firm's employees in the Cornish mines, the partners decided to launch a criminal prosecution against him.

It was in this somewhat strained atmosphere that Heinrich Friedrich Carl vom Stein (a genuine baron and the future reform statesman of Prussia) arrived in England and let it be known that he would like to make a tour of Soho. But again, reports went ahead that he was planning to visit using an alias. Charles Greville, the second son of the Earl of Warwick whom the baron had hoped to enlist as an intermediary, would acknowledge that Stein had initially assumed the title of Count Vidi – prompting Boulton to observe 'I say you would not have treated or considered any persons as Gentl.[m] who passed under two names.'[13] Understandably, therefore, the partners adopted a distinctly lukewarm demeanour. Still smarting from his experiences at the hands of Bückling, Watt advised that business relations be terminated the moment there were grounds for believing that Stein's draughtsman had made sketches of their engines. Alluding in veiled terms to the Bückling affair, Boulton spelt out the situation to Stein in as civil a manner as he could muster: 'I observe that you propose to make a journey to Cornwall soon. If I can promote your views as a Natural Philosopher, as a Mineralogist or as a Gentleman I shall be happy, But [sic] as a Mechanick & as an Engineer you must pardon me if I throw obstructions in your way.'[14] In the event instead of going to Cornwall the thick-skinned Prussian nobleman headed straight to Birmingham, for we know that he and his entourage dined at Soho on 3 April 1787, less than a month after Boulton's letter was drafted. It seems unlikely that he saw very much during his visit to England, for by now all of the West Midlands manufacturers were on the lookout for him. Stein would later describe his 1786–7 sojourn in England as a 'wasted year.'[15] Matthew Boulton, for his part, drew a more optimistic conclusion. In a letter to his son who was studying in Germany, he related how 'we have had many philosophical & mechanical robbers, who have come here under false names and pretences', adding the observation 'they have all fail'd in making good machines upon returning to their own countries, after laying out a much greater sum than they

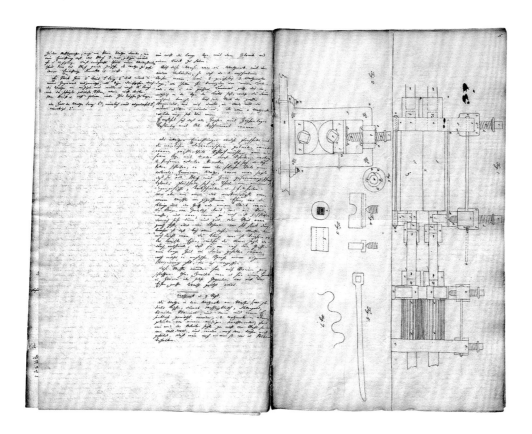

56 J. M. Ljungberg, Soho Manufactory rolling mill, 1788. Drawing, sketchbook size approx 42 × 30 cm. National Archives, Stockholm.

57 G. von Reichenbach, Soho Manufactory 'lap' engine, 1791. Ink on paper with additional colour, 12 × 19 cm. Deutsches Museum, Munich.

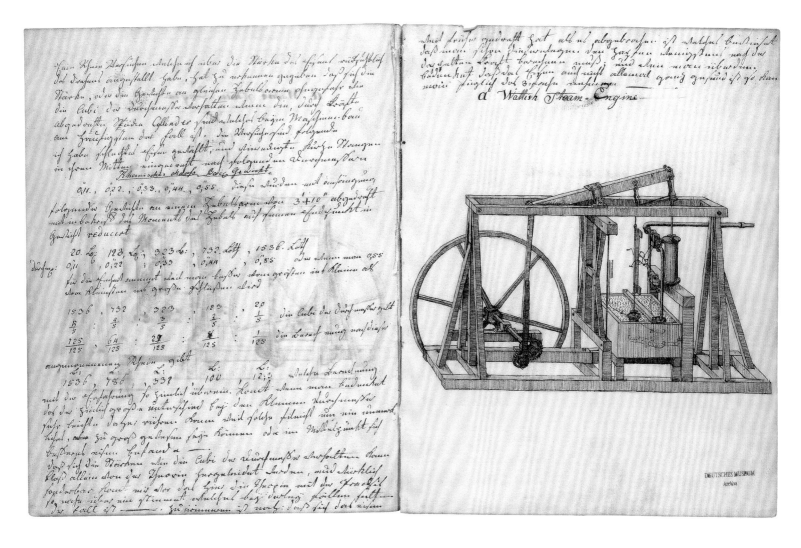

might have purchased of us perfect machines that would have burnt much less coals than those they have erected …'[16]

Jøns Matthias Ljungberg, a Swedish traveller in Danish service, adopted a more skilful approach. In fact it was men like Ljungberg, combining the personas of itinerant natural philosopher and government agent, who posed the biggest challenge. The scale of his intelligence-gathering activities would only come to light by accident. By all accounts he had been travelling in Britain since the late-1770s; no doubt he had discovered that his Chair in mathematics and astronomy at the University of Kiel opened doors that would otherwise have remained closed. It is probable that he settled in Birmingham for the first time around 1784 and secured easy admission to the local community of natural philosophers. Although everyone seems to have known that he was in the pay of the Danish Crown, this does not appear to have been viewed as a problem. Matthew Boulton entertained him at Soho, and received fulsome thanks for his hospitality when Ljungberg finally packed to go home. However, when four heavy crates accompanying his luggage were opened in the London Customs House in August 1789, it became immediately apparent that he was shipping out the fruits of more than a decade of successful industrial espionage carried out in defiance of the Tools Acts of 1785–6. Josiah Wedgwood of Etruria expressed the greatest indignation since inspection of the boxes uncovered evidence that Ljungberg had been trying to access his kiln technology by bribing his workers, as well as collecting specimens of his clays. Matthew Boulton and James Watt joined in the condemnation, for Ljungberg's sketchbook also contained many drawings of machines, tools and products seen in Birmingham and particularly in Soho (Fig. 56).

These cases form the backdrop to Matthew Boulton's increasingly testy attitude towards visitors as the 1780s turned into the 1790s. Certainly promotion remained necessary, but it now focused more on his steam-operated mint than on enlarging the market for objects of taste. At best visitors were interruptions; at worst they were potential engine pirates and 'kidnappers' of useful knowledge. It is in this light that we should view a final example of eighteenth-century industrial espionage: the case of twenty-year-old Georg Friedrich von Reichenbach. A banal episode of spying, it differs little from half a dozen other attempts to filch Watt's engine secrets, but for the fact that it is unusually well-documented from the perspective of the would-be intelligence gatherer. In 1791 the Elector of Bavaria ordered from Soho a small steam engine for the purpose of pumping water into the Mannheim city reservoir. Whether the order was anything more than a pretext devised by Sir Benjamin Thompson, the Elector's general factotum, to send a man out to Soho, is hard to fathom. At any event, a capable young mechanic by the name of Reichenbach appeared at the gates of the Manufactory on 10 July of that year. He was escorted by Joseph von Baader, an English-speaking Bavarian engineer whom Boulton knew all too well. After a couple of days in Birmingham, Baader departed for the north of England leaving the monolingual Reichenbach to cope as best he could.

At this point we can pick up the story in the young mechanic's notebook-cum-diary. Matthew Boulton, we learn without surprise, was not very pleased to see the pair. He could scarcely refuse them hospitality, yet they were not offered a bed for the night at Soho House. Once Baader had moved on, Reichenbach had to make the return visit to Soho from his inn in Birmingham on his own, and on foot. He soon became lost on Birmingham Heath since he was not capable of asking for directions. After a crisis of doubt about the whole undertaking, he pulled himself together and, as he put it in his diary:

> I endeavoured as speedily as possible to reconcile myself to this unpleasant position: and I soon observed it had its advantageous side, for I was able by giving a few small tips to obtain the opportunity, despite the secrecy of Mr Watt and Mr Boulton, thoroughly to study the mechanism of the fire – or steam engine. I worked at my drawings for six weeks, for I had to maintain secrecy not only against Mr Boulton but also against all the workers who were there. For this reason, this work cost me indescribable labour, for not only could I ask no questions of anybody, but also might not for fear of arousing suspicion; so on the other hand I was only allowed to look at them at certain times.[17]

He went on to describe in great detail, and to sketch (Fig. 57), the workings of the Soho 'lap' engine, that is to say, Watt's double-acting steam engine fitted with a rotary drive. This same engine can be seen today in the London Science Museum. The Soho assignment must have been a character-building experience for a twenty-year-old, for we know that Reichenbach was disdained and suspected throughout the Manufactory even though Boulton had apparently agreed to train him as part of the transaction over the Mannheim engine.

Would Matthew Boulton ever have reached the point of denying visitors access to his showcase Manufactory? Fellow manufacturers, after all, were moving in this direction in the more competitive international climate of the late 1780s. In his popular tourist guide of England published in 1789, the Frenchman Louis Dutens[18] felt it necessary to advise travellers that it was no longer a straightforward matter to gain access to Birmingham's workshops. Yet it seems unlikely that Boulton would have taken this step of his own volition, for he liked to think of himself as a man of the Enlightenment whose *raison d'être* was not to be circumscribed by mere questions of commercial advantage, as this essay has sought to make clear. The pressure to seal off Soho from visitors and would-be spies probably came from the next generation: Matthew Boulton and James Watt's sons. They were men who had not been cast in the Enlightenment mould, and for whom the machine-building business was just that – a business. Once a decision had been taken to build an ironworks and integrated engine assembly

58 Matthew Boulton's draft of the intended notice curtailing visits to Soho. MS 3782/12/108/86, Notebook 35, 1800 (Birmingham Archives & Heritage).

59 Matthew Boulton's diary showing his entry for 21 Sep. 1802, 'Lord Nelson here'. MS 3782/12/07/30 (Birmingham Archives & Heritage).

plant on a greenfield site not far distant from the old Manufactory in 1795, the step of banning visitors followed almost inexorably. Matthew Boulton inclined with considerable reluctance to the protectionist arguments of his young partners, and from 1800 onwards notices were inserted in the London and provincial newspapers advising travellers that both of the Soho industrial sites, the Manufactory and the Foundry, were now out of bounds (Fig. 58).

Yet visiting did not altogether stop, and the interdict had several times to be reissued. For Boulton had retained ownership and undivided management control of Soho Mint – the last and most satisfying of his numerous speculative ventures. Many a time he compensated disappointed Birmingham-bound travellers with an invitation to inspect instead his steam-driven coining presses. Nevertheless the denial of access to the Manufactory and the Foundry was regarded by some as an unpardonable breach of Enlightenment civility, particularly when the policy was applied in a manner that was visibly lacking in even-handedness. In his published travelogue, the German tourist Christian Goede[19] grumbled that when he had passed through Birmingham towards the end of 1802, Soho was off-limits, whereas Rear Admiral Horatio Nelson and the Hamiltons had been admitted only a short time earlier. To be fair to Boulton, he was acutely aware of the risk to reputation that such inconsistencies posed. When Sir William Hamilton first wrote on 24 August to announce the impending arrival of Britain's Hero of the Nile, Boulton was thrown into agonies of embarrassment. Pleading ill-health, which was real enough, he answered 'it is not within my Power to wait upon you,'[20] before wheeling out the prohibition on visitors, and reciting the names of all the illustrious families whose members had been refused access to Soho over the previous two years. Yet something must have happened to alter his resolve, or rather that of his young partners. Presumably there was an outcry in the streets of Birmingham as the news leaked out that the Hamilton travelling party would bypass the town and return to London by way of Worcester and Warwick instead. At all events, Nelson made a triumphal *entrée* into Birmingham in September and the whole party would also be received in Soho by the bed-bound Boulton (Fig. 59). Emma Hamilton was even invited to strike off specimen coins from the new Mint presses.

By this date, however, it was the chance to meet and converse with Matthew Boulton that visitors appear to have prized above all else. Sedulous self-promotion as England's greatest metalwares manufacturer had resulted in his becoming something of a legend in his own lifetime. For Boulton, too, the arrival of visitors was an opportunity to practise the art of conversation, even if the timing was not always to his liking. The consequences of this single-minded addiction to the cultural practices of the Enlightenment could sometimes border on the surreal: 'our house has been like that of Babel', he once reported to James Watt following the visit of six assorted foreigners, 'by bad English, French, Italian, German, and bad translations.'[21] Yet there can be no reason to doubt Matthew Boulton's powers of attraction. Tourists both domestic and foreign 'seem much disappointed when they cannot see you,'[22] his silver and plated goods manager would note. For some of the visitors from overseas, in particular, the tour of the Manufactory must have resembled a veritable voyage into a world unknown – a voyage in which Boulton played the part of the great navigator. 'Of all the interesting things to be seen', confided Fanny De Luc after a pilgrimage to Soho in 1800:

> Mr Boulton himself is the most interesting. We passed an evening listening to him, & we wished he had never ceased to speak; we could have heard him constantly without having another wish; he is an inexaustible [*sic*] mine of precious knowledge, & of every thing that is amiable; he treated us with the utmost Kindness & made us travel through a Country of Wonders![23]

Imitation is one thing
Counterfeiting another

The finest & most difficult Coin
that is possible to make may be
imitated though not Counterfeited

Perfection in Engraving is
undoubtedly desireable both
for the beauty of the Coin as well
as one of the difficulties of Counterfeiting
but that alone is a very feeble
security because that extream
delicacy & beauty cannot be
preserved in the gross
particularly in Copper
Coinage of Millions, as it is
in Specimens struck from new
dies — Moreover if the
hand of or the first Engraver
that ever existed was to produce
a die in the happiest moment
of Genius I say such die
may be so closely copyed by
Mechanical means that
the Publick w.d not be

able to discriminate the
true from of false Coin &
even the artist himself
would pause for a few
Moments whether it was of
work of his own hand or not

Perfections in N.B Coin
Perfect round
D.o = Diam.rs
polished Ground
applicable to a gage
Durability in Letters
compact stacking
less liable to wear
Concave Millies
Striking D.o or inscription on the
Edge at the same blow

10: Matthew Boulton's Mints: Copper to Customer[1]

Sue Tungate

Matthew Boulton is well known for building what was, in its time, the most famous manufacturing business in the world at Soho, near Birmingham; he then went on to make a practical reality of James Watt's steam engine including the rotary motion engine developed in 1781. Finally in the latter part of his life, he turned his talents to the nation's coinage. The productions of the Soho Mint, including the 1797 regal coinage, were among his proudest achievements and involved no other partner. James Watt said in Boulton's obituary:

> Had Mr B done nothing more in the world than what he has done in improving the coinage, his fame would have deserved to be immortalised; and if it is considered that this was done in the midst of various other important avocations, & at an enormous expense for which he could have no certainty of an adequate return, we shall be at a loss as to whether to admire most his ingenuity, his perseverance or his munificence.[2]

With his development of steam-driven coining presses Boulton virtually invented modern, high-quality coinage as we know it today, and proceeded to sell his coins, tokens and medals, and the minting process, to customers all over the world.

The Mint at Soho was the first in the world to be powered by steam. In order to produce his coin, Boulton had to develop the technology and build the equipment needed; obtain metal of a suitable quantity and quality; and train suitable workers for the Mint, including engineers, workmen, designers, engravers and salesmen. He had also to acquire orders and make suitable designs for his customers, organise the finances, and finally transport both copper and coin. Boulton was involved in every aspect of the Mint's growth, giving detailed instructions which are documented in the Archives of Soho at Birmingham Reference Library.

In the eighteenth century coins of small value were in short supply, and their quality was a national disgrace. As early as 1771 there were complaints that 'The scarcity of cash in this part & for many miles round us has been for some time past greater than I ever remember'.[3] The lack of copper money led to much counterfeiting of coins, often in backstreet workshops in Birmingham.[4] Boulton had been involved in the Royal Mint's re-coinage of 1773 as an agent to receive worn gold sovereigns (the Royal Mint collected such coins via trusted agents, who returned them to the Mint for restriking). But where copper coin was concerned, apart from a small issue in 1775, and despite protests to parliament and the king, no small change was provided until Soho Mint produced the first copper penny and twopence coins. No copper coinage was made by the Royal Mint until 1821.

Boulton was concerned about workmen being cheated, and wanted to provide a coinage with intrinsic value of metal, of constant diameter, thickness and weight, and to use steam power to produce it cheaply so that it would not be worth counterfeiting (Fig. 60). He wrote to Sir Joseph Banks in October 1789: 'I took up the subject because I thought it would be a publick good, and because Mr. Pitt had express'd a wish to me of seeing something done to put an end to counterfeiting the copper coin'.[5] Also, Boulton was always interested in developing new schemes. His friend James Keir said in a memoir, 'it was always in Mr B's mind to convert such trades as were usually carried on by individuals into Great Manufactures by the help of machinery, which might enable the article to be made with greater precision, and cheaper than those commonly sold'.[6] So it was natural that he would start to think of producing coins en masse, when the Royal Mint was reluctant to coin copper. By 1789 Boulton claimed that he could make one hundred million halfpence per year, 'whereas the officers of the English Mint ….are very much hurried to make 3½ million of Guineas per Year'.[7]

The links with the mining industry which Boulton developed through supplying steam engines were another important factor in the Mint development. By 1780 twenty Boulton & Watt engines were operational in Cornwall to pump water out of copper mines so that more ore could be mined. Boulton became part-owner in several mines, and started selling copper as well as using it. In 1782 he was corresponding with Samuel Garbett, who was reviewing the Royal Mint, and thought that coining was one way to use up the copper surpluses.[8]

Despite encouragement from Pitt's Government, it took until

60 Matthew Boulton's thoughts on how to eradicate counterfeiting of coinage, from his notebook for 1799. (Birmingham Archives & Heritage, MS 3782/12/108/83)

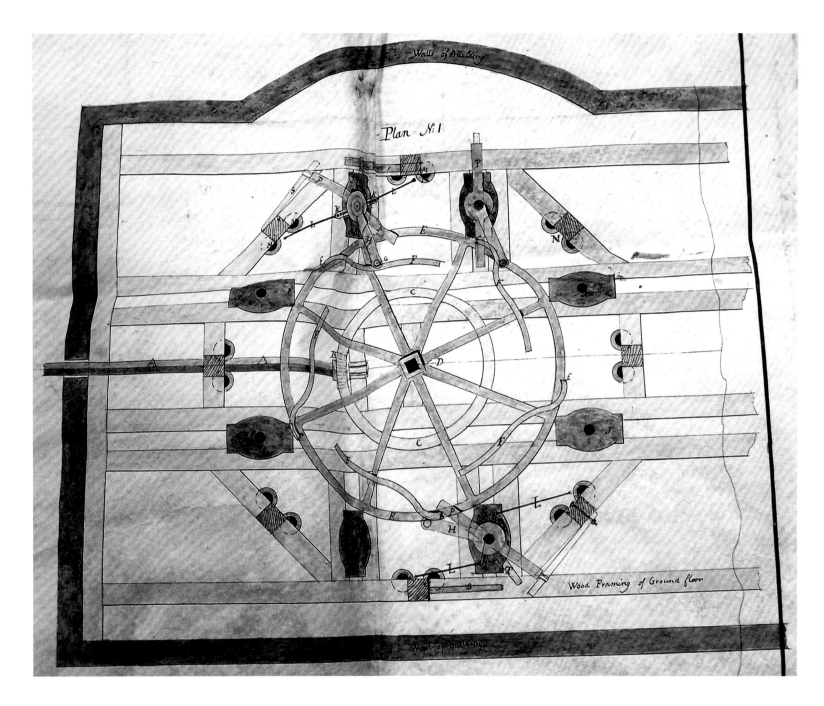

61 Diagram of coining presses, from the 1790 Patent Document. Parchment, overall size 36.6 cm high, 54 cm wide. (Birmingham Archives & Heritage MS 3782/17/2)

1797 before Boulton was finally allowed to provide a new regal coinage for Britain. In the meantime Soho Mint coins, medals and tokens were ordered by a variety of customers, and some remarkable works of artistic merit were fashioned. These were some of the first mass-produced pieces which the general public could own, not only for their intrinsic value, but also as symbols of industry, culture and of Britain's rising international reputation.

Mint technology

Normally coins, with designs in high relief, were struck by blows of a die on metal, using a press powered by several men.[9]

Boulton's first coining orders were done this way in 1786 and 1787 for the Honourable East India Company for their colony at Bencoolen in Sumatra. This commission was struck at a makeshift mint in London, with blanks made at Soho, using similar techniques to those used in Boulton's button-manufacturing business. The experience made Boulton decide to apply steam power to coining. At Soho all the coining operations were concentrated on one site.[10] The Mint was placed about a hundred yards from the principal buildings of the Manufactory in order to maintain the secrecy of the new technology. The rebuilding of the Soho water mill in 1785 had significantly increased rolling capacity, so that sufficient copper could be flattened into sheets. New steam-powered presses were designed for coining and blank cutting. Special rooms for annealing and cutting of blanks, and forging and multiplying dies, were also installed; and a new warehouse was built in Livery Street, Birmingham, in 1787.[11]

In February 1788, Boulton was encouraged to think that a regal coinage contract was imminent. He wrote to his son Matthew Robinson Boulton with enthusiasm:

> I was sent for to Town by Mr. Pitt and the Privy Council about a new copper coinage which I have agreed for, but at a very low price, yet nevertheless it shall be the best Copper Coin that ever was made. I am building a Mint & new Manufacture for it in my Farm Yard behind the Menagery at Soho where I shall be close engaged for 1½ Year.[12]

Many experiments to impart motion to work the coining presses were tried. In January 1788 the presses were in two rows linked to the steam engine at the east end by sliding rods; by November 1788 they were arranged in a circle and powered by bars from a horizontal wheel turning above (Fig. 61) This imparted the force to the screw presses. Each press had an air pump to provide the return stroke, and an automatic fork, or layer-in, which fed the blanks under the die and removed the completed coin. Inventories taken of the Soho Mint mention the great presses for coining, the fire [steam] engine, and a variety of workshops including the multiplication shop and cutting shop.[13] Vacuum pumps were introduced later and a new coining room built in 1798. Boulton wrote detailed lists of instructions, including how the rooms should be arranged, how the metal should be rolled and cleaned, how the dies should be forged and so on.[14]

In addition to sorting out the steam-coining presses, Boulton set his mind to improving other areas such as die making. For a coin, medal or token, two hardened steel dies, each engraved with a unique design, are used, one for the obverse (front side, or 'heads') and one for the reverse (or 'tails'). In order to speed up coining, the dies needed to be made with shallower images, which extended their life due to the reduced forces used in striking. Boulton also introduced a third die, the collar. This ring-shaped piece of metal surrounds the metal blank, to make the edges straight and vertical and enables the coins to be struck consistently the same size, which, as Stebbing Shaw noted in 1798, 'is not the case with any other national money ever put in circulation'.[15] Initially the sexpartite collar invented by Jean-Pierre Droz (Fig. 62) was tried, but later a one-piece collar proved more successful. Others made attempts to coin in a similar fashion, but as Boulton wrote to Joseph Banks in 1789, no one else could do it so cheaply:

> I have also heard of an attempt to strike crown pieces at the Tower in collers, but it was found so troublesome and the collers so hazardous that I believe there never was half a dozen of them struck, and if such a thing had been proposed to the moneyers they would have concluded that it would be worth a peny [sic] at least to make a half peny.[16]

62 The Droz sexpartite collar, designed to produce straight edges to coins. Steel, external diameter 16.5 cm (Birmingham Assay Office)

Dies were usually made from soft steel; then after engraving they were hardened (heat-treated to make the steel hard and durable). Hundreds of working dies were made for a large coinage as they needed to be changed regularly. Boulton wrote in 1791: 'Pray request Mr Lawson to consult with Nichols about supplying the mint constantly with perfect neat sharp well polished dies which I think should be changed after striking every 30 or 40 thousand pieces'.[17] The finished dies were kept in a special locked mahogany cabinet.

In Boulton's Mint notebooks there are detailed descriptions of the various processes of die-making. For example: hardening was done by placing the die in a cast-iron pot, completely embedded in animal charcoal, chiefly made from leather. The pot was placed in an air furnace, in which coke was burned to give an even heat. Once the die reached the correct heat, judged generally by experience and the colour of the flame and metal, it was immersed in a large cistern of water, which was kept at a constant temperature by a continuous flow of cold water.[18] At this point the dies often cracked and the work of the engraver was lost, as happened with some of Boulton's earlier medals. Boulton was very particular about the steel used for dies and had a variety of suppliers, including John Rennie, in 1791. He was concerned with getting the best rather than worrying about the price.[19] He wrote to Benjamin Huntsman in 1797 to order twelve dozen two-and-a-half- inch diameter dies:

> These dies are steeled quite through & take from 1½ lb to 1 lb 10 oz of steel. They are for striking penny pieces of the size [halfpenny] I have sent with the Die but as they are struck in Collers they may require a very hard blow, and if not steel'd through, the dies would sink in the middle.[20]

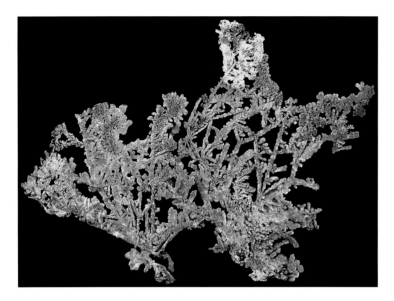

63 Specimen of native copper from Cornwall, from the collection of Boulton & Watt's engineer William Murdock. Width 20 cm, height 15 cm [Lapworth Museum of Geology, University of Birmingham]

The use of good-quality metal was also important for coining blanks. Copper was delivered to the Soho Mint from suppliers as sheet or cake copper, and, during the early rolling processes, scale (copper oxide) was removed by scraping, and by using urine. The metal was rolled several times to precise thicknesses.[21] Boulton noted that 'no flame or black smoak should pass over it at time of scaleing' and that the workers should 'Take care in the finish rolling not to handle the surface with dirty greasy fingers as it rolls in that dirt & tarnishes the metal'.[22] Gold and silver might be melted down on the premises, assayed and formed into strips, but were less commonly used.

After the strips were rolled, a steam-powered blanking press punched out discs of metal of a specified size-to-weight ratio with so many per pound avoirdupois. Because of the shearing action of the punches, the blanks had rough edges (burrs) which were removed by abrasion in shaking bags, again powered by the steam engine. The blanks were then cleaned again, sometimes with vitriol (sulphuric acid) or whiting (lime) and dried with sawdust.[23] Boulton encouraged innovation in his team of workers; his Mint engineer, James Lawson, reported in June 1789:

> The principal improvement is in Burnishing the Blanks; Peter Ewart has the merit of it. It is by putting them between two Brushes at a small angle so that the pieces, sand and water being put in at one end, … come out perfectly well brushed both sides and edges.[24]

The blanks were next sorted over a gauge to eliminate defective pieces, which were remelted. The scrupulously cleaned, dried and polished blanks, now known as planchets, were then ready to be struck on the Mint's coining presses. They were fed by gravity in a stack or rouleaux (rolls) of twenty or so planchets into a vertical tube attached to the coining press. The lowest planchet was pushed forward by the metal fork, known as a layer-in, to drop into a circular steel plate, the collar, in which the lower die is recessed. This layer-in enabled coining to be much faster than the normal method of locating blanks by hand.[25]

After striking, the pieces were checked, counted, weighed and packed. Packaging had to be sufficient to enable products of the Soho Mint to reach the customer without damage or tarnishing.[26] This could involve travel by horse and cart or wagon, canal boat, or ship to London, and then vast distances by sea to places as far afield as India, America and Sumatra. To prevent damage to the finish, the pieces were wrapped in paper, and then placed in specially made casks.[27] Medals in silver and occasionally gold were dispatched by coach in individual display boxes.

Metal

The supply of metal was vital to the organisation of the Soho Mint. The Manufactory already produced a wide range of metal products from buttons to steam engines, for a variety of markets, and thus there was an extensive array of metal-working skills in the existing workforce. Collectors of Soho coins, medals and tokens were offered a range of metal options, from cheap versions in tin, copper, or white metal, to special orders in gilt copper, silver and gold, or 'bronzed'. Examples in other metals can be found in the collection at Birmingham Museum and Art Gallery, such as the early brass Otaheite medal of 1772 (see Chapter 11, Fig. 67) and trial strikes in lead. Later, after the discovery of aluminium in 1824, there were also restrikes in that metal.

Gold coins were produced by the Royal Mint and Boulton was very clear that he would not interfere with that privilege. But he did use gold in the production of medals for special customers such as George III. Silver had been used in world coinage since antiquity; Boulton struck a silver coinage for Sierra Leone, and also overstruck silver tokens for the Bank of England and Bank of Ireland.

Boulton also produced bronzed and gilt pieces. Bronzed copper blanks were formed by baking in bronzing powder to produce tones varying from yellow to dark chocolate. This powder was readily available in London. Soho Mint struck prepatinated bronze blanks so that the dies could impart a brilliant mirror-like finish.[28] Gilding involved the use of poisonous mercury and nitric acid as well as gold, and Boulton tried to avoid harm to his workers wherever possible. He wrote to a correspondent in 1802:

> If the Gilding Chimney doth not properly draw off the Mercurial Vapour the party will very soon grow paralitick… I have contrived a Gilding hearth or Stove which effectually draws off all the fumes & condenses them in an Iron Vessel by which the Whole of the Mercury is saved & the Gilder is never injured.[29]

However most of his coins and tokens were produced in copper including the bulk orders for the Government and for the East India Company.

Copper was used for many products at Soho, and originally Boulton may have obtained it from Bristol and Cheadle. Here, under the leadership of the Bristol Brass Company, a price cartel controlling the price of copper ore had been established.[30] It was not very economic to smelt the copper ore locally in Cornwall due to fuel costs, so the ore was shipped from there to Swansea where smelters, not miners, profited most (Fig. 63).[31] Boulton became very influential in the Cornish copper industry. Thomas Williams of the Anglesey Copper Company, was not part of the smelting cartel, and he wrote to Boulton in June 1781 to confer on the 'Copper Trade of this country in general & the Mineral Concerns of the counties of Cornwall & Anglesey in particular'.[32] He wanted to set up independent smelting works. Unfortunately the Cornish miners would not initially agree to cooperate with Williams and he was able to undercut their copper ore prices.[33]

Both Boulton and Williams wanted a new regal coinage to use up the surplus supplies of copper profitably. They had been cooperating on coinage contracts such as that for the East India Company in 1786.[34] They agreed in October 1787 for Williams to provide the copper and Boulton the coining, but later Williams' virtual monopoly on the supply of copper became a serious problem.[35] In 1788, when Boulton originally expected the regal coinage contract, copper prices were falling, to around £73 per ton.[36] However, by the time the contract was finally agreed in 1797, Williams' Parys Mine in Anglesey was declining, and there was a lack of copper at a reasonable price.

For the first regal coinage Boulton paid around £108 per ton for copper, but for the 1799 contract the price was £121, and by 1805 it had risen to £169.[37] The fluctuating price of copper caused difficulties throughout the life of the Soho Mint. Boulton kept track of copper prices in places as far away as Calcutta, India and Bussorah (Basra) in present-day Iraq. The highest price in Calcutta was in 1793 when the cost had risen to £115 from a previous low of £77 in 1788.[38]

Distribution

To fulfill an order, bulky consignments of copper were shipped from Cornwall via Swansea smelters to Soho, and tons of valuable coins, medals, or tokens were transported from Soho to customers (Fig. 64). Freight and insurance, wharfage, customs duties and so on, had to be included in the price as well as the cost of coining and copper. For foreign markets there were exchange rates to consider, as well as the credit-worthiness of foreign agents and factors (merchants working on commission). Most of the Mint production was made during the Revolutionary and Napoleonic Wars, which added further complications. Capital was tied up in copper stocks, and payments for completed Mint orders were not always easily obtained.

Roads improved throughout the eighteenth century and by the 1790s it was possible to receive post daily from London, which took around sixteen hours per trip, but the cost of land transport remained high. Water transport was slower but much cheaper. Boulton used a variety of routes to deliver to his customers depending on how important the order was, and where they were located. For fast bulky deliveries to London, Boulton used one of his regular carriers, 'Sherratt's Flying Waggon'; small orders for medals were sent by coach.[39] Copper generally came up the River Severn to Stourport and then along the canal to Birmingham. Boulton also used the canals to send completed orders northwest to Liverpool or northeast to Hull. For America, orders could be sent to Bristol along the Severn or to Liverpool depending on port charges and sailings.[40] His usual waterways route to London and the Continent was via Hull, but sometimes it was via Oxford and then the Thames.[41] As early as 1769 Boulton had decided that it was better to ship goods in casks despite the cost, rather than have metal or ore shipped loose, as consignments were more secure and losses minimised.[42] Each cask could be individually weighed and labelled, and consignments could be traced.

Unfortunately transport was dependent on the weather. During the correct season the East India Company's coin orders had to reach St Botolph's Wharf, London, in time for regular sailings to the Far East on Company ships. Matthew Robinson Boulton wrote in 1795:

> I am afraid it will be some time before you receive advice of the last Ton of E.I Coins; there is yet 8 Tons to strike & they cannot work overtime for want of Metal, the floods having put a stop to Land as well as Water Carriage, the copper coming in the wagons have been detained upon the Road & this disappointment in the supply of Metal will much retard the coinage.[43]

Frosts were also a problem. Delays could mean that it might take up to two years for the round trip to places such as Ceylon (Sri Lanka).

In the distribution of the regal coinage of 1797 transport expenses were a significant factor. It could cost more than £10 per ton to get coin by road to Newcastle, for example, but less than £2 by water. Boulton's agent John Southern advised that 'your attention should be directed to those parts of the kingdom *only* where water carriage can be employed'.[44] According to the original coinage license Boulton was to be allowed seventeen shillings for delivery no matter how near or far. This was increased to £4 for the 1799 contract.[45]

Workers

Boulton already had experienced workers at the Soho Manufactory to develop the equipment needed for the Soho Mint, and he was fortunate in having able assistants such as James Lawson and John Southern, and his son Matthew

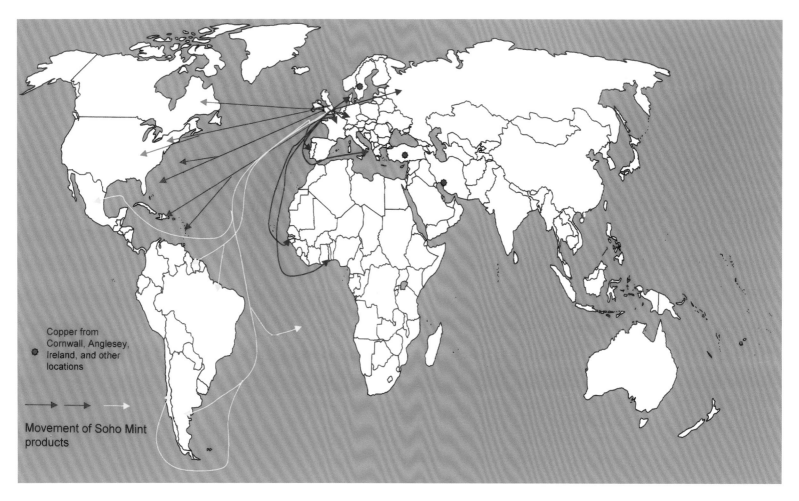

64 Map showing worldwide distribution of coin from the Soho Mint. Green and yellow arrows indicate the Americas, and purple arrows indicate Europe and W. Africa.

Robinson Boulton. The Soho rolling mill run by John Kellet was able to expand to roll for the Mint, and there were on-site specialists in die forging, cutting and annealing such as John Busch, John Peploe and Joseph Harrison.[46] Boulton was keen to retain key workers for the Mint and he used them in other areas of the Manufactory when coining work was scarce. For example, Nelson and his team prepared gilt blanks as well as gilding plated wares.[47] Workers were encouraged to put forward their own ideas; John Middlehurst was employed to bronze blanks, and Boulton wrote: 'It would be well to have such a hearth as he might approve built up in the room over the Shakeing shop'. He added 'I presume an Iron plate will not answer so well as a Coal fire, but *please to consult* John about the best means of Bronzing 40 gross per day'.[48]

Soho also boasted a pool of talented artists, skilled in producing articles for the fashionable world. Some of Boulton's long-term employees had been apprenticed there, and he trained others who were later involved in token or medal making, such as Edward Thomason, John Gregory Hancock and Thomas Wyon, who multiplied dies for early tokens and also made the 1791 guinea die and a pattern die for Barbados.[49] John Westwood senior engraved some of the earlier medals from 1772, and Hancock provided the dies for Anglesey and Wilkinson tokens (see Cats 259, 260).

In late 1786 Boulton wanted Jean-Pierre Droz, who was employed by the Paris Mint, to engrave a bust of George III for his projected regal coinage. In addition to being a fine engraver, Droz also claimed to have produced improvements in mint design. He came to Soho Mint in 1788 but had left by July 1791. Boulton was disappointed with him, 'the most ungrateful, most ungenerous, & basest man I ever had any concern with'.[50] Boulton later employed Rambert Dumarest from August 1790 until June 1791, and Noel-Alexandre Ponthon from August 1791 to September 1795, and they engraved many of the tokens that Boulton issued. Richard Phillips was used as a specialist letter engraver.[51]

Boulton's best-known engraver was Conrad Heinrich Küchler, who from 1793 to 1810 produced dies for the British regal coinage and a series of beautiful medals. John Phillp came to work as an apprentice draughtsman at the age of fourteen from March 1793.

65 Obverse and reverse of the medal struck to celebrate the victory of Marquis Cornwallis over Tippoo Sahib, the Sultan of Mysore in India in 1792. Diameter 48 mm. (Cat. 272) [BMAG 1974 N27]

66 Obverse and reverse of the medal commemorating 'The Glorious First of June', a naval battle in the Atlantic in 1794 between the British and French fleets during the French Revolutionary War. Commanding the British fleet was Admiral Richard Howe. His aim was to intercept a grain convoy heading from America to France to relieve the near-starvation of the French people. Howe's ships captured six French ships, sank one and damaged many more and their gallantry was greeted with enthusiasm in Britain, but the French grain convoy still got through safely to the port of Brest. Diameter 47 mm. (Cat. 275) [BMAG 1885 N 1536.10]

He produced many drawings of the Soho Manufactory and Soho Mint, and also designs for medals and tokens.[52] Some medals had the contribution of several artists, such as the Monneron medals with input from Ponthon, Dumarest and Dupré.[53] Conrad Heinrich Küchler and Carl Frederic von Breda (who painted Boulton's portrait in 1792) collaborated on a medal of Gustavus III of Sweden in 1793.[54] Sir William Beechey was also pressed into service with the modeller Peter Rouw to produce a notable medal of Boulton.[55]

Design

Why were some of Boulton's coins so elaborately made? He did produce simple coins and tokens (as for the East India Company) but many of his pieces are beautifully designed and executed. The complexity of the design of each piece depended on a variety of factors, such as the subject matter, the intended purpose, and the budget priorities. Time factors were often important, too, for example, George III ordered two hundred silver medals to celebrate Queen Charlotte's birthday and gave Boulton just a few days' notice. It was a matter of pride and reputation to complete the order on time. Boulton wrote, 'the King was told it was impossible to which he replied "that if Boulton could be favour'd he was sure he w^d do them as nothing was impossible wth him".'[56] Other orders took years to complete. Major-General Claude Martin sent his first design for a coin from India to Soho in 1794, but the order was not delivered until just before his death in 1799.[57]

Boulton was very concerned to maintain the name of Soho as a 'hallmark of excellence'.[58] He had a large library of books which he used as design sources, such as *Complette de toutes Les Médailles de Chevalier Jean Charles Hedlinger*, with 132 illustrations of medals, and he bought important books of prints and engravings from his bookseller Peter Elmsley.[59] Boulton's collection of sculptures and models, for example from Flaxman, were also used.[60] Boulton was an avid collector of medals and in 1797 was sent a complete set of Russian medals by Emperor Paul of Russia.[61]

A design produced for the Soho Mint needed to please the aesthetic sense, but also to be practical for mass production. The image was the pictorial representation of the issuing authority, and some commissioners of tokens wanted to show their industry and importance by employing devices such as a coat of arms, or their own profile. On regal coins, ships were exploited as symbols of British naval power in a time of war. Allegorical figures were used, but also images from industry and commerce, and scenes from contemporary life. The inscription had a functional use, for example to show the monetary value of the coin; or to instill some message, whether mundane such as 'SELLS WHOLESALE WOOLEN & LINEN DRAPERY GOODS … CHEAP' as on the Dundee token, or moralistic: 'FAS SIT PARCERE HOSTI' (Let it be the right thing to spare an enemy) on the Cornwallis medal celebrating a victory in India (Fig. 65; Cat. 272).

Sometimes Boulton's engravers were given a specific image and exact specifications by the customer. His first Sumatra coinage in 1786 consisted of keping coins with a simple design, the so-called balemark, or trademark, of the East India Company, and an Arabic inscription of value. This design had been established in 1783 by John Marsden, the elder brother of the orientalist William Marsden, who supplied the denominations, their proportions and the inscriptions to Boulton (Cats 280–82).[62]

Some medals were produced as speculative products with no specific customer in mind. In such cases the imagination of the engraver was given relatively free rein but Boulton often had some input and took advice from scholars. An example is the early medal for 'the Glorious First of June' 1794 (Fig. 66; Cat. 275). It was based on a print of an original picture of Lord Howe, painted by Copley, and bought by Boulton. He suggested that on the reverse showing warships 'you should distinguish the 3 coloured Flag in the French ship, and suppose you may see one of their flags at St Paul's'.[63] He also pointed out that the ships should be less perpendicular due to the firing of broadsides. The inscription on the medal was corrected by James Watt junior in a letter of June 1795.[64]

There are many examples of initial ideas for designs in Boulton's letters. More accurate designs were done on paper, as with the Board of Agriculture Medal found in the Timmins album.[65] There are also many trial strikes of coins, medals and tokens in the Birmingham Museum & Art Gallery collection which give an indication of the various ideas tried. Many pattern coins were produced which did not go into full production.

Conclusion

Boulton does not seem to have gone into coining purely to make a profit. His primary motive seems to have been the enjoyment of innovation, the desire to improve the supply of money and the enhancement of his reputation. Boulton's coins were appreciated at the time as a means of protecting against counterfeiting, and for their artistic merit.[66] A recipient of coin wrote: 'The specimens of the intended copper pence are, like all your productions, strong marks of the superior excellence of the English artists over those of any other country'.[67] The coins, medals and tokens that Boulton produced are still admired.

The improvements made by Matthew Boulton at the Soho Mint represented a step change from what had gone before. He made the first recognisably modern coinage, improving coining technology, overcoming the problems of obtaining copper and distribution to customers. Many of the pieces show artistic merit in their design as well as superior manufacture, and provide a valuable historical record of the eighteenth century. The Soho Mint was his greatest adventure and eventually produced coins, medals and tokens for customers all over the world.

11: 'Bringing to Perfection the Art of Coining': what did they make at the Soho Mint?

David Symons

The Soho Mint's output of coins, medals and tokens was so large that it is impossible to give more than an overview here. In particular I will not consider here the wide range of patterns, proofs, trial pieces and specimens which were made, including those relating to issues that for one reason or another did not go into full production. Nor will I discuss the problem posed by the Soho restrikes, pieces made in the later nineteenth century using original Soho Mint dies obtained after the mint was closed in 1850. Readers who wish to investigate these areas, or simply to find out in more detail about Soho's products, should consult David Vice's paper, soon to be published by the British Numismatic Society, which provides a complete account of Soho's numismatic products from the 1770s down to 1850.[1]

Matthew Boulton's numismatic roots stretched back to 1772, well before the establishment of the Soho Mint proper. In that year he was commissioned by the Admiralty to make two thousand medals for Captain Cook to take with him on his second voyage to the Pacific (1772–5) as gifts to the natives. The result was the 'Otaheite', or 'Resolution and Adventure', medal (Fig. 67; Cat. 366), probably struck from dies cut by the Birmingham engraver John Westwood senior.[2] These medals, and everything else that Boulton made until the late 1780s, were struck by hand using the traditional technology of the period. In addition to the medals ordered by the Admiralty, Boulton struck a further 106 specimens in silver and two in gold as a private commission for Joseph Banks, who distributed them as gifts.[3] Boulton also struck a few more examples in silver for sale in the 'toyshops' of Birmingham and London, as a speculative venture on his own account.

Boulton made further occasional forays into minting in the following years – in 1774 award medals for the 37th Foot, and in 1781 a medal marking the capture by Admiral Rodney of the island of St Eustatius in the West Indies[4] – but it was not until 1786 that a really significant business opportunity arose. In that year the East India Company decided to produce a copper coinage for use in Bencoolen (modern-day Benkulen), one of its possessions on Sumatra, and commissioned Boulton to carry out the work. At this stage he did not have a mint of his own, so it was agreed that the Company would deliver the required copper to the Soho Manufactory, where it would be rolled to the correct thickness and the coin blanks cut.[5] The blanks would then be shipped to London, where they would be struck into coins using (hand-operated) coin presses and a milling machine that Boulton would install in a warehouse owned by the Company. In this somewhat cumbersome manner just over forty-eight tons of copper were converted into one-, two- and three-keping coins in 1786 and 1787 (Fig. 68; Cats 280–82).[6] This was to be the start of a long, and very fruitful, relationship between Boulton and the East India Company.

Boulton encountered a number of problems with these Sumatran coins which seem to have convinced him that it would be much better to set up his own mint at Soho, where all the various stages of the process could be kept firmly under his control. He also decided that he could strike coins better, faster and cheaper if the machines in that mint were to be powered by steam. This was to be the genesis of the Soho Mint in 1788. It is hard to say when he first thought about using steam to make coins. After Boulton's death in 1809 James Watt possibly implied that steam was considered as far back as 1774, but other evidence suggests the early 1780s as more likely.[7] Boulton was encouraged to take the plunge and create his mint because he had high hopes of winning a contract for a new copper coinage from the British government (rumours of such a contract were circulating by at least October 1786).

By the later 1780s Britain's copper coinage was in a deplorable state. Very few copper coins (halfpennies and farthings) had been issued since 1754, just when the growth of an industrial workforce, which relied on small change to buy its everyday necessities, meant that more such coins were needed. As a result people found themselves forced to use a mixture of old, worn regal coins, low-weight Irish coins, 'evasive' halfpennies, out-and-out forgeries, and even blank copper discs (Fig. 69; Cats 294, 297, 299, 304).[8] Some contemporary estimates suggested that as few as eight per cent of the halfpennies in circulation bore even a 'tolerable resemblance' to regal coins.[9] Understandably there were loud calls for a reform of the coinage, and Boulton hoped to be able to profit from this situation.[10]

As so often with Boulton, his interest and enthusiasm got the better of his business sense, and from April 1787 he forged ahead with his plans without any guarantees that a new contract would even be placed, let alone that he would get the work. By 1789 he had spent between £7,000 and £8,000 building a new mint

67 Silver Otaheite medal, 1772, diameter 43.5 mm (Cat.366–7) [BMAG 1981 N 17]

68 Copper proof coins for Sumatra (Bencoolen), 1786: one keping, 20 mm [BMAG 1885 N 1541.183]; two keping, 25mm [BMAG 1885 N 1541.184]; three keping, 27.5 mm. (Cats. 280–82.) [BMAG 2007.1261]

equipped with steam-powered coin presses initially at least based on the design of Jean-Pierre Droz, a Swiss die-engraver. At great expense, and after much delay, Boulton even lured Droz over to work at the Soho Mint as his chief engraver in 1788. Boulton also spent over £2,000 in upgrading the rolling mill that he had already modernised in 1785, so that it would be able to deal with the hundreds of tons of copper that a government contract would bring in.[11]

Droz engraved dies for a 1788 pattern halfpenny that Boulton provided to the Privy Council committee considering coinage reform. This was undoubtedly much superior to that produced by the Royal Mint at the same time (Fig. 70; Cats 307, 310), but Boulton found it very hard to get much more work out of him and their relationship eventually deteriorated beyond repair, with Droz leaving England in 1791.[12]

Unfortunately for Boulton the government repeatedly delayed a decision on the new coinage and finally, in June 1790, decided to postpone the idea indefinitely. In effect he was left with a powerful, modern mint, but had nothing to make in it. To keep it in business, and to keep all his key personnel employed, he therefore took on whatever work he could find and in consequence the Soho Mint was to produce a very varied output over the coming years.

Given that he had cherished hopes of producing a reformed copper coinage it is somewhat ironic that the first things made at the new mint were trade tokens. These were the latest addition to the circulating currency – effectively they were unofficial money (mainly halfpennies) issued by a variety of industrialists, merchants and towns to help facilitate trade. The first were issued in 1787 by Thomas Williams, known as the 'Copper King', who ran the Parys Mines Company, Anglesey, but many others soon followed his lead and began to issue their own tokens.

'BRINGING TO PERFECTION THE ART OF COINING'

69 The British copper currency of the late eighteenth century, from top: a halfpenny as it left the Royal Mint, 1771, 28.5 mm [BMAG 1969 N 602]; a worn halfpenny of George II (1727–60), no date visible, typical of what was actually in circulation, 27 mm [BMAG 2008.1413]; a forged halfpenny, '1775', 27 mm [BMAG 1932 N 107.13]; an 'evasive', no date, based on an Irish halfpenny, GOD SAVE THE KING / NORTH WALES, 26.5 mm [BMAG 1885 N 1526.258] (Cats. 294, 297, 304, 299).

The first tokens struck at Soho were copper halfpennies for the Associated Irish Mine Company of Cronebane, County Wicklow, Ireland, produced in 1789 (Fig. 71; Cat. 261).[13] They were soon followed by others for Williams's Parys Mines Company, also in 1789, and for John Wilkinson 'Iron Master', in 1790. These tokens were all intended for general circulation and were struck in substantial numbers: 1,700,000 Cronebane tokens in this one order; 2,150,000 Parys halfpennies and 34,000 pennies in two orders covering 1789–92; and 672,000 Wilkinson halfpennies in a series of four orders spread across 1790–95.[14]

In the next few years Soho produced smaller numbers of trade tokens for various other issuers as well, generally merchants or businesses based in the town or county named on the tokens, which were intended for local circulation. They comprise the following (all copper halfpennies):[15]

1791	Cornwall	76,000 tokens
1791	Glasgow	483,000 tokens (Fig. 5; Cat. 262)
1791, 1792	Southampton	193,000 tokens
1793	Leeds	179,000 tokens
1793, 1794, 1795, 1796	Inverness	377,000 tokens
1794	Lancaster	104,000 tokens
1795, 1796	Dundee	53,000 tokens

70 Copper pattern halfpennies, 1788, (top) engraved by Jean-Pierre Droz for Soho Mint, 36 mm [BMAG 1969 N 637]; (bottom) by Lewis Pingo for the Royal Mint, 37 mm. (Cats. 307, 310) [BMAG 1969 N 624]

71 Late eighteenth-century trade tokens, from left to right: Cronebane halfpenny, 1789, gilt proof, 29 mm [BMAG 1885 N 1536.94]; Glasgow halfpenny, 1791, copper, 28.5 mm [BMAG 1885 N 1536.87]; Bishop's Stortford halfpenny, 1795, gilt proof, 29 mm (Cats. 261, 262, 265.) [BMAG 1885 N 1536.90]

72 Copper Monneron tokens and medal, 1791–92, from left to right: two sols token, 32 mm [BMAG 1885 N 1536.97]; five sols token, 38.5 mm [BMAG 1885 N 1541.4]; medal depicting Jean-Jacques Rousseau, 34 mm (Cat. 343, 342, 344.) [BMAG 1885 N 1541.10]

Other issues, however, were on a much more restricted scale and may at least partly be regarded as a kind of 'vanity publishing' on the part of the issuers. Among these we may single out the Bishop's Stortford halfpenny (Fig. 71; Cat. 265), struck for Sir George Jackson in 1796 (although dated 1795) and especially noteworthy for the remarkable detail in the landscape on the reverse (the die for which was engraved by Boulton's new engraver, Conrad Heinrich Küchler). Some 24,000 of these were produced. However only 10,563 specimens were struck of a halfpenny token made in 1795 for George Cotton of Hornchurch in Essex, which represents a face value of just £22 0s. 1½d. for the entire issue.[16] Soho's output of trade tokens virtually came to an end in 1796, although there was to be one last flourish in 1800–1 when no fewer than 655,000 halfpennies were ordered by Woodcock's Bank of Enniscorthy, County Wicklow, Ireland.[17]

Another important customer in the early days of the Soho Mint was Monneron Frères of Paris, a business run by the brothers Pierre and Augustin Monneron. Working in close collaboration with Boulton, the Monnerons decided to address a shortage of small change in France by issuing tokens like those in circulation in Britain. As a result, during 1791 and 1792 something near to two hundred tons of copper were struck into two sols and five sols tokens (Fig. 72; Cats 343, 342). Concurrent with the tokens Boulton also struck a number of medals for the Monnerons depicting important scenes and persons of the French Revolution (Fig. 72; Cat. 344). The Monnerons' financial difficulties and the deteriorating political situation in France caused problems for their business, which was eventually killed off by a decree of the French National Assembly on 3 September 1792 forbidding the import and use of such private tokens. Nevertheless Boulton still optimistically sent specimens of his medals to the Monnerons until August of the following year, well after France had declared war on Britain on 1 February.[18]

The Monneron tokens also nicely illustrate one of the major problems that Boulton faced – how best to send the coins and tokens that he had struck to his customers given the state of communications in the late eighteenth century? The standard method evolved at Soho was to wrap a specified number of pieces in a paper wrapper and then to put the resulting rouleaux into a cask or small barrel. In the case of a shipment of Monneron five-sols tokens, 140 rouleaux each containing forty tokens went into each cask, which had a total weight of about 340 pounds when full. Wherever possible the casks were shipped by water, which was much cheaper than road transport. The Monneron tokens, for example, were sent by canal to the River Trent, down the Trent to Hull and were then sent on by sea to France. One such shipment was tampered with by the bargemen, who opened three of the casks, took out some of the tokens and filled up the space left with cow dung and hay. They then used the tokens to buy food from shopkeepers in Gainsborough, Lincolnshire.[19]

73 Coins minted at Soho, from top: Bombay, 1804, two pice, bronzed proof, 30 mm [BMAG 1885 N 1541.193]; Sumatra, 1804, four keping, copper proof, 30 mm [BMAG 2003.0035.3.1.]; Sierra Leone, 1791, 50 cents, bronzed proof, 31 mm [BMAG 1885 1541.173]; Isle of Man, 1798, penny, gilt proof, 38 mm (Cat. 285, 283, 286, 290.) [BMAG 1885 N 1541.211]

Also important to Boulton was the business that came his way from the East India Company, which, following on from its order of coins for Sumatra in 1786, was to remain a major customer of the Soho Mint until his death and indeed beyond. Over the years the directors of the Company placed the following orders for copper coins:[20]

1791	for Bombay Presidency, India	17,200,000 coins	½, 1, 1½, 2 pice
1794	intended for the Northern Circars, a region of the Madras Presidency, India (but many ended up in use in Ceylon)	13,558,000 coins	1/96th, 1/48th rupee
1794	for Bombay Presidency, India	8,650,000 coins	½, 1, 2 pice
1797	for the Northern Circars, Madras Presidency, India	16,530,000 coins	1/96th, 1/48th rupee
1798	for Bencoolen, Sumatra	2,560,000 coins	1, 2, 3 kepings
1803	for Madras Presidency, India	37,936,000 coins	1, 5, 10, 20 cash
1804	for Bombay Presidency, India	12,240,000 coins	½, 1, 2 pice (Fig. 73; Cat. 285)
1804	for Bencoolen, Sumatra	9,775,000 coins	1, 2, 4 kepings (Fig. 73; Cat. 283)
1808	for Madras Presidency, India	86,515,000 coins	10, 20 cash

Not all orders were as substantial as these, however. For example, in 1792–3 Soho struck silver and copper coins for the Sierra Leone Company, which had been founded in 1791 (the date that appears on the coins) to settle freed slaves in Africa. Denominated at first in dollars and pence, and then in dollars and cents, some 790,000 coins were made in total and shipped to West Africa. Appropriately, since Sierra Leone is the Spanish for 'Mountain of the Lion', the obverse ('heads' side) of all the coins showed a lion standing on a mountain (Fig. 73; Cat. 286).[21] Much closer to home, in 1798–99 Boulton produced 193,000 halfpennies and 95,000 pennies for the Isle of Man (Fig. 73; Cat. 290). The design of these pieces was closely based on that of the 'Cartwheel' coins that Soho had just struck for use in Britain and which we will consider below.[22]

As well as finished coins the Soho Mint could also supply coin blanks to be minted into coins elsewhere, as Boulton had already done for the Sumatran issue of 1786. The main customer for this service was to be the newly-independent United States of America. Boulton had entertained hopes of winning a contract to strike copper coins for them, but for reasons of prestige and economic security the Americans decided that they should have their own mint. Nevertheless, between 1797 and 1807 no fewer than 16,000,000 blanks for cents and 4,886,000 blanks for half cents were made in Birmingham and shipped across the Atlantic to Philadelphia.[23]

The real prize for Boulton, however, remained a contract to strike a new copper coinage for Britain, and he had to wait until 1797 to achieve this. When the time came things happened very rapidly indeed, and Graham Dyer has shown how the decision to produce new copper coins was an integral part of the government's response to a currency crisis that struck Britain in late February of 1797.[24]

Under the pressure of financing the war against France, the Bank of England's reserves of gold were falling very rapidly and, on 26 February, the Privy Council was forced to tell the Bank to stop redeeming its banknotes for gold. Faced with a potential collapse of confidence in the entire financial system, the government moved with remarkable speed. Over the next few days the Bank of England and the 'country' banks in England and Wales were authorised to issue banknotes of less than £5 value, which they had previously been forbidden to do by law. The government also decided, as an emergency measure, to countermark the large number of Spanish silver dollars held by the Bank with a stamp bearing the king's head and to put them into circulation with a face value of 4s. 9d.[25] These measures were intended to ease the problems resulting from a shortage of gold coins in circulation. In the same way, the new copper coins were clearly intended to alleviate the shortage of lower-value silver coins in circulation: hitherto the government had only ever minted copper farthings and halfpennies, but the new issue was to be of pennies and twopences.[26]

The first Boulton knew of all this was an invitation from Lord Liverpool dated 3 March to come to London to discuss the matter. He must have been highly gratified to read in it that:

> it has been suggested, that it may be proper in such a Moment, to have a new Copper Coinage – There is no Man who can better judge of the Propriety of the Measure, and of the Plan that ought to be adopted, in issuing a Coinage of this Nature, than yourself: and no one will execute it with more Accuracy and Expedition.[27]

Full of hope he left Birmingham on 5 March, reached London on the morning of 6 March, attended a meeting of the Privy Council on 7 March, knew the job was his by the end of the month, and had the first of the new coins in circulation by August. Over the next eighteen months Soho Mint turned a total of some 1,250 tons of copper into 43,970,000 pennies and 722,000 twopences. These are the famous 'Cartwheel' coins, which got their nickname from the prominent raised rim which was intended to protect the rest of the design from wear (Fig. 74; Cats 317, 315).[28] They are also notable as being the first coins on which the figure of Britannia carries a trident rather than a spear.[29]

Further contracts for British copper coins followed:[30]

1799		42,480,000 halfpennies	4,224,000 farthings
1806–7	30,645,000 pennies (Fig. 74; Cat. 329)	129,288,000 halfpennies	5,909,000 farthings

Boulton also helped the government to address the continuing shortage of silver coins. The emergency countermarking of Spanish dollars in 1797 had not been a great success, as it proved too easy to forge both the coins and the countermark. In 1804 the expedient was repeated, with the same results. Boulton proposed instead that Soho should restrike the Spanish coins with British designs and in due course persuaded the Bank of England to commission him to do so. Just over 1,000,000 of what Boulton called these 'regenerated dollars' were produced in 1804. A further 3,490,000 followed in 1809–11, although they all still carried the date 1804 (Fig. 74).[31] While producing these Bank of England dollars, Boulton was also making similarly overstruck dollars for the Bank of Ireland, striking 790,000 in 1804.[32] In the following year he also won a contract to supply Ireland with a new copper coinage, striking 8,788,000 pennies, 49,795,000 halfpennies and 4,997,000 farthings.[33]

Alongside its coins and tokens, Soho is also well known for the production of medals. The earliest one struck at the Soho Mint proper was produced in 1789 and was intended to celebrate King George III's recovery from a serious illness (Fig. 75; Cat. 271). April 23 was appointed a day of national thanksgiving and Boulton decided to make a commemorative medal for sale to mark the event. Unfortunately Droz was responsible for engraving the dies and the experience was not to be a happy one for Boulton. Although he was adapting the obverse die from a pre-existing punch and had only to engrave a new reverse die from scratch, Droz did not finish the work in time and the first medals were not ready until 26 April, three days too late. (In fact the bulk of the medals were delayed for another month because the first obverse die developed a crack during use and Droz had to cut another.)[34]

Nothing daunted, Boulton continued to strike medals, many of them made for sale to collectors as collaborative speculative ventures between him and Küchler, Droz's successor as engraver at the Soho Mint.[35] Just over a dozen medals were produced on this basis between 1792 and 1802. They include one to mark the assassination of King Gustavus III of Sweden on 26 March 1792,

74 Cartwheel and later coins minted at Soho, from top: copper Cartwheel penny, 1797, 35.5 mm [BMAG 1969 N 775]; copper Cartwheel twopence, 1797, 41 mm [BMAG 1969 N 736]; Bank of England silver dollar, overstruck on a Spanish coin, 1804, 40.5 mm [BMAG 1932 N 285.434]; copper penny, 1806, 34 mm. [BMAG 1969 N 927] (Cats. 317, 315, 337, 329.)

75 Medals minted at Soho, from top: Trafalgar medal, 1806, silver, 48 mm [BMAG 1885 N 1536.28]; restoration of the King of Naples, 1799, gilt, 48 mm [BMAG 1885 N 1541.58]; recovery of the King's health, by Jean-Pierre Droz, 1789, silver with gilt edge, 34 mm [BMAG 2003.0035.1.1.]; execution of Marie Antoinette, 1793, copper, 48 mm [BMAG 2000 N 7.2] (Cats. 236, 276, 271, 274.)

another to celebrate the victory of Marquis Cornwallis over Tippoo Sahib, the Sultan of Mysore in India also in 1792 (see p. 87, Fig. 65), and a trio of medals that depict Louis XVI's farewell to his children, his execution (see Cat. 273), and the execution of his wife, Marie Antoinette (Fig. 75; Cat. 274), on 21 January and 16 October 1793 respectively.[36] Others record the victory of Lord Howe over the French fleet on the Glorious First of June (1794) (see p. 87, Fig. 66), the restoration to his kingdom of Ferdinand IV, King of Naples, by Lord Nelson in July 1799 (Fig. 75; Cat. 276), and the Union of Ireland with Britain in 1801.[37]

These medals were struck in a variety of metals, normally grain tin (often today described as 'white metal'), bronzed copper, copper gilt and silver. Because the market for what were essentially miniature works of art collected by the middle and upper classes was quite limited, each medal was produced in what may seem to be surprisingly small numbers. For example, Soho Mint records show that just 429 of the Marie Antoinette medal, in all metals, had been sold by 1800, and this is a quite typical figure.[38]

Business did not always run smoothly. In January 1795, Küchler suggested to Boulton that they produce a medal to mark the marriage of George, Prince of Wales (the future George IV) to Caroline of Brunswick (Fig. 76; Cat. 268), which would take place on 8 April. Boulton agreed and on 12 February he told his son in a letter that 'The day before yesterday I was wth the Prince of Wales near an hour he shewed me his princesses picture & has consented to sit for a drawing of his profiel [sic].'[39] Unfortunately this time it was Küchler who was late completing the dies, and on 12 April, four days after the wedding had taken place, Boulton was still commenting unfavourably on the specimen medals he had been sent. The dies were not finally ready until 10 May and Boulton decided against striking any medals so long after the event. It is ironic that the project to produce this medal ended in fiasco, as the marriage itself proved a catastrophic failure and the royal couple separated the following year. However, Matthew Robinson Boulton did manage to recoup something from the project as it seems that examples were struck later, in about 1810. The obverse of the medal is a miniature masterpiece. Remarkably the original die has survived and is now in the Birmingham Museum and Art Gallery collection (Fig. 77; Cat. 269). Given all the attention that was lavished on this medal, it seems incredible that apparently no one spotted that the wrong date for the wedding appeared on the reverse (1797 instead of the correct 1795).[40]

Other medals were commissioned from the Soho Mint by outside buyers. For example, in 1797 the American Minister to London ordered medals which seem to have been intended for distribution as Peace Medals to Native American leaders. There are three different obverse designs – a farmer sowing grain, a woman spinning, and a shepherd with his flock – each combined with a common reverse commemorating George Washington's second term as president and dated 1796. About 240 of each version was struck, a quarter of them in copper and the rest in silver.[41]

76 Medal, struck to commemorate the marriage of the Prince of Wales to Princess Caroline of Brunswick, formerly in the possession of the Watt family, with its original shell [case] and paper wrapper. Diameter 48 mm. (Cat. 268) [BMAG 2003.0035.2.1-3]

77 Obverse steel die for medal struck to commemorate the marriage of the Prince of Wales to Princess Caroline of Brunswick. The marriage took place in 1795, but not only did the medal bear the wrong date (1797), the die was so late being delivered by Küchler that Boulton abandoned the plan and the medals were not actually struck until about 1810. (Cat. 269) [BMAG 2004.0182.1]

Another commission came from Alexander Davison, Lord Nelson's prize agent, who ordered medals for distribution to those who had taken part in the Battle of the Nile in 1798, the medals being paid for out of the profit he had made selling the French ships captured there on Nelson's behalf. These medals were produced in bronzed copper for the seamen and marines (6,531), copper gilt for petty officers (506), silver for lieutenants and warrant officers (154), and gold for Nelson and his captains (twenty-five). One hundred additional examples in bronzed copper were struck and kept in stock at Soho.[42]

78 Bronzed Hafod Friendly Society medal, 1798, 40mm [BMAG 1885 N 1541.105], and John Phillp's corrected design drawing. (Cats. 279, 278.) [Birmingham Museum & Art Gallery]

Probably prompted by Davison's example, Boulton decided to mark the victory at Trafalgar in 1805 and the death of Nelson (who had visited the Soho Mint and Manufactory on 1 September 1802) by striking a commemorative medal at his own expense for distribution to the common seamen and marines who took part in the battle (Fig. 75; Cat. 236). The medals were ready for distribution in October 1806. Their production was no small gesture on Boulton's part as over fourteen thousand grain tin medals were eventually distributed at a cost of about £400. (Typically though he also produced examples in bronzed copper, gilt copper, silver and silver gilt for presentation to his friends and various persons of influence, so his generosity would not go unnoticed.)[43]

Finally we should mention some of the perhaps less well-known items that were made at the Soho Mint over the years. These range from agricultural prize medals for the Board of Agriculture and groups like the Essex Agricultural Society, all struck in very small numbers in precious metals, to the four thousand copper theatre passes made for the Ipswich Theatre in 1802 and the two hundred copper medals ordered in 1803 by the Dowager Lady Spencer for the wonderfully named St Albans Female Friendly Society.[44] We may end with a rare instance where an original design drawing seems to have survived. This is preserved in the Phillp Album at Birmingham Museum and Art Gallery, and seems to be a drawing by John Phillp, one of Boulton's protégés, for the medals ordered by Colonel Thomas Johnes for the Hafod Friendly Society in Cardiganshire in 1798. The drawing shows that Phillp had to correct his original spelling of the Welsh inscription on the obverse, which translates as 'Temperance, Industry and Brotherly Love' (Fig. 78; Cat. 278).[45]

The Soho Mint was probably the one of his many creations that was dearest to Boulton's heart. When his health failed him in his last years and he was forced to hand over the running of his businesses to his son, he still loved to go and visit the mint and watch the presses in operation, even though he had to be carried down the hill from Soho House by his servants.[46] As he himself said: 'Of all the mechanical subjects I ever entered upon, there is none in which I ever engaged with so much ardour as that of bringing to perfection the art of coining …'[47]

12: A Walking Tour of the Three Sohos

George Demidowicz

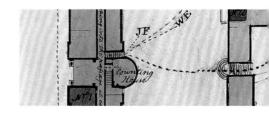

There were many visitors to the various Soho works, and in particular the Manufactory in Handsworth. They have left descriptions, which provide an important source for understanding the layout, function and contents of the buildings.[1] The accounts are, however, understandably biased, reflecting the preferences, tastes and expertise of the visitors. Furthermore the tours around the buildings were necessarily selective, both representing the personal inclinations of the visitors and the guiding hand of the Soho personnel. These were usually the managers, including Matthew Boulton himself, as there were no dedicated guides. They found themselves torn between the good publicity and marketing that visitors could generate and the time taken from their own work. They were also conscious that Soho trade secrets could lead out when more technically aware visitors observed new processes and technologies. Certain areas therefore remained out of bounds.

To help distinguish the Soho Manufactory, the Soho Mint and the Soho Foundry, we shall accompany a small group of imaginary visitors touring all three locations on three consecutive days in 1805. In fact, this would have been highly unusual at the time, since, as Peter Jones has also shown (p. 78), visitors had been banned from the Manufactory in 1800 due to concerns about industrial espionage and disruption, and the 'learned, ingenious, inquisitive, the young, the gay, the beautiful' were no longer welcomed.[2]

The tour is based on detailed research carried out over fifteen years, initially on the Manufactory and Mint and then on the Foundry, which was threatened with demolition in the mid-1990s.[3] The documentary material was sufficient to produce a large-scale three-dimensional (axonometric) reconstruction of the Manufactory and Mint, which was used as the basis for a colour poster on sale at Soho House and is reproduced here (Fig. 79). This will help in visualising the lost buildings and tracing the route taken.

The Manufactory

The tour of the Manufactory takes place four years before Matthew Boulton's death in 1809. At this time the buildings have more or less reached their maximum extent, following a recent reorganisation of the engine works; they will not now change significantly before their demolition between the early 1850s and 1863. There are detailed internal plans of the principal building, dated 1858, but individual rooms are unfortunately not identified (Fig. 80). The function of some of those entered during the tour cannot, therefore, be given with any certainty.

*

About two miles from Birmingham, carriages heading for the Manufactory turn sharp left from the turnpike and descend a narrower road across the former heath towards the Palladian principal building (Fig. 82). Turning left through the gates, visitors arrive at the broad terrace in front of the mansion-like building, which now, in 1805, is occupied as the headquarters and main workshop of Matthew Boulton's Plate Company, managed by John Hodges. Opposite the gates lies the large original mill pool, constructed in 1757, from which water flows under the road and into the canal that runs alongside the forecourt terrace. This canal symbolises the dual purpose of much that can be seen at the Manufactory, as it takes water to feed the water mill but also provides an important aesthetic feature at the boundary of Boulton's park and garden. To the front and right of the terrace the park rises towards Soho House, hidden amongst greenery at the summit of the hill. From the edge of the terrace water can be seen flowing into a culvert, which runs beneath the terrace and then under the principal building. This is the mill head race.

The principal building has generous, symmetrical proportions. It is mostly three storeys in height, constructed of red brick with a slate roof. Vertically sliding sash windows rhythmically punctuate the main and side façades. The last three bays of windows at either end of the building stand slightly forward, emphasised by pediments. The entrance lies within a projecting central bay, four storeys high, topped by a pyramidal octagonal roof and lantern with four small domes at each corner. A Diocletian window surmounts the main door which is flanked by single Doric columns. Inside, the entrance hall is an unpretentious space and fairly dark. Narrow corridors at the far end run left and right down the spine of the building, and a modest staircase ascends to the floor above. All the doors are numbered. A glance down the corridors from the bottom of the stairs confirms that they do not penetrate the two projecting wings of the building, but stop at internal windows. These borrow light from passages leading to external doorways at both ends of the building, where living accommodation is provided for some of the managers and clerks.

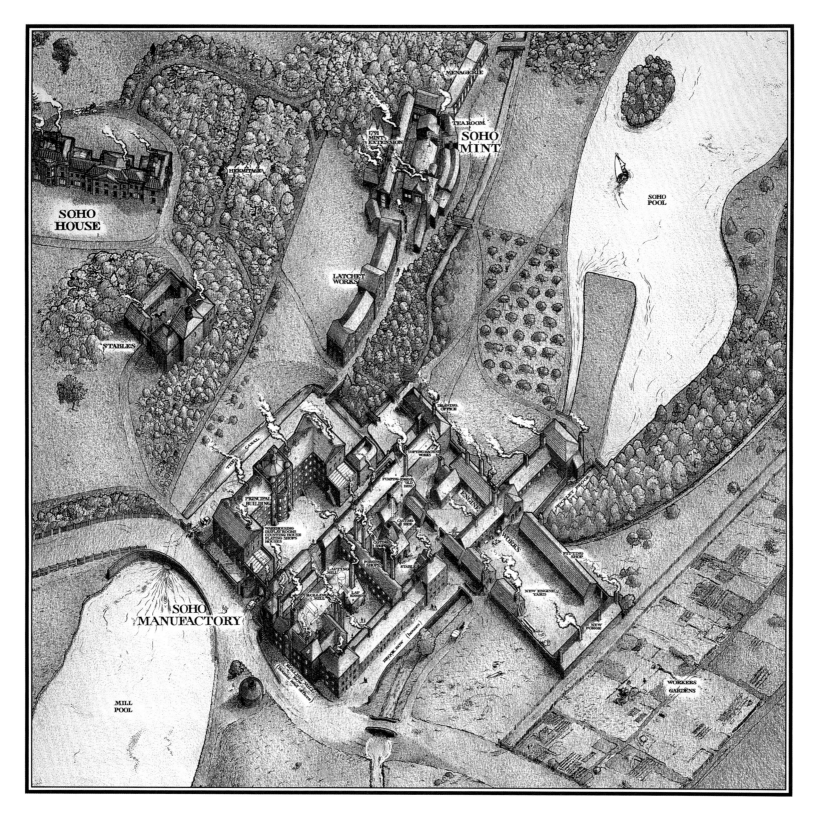

79 Axonometric projection of the Soho Manufactory and Mint, based on archive evidence and contemporary views. Although all such views show the latchet building (centre, between the main Manufactory and the Mint) as complete, research since this drawing was made has shown that its north wing was not completed until 1825–6. (© Bremner & Orr Design Consultants Ltd., 2006, based on a drawing by George Demidowicz.)

This is why curtains can be seen in some of the windows. John Hodges leads the group of visitors to a door on the left-hand side of the hall, from which a few steps descend into a long, tall room partly below ground level. This is probably a warehouse where non-perishable materials are stored. Immediately to the left of the stair is a door leading to a room with three windows in a semicircular bay at the rear of the building, used as a counting house. The room is lined with shelves on which ledgers are stacked. Looking out of the sash windows, it is obvious that these are restricted to the projecting bay and that the rest of the openings on the rear elevation are broader, with fixed workshop windows. These windows reflect the twin nature of Soho, which was so aptly expressed by a previous visitor.[4] Adapting his words, the architecture of the *dulce* graces the front whilst the *utile* is sufficient for the rear. The back of the building is four storeys high with an extra 'undercroft' below the ground floor, so that the rear yard lies lower than the front terrace.

The cramped, dark staircase next to the counting house leads up to another pair of narrow corridors on the first floor. Behind their closed doors beautiful silver and plated wares are being painstakingly fashioned by a combination of relentless innovation and traditional skills and craftsmanship. The visitors step into only one of the workshops to watch the work for a few moments, before being directed into a room at the front, opposite the staircase. This is architecturally the grandest room in the building, rising through two storeys, and is probably the showroom where the finest of Soho products are displayed for admiration, followed, it is hoped, by purchase. The room is lit by the Diocletian window and two circular windows, and is dominated on the opposite wall by a single engaged Doric column, which ingeniously disguises a stove for heating. Next to this room is another two-storey room, the length of the half-cellar warehouse below, which may serve as storage space. To the left of the stairs, as on the floor below, a door leads perhaps to another counting house in the semicircular bay at the rear. From a third floor viewpoint the true extent of the Manufactory can be appreciated for the first time. Chimneys of all heights and sizes rise out of the workshops below to confuse the panorama and obscure it with smoke.

The visitors now descend and leave the principal building to explore the myriad industrial activities taking place within the maze of workshops and courts just observed from above. The rear yard is enclosed by the principal building on three sides, but the fourth side appears to be a separate workshop only one storey in height. A quick detour to the right reveals the partly-exposed channel of the head race that is the continuation of the culvert leading to the water mill, still not visible (Fig. 81). The single-storey workshop runs the whole length of the Manufactory, but is intercepted by steps, semicircular at the top, descending sixteen feet to the yard below. During the descent it becomes apparent that the workshop that appeared to be only one storey high in the direction of the principal building, has an opposite façade

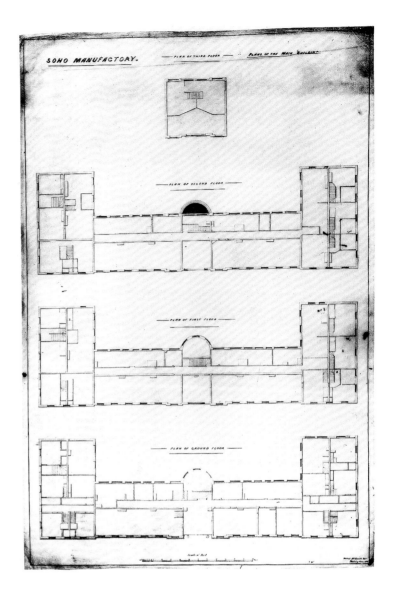

80 Robert McLeish junior: *Plans of the 'Principal Building*, 1858. (Photo: Birmingham Museum & Art Gallery)

of three storeys facing the water mill, which dominates the scene immediately ahead. The asymmetrical building is explained by the fact that it straddles a steep bank ('the great bank'), which was originally exploited by the mill for its head of water.[5] The bank forms part of the modified slope down to the Hockley Brook. To the right a timber trough emerges from this building above head-height and carries the head-race water on timber stilts into the mill.

Thomas Kellet, the rolling mill manager, now takes over the tour. He explains how the mill, a broad two-storey building, was entirely rebuilt in 1785, when it was turned round ninety degrees, and then extended in 1788. The section to the right is now attached to buildings that face the steep road descending from the mill pool, and contains the rolling mill. Kellet directs the visitors into this broad, high room open to the roof, but with a gallery running on three sides. Four trains of rolls for rolling

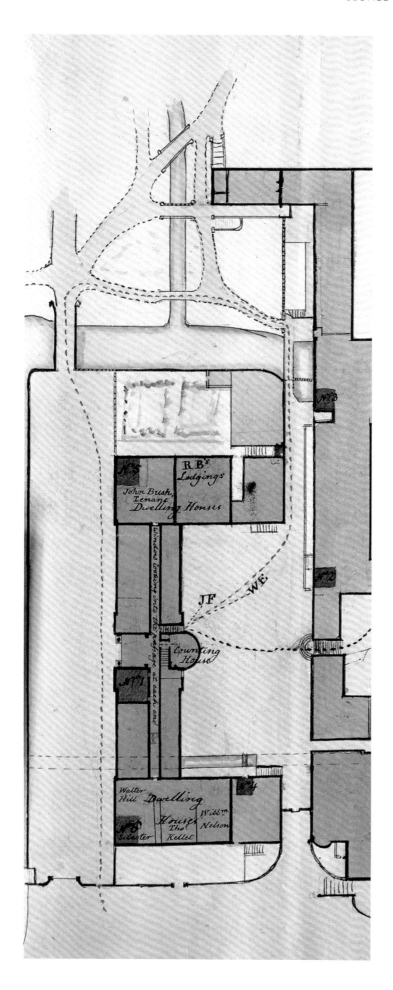

metal out into thin sheet can be seen connected to large metal cogs, which in turn are powered by a twenty-foot diameter water wheel in its adjacent, separate compartment. A great vibration can be felt in the floor as one pair of rolls, screaming resistance, reluctantly gives passage to a piece of metal, while the heavy water wheel turns slowly in its pit. Most of the metal rolled is copper for Matthew Boulton's coining business at the Mint, but other metals are flattened for the Plate Company. Up on the gallery the timber trough, or 'foreboy', can be seen disgorging water at the twelve o'clock position to the giant overshot mill wheel.

The next room is the lap shop, where two lines of 'laps', or polishing discs, are capable of being set in motion by leather belts emerging through the floor. The belts wrap around two revolving timber drums in the room below. The latter were originally powered by the water wheel, but both rolling and lapping placed too heavy a demand on the wheel, and in 1788 a new steam engine was installed to drive the laps directly, hence its name – the lap engine. The button business, a former principal user of the laps, has been in decline lately, and some laps were changed to glass-cutting wheels used by the Plate and Latchet Company. The latchet business, making flexible detachable buckles, has itself almost ceased trading; its underused premises will be seen later.

There is one final room to be visited at this level, where the ingenious cutting-out machinery is housed. The lap engine is additionally employed to power these presses, of which there are eight arranged in a circular wooden frame. The workers, usually women, pass copper strip through the presses, which cut out coin blanks of various sizes to be processed in the separate Mint situated in the park above the Manufactory. The tight fit at this end of the building is not helped by the beam of the steam engine below, which penetrates the floor adjacent to the presses. The visitors get a glimpse of the lap engine itself by leaving the mill and entering the engine house through a separate door at ground level. The Boulton & Watt engine crank is described as a 'sun-and-planet'; it transforms vertical motion to rotary action.

At right angles to the mill across a narrow space is a long range of workshops through which a narrow passage leads into the next yard. Through an open door can be seen the smithy containing a series of hearths where metal is heated, worked and skilfully shaped into various objects. The visitors cross the second yard, dominated by the brick furnace chimney, with attached lean-to sheds where metal castings are made, and enter a tall building with a chimney. This houses a large pumping engine, affectionately known as 'Old Bess,' which soars above their heads to a massive pivoting timber beam. This is the first working Boulton & Watt steam engine in the world, and still in its original

81 Plan of Soho Manufactory, one of three drawings prepared as evidence in the trial of the Soho robbers, 1801. Ink and watercolour on paper. The plan shows the internal layout of the principal building at ground-floor level and the head race channel in the rear yard. (Birmingham Archives & Heritage, MS 3069)

Soho, Staffordshire.

82 J. Walker, engraving from *The Copper Plate Magazine*, 1798–9, later republished in *The Itinerant: A select collection of interesting and picturesque views, in Great Britain and Ireland* (1798), showing fashionable visitors leaving the Soho Manufactory. The building on the left, the crescent building or latchet works, never had a dome but was built instead with a series of pediments. The wing extending towards the principal building was not constructed until 1825–6. (Cat. 360) [BMAG 1965 V 221.80]

position. To the left a great metal tube ascends through the engine house. Hidden within the tube, the up-and-down motion of the piston pulls water to a height of twenty-four feet from the bottom of the culvert to the level of the canal at the front of the Manufactory. Back in 1774, instead of the water from the mill merely entering the next pool below, known as Soho Pool, a culvert was taken off the tail race channel. It ran along the edge of the Manufactory site at that time to the 'great bank'.[6] Watt's engine was installed here to lift the diverted tail-race water and return it to the head race so that it could be recycled over the water wheel. A new section of channel was constructed along the side of the principal building to link with the existing canal. In truth, the first working Boulton & Watt steam engine was erected to aid a traditional source of power – the water mill. The visitors are awestruck by its sheer size and capability, but they are about to see how steam-engine technology has advanced in thirty years.

At the gates to the left of Old Bess's house the engine manufactory superintendent, William Brunton, takes over the tour, first pointing out the three-storey building to the left perched on the 'great bank'. This part is mostly occupied by the document-copying machine works; here an ingenious device, invented by James Watt in 1779 to enable copies to be made of the firm's voluminous correspondence and engine drawings, is manufactured (see pp. 67–8). The magnificently located engine-drawing office, whose windows overlook Boulton's garden and Soho Pool, is attached to the far side of the copying-machine workshops and is headed by John Southern. Brunton's counting house for the engine works is also in the same building.

Looking in the opposite direction into the engine yard, the first buildings in view were the earliest to be constructed many years previously, and the visitors are escorted to the junction between four linked workshops, the hub of the engine manufactory. Here they enter the engine house, which, unlike the tall edifice of the pumping engine behind them, is a much squatter structure. William Murdock has recently invented the three small engines here, of only 1H, 3H and 6H, which are ideally suited to power the various drills and lathes.[7] The compactness of these engines is due to the lack of a rocking beam; this has the additional advantage that smaller buildings are needed to house them. Matthew Boulton's son, Matthew Robinson Boulton, has only recently reorganised and extended the engine works (1801–4). A new yard was laid crossing the Hockley Brook onto

Birmingham Heath, around which were erected new fitting shops and a forge. This investment will help increase the production of steam engines, including Murdock's new compact model. Previously most have been made at the Soho Foundry in Smethwick, founded in 1795 under the direction of James Watt junior. The second generation Boulton & Watt have now taken charge of the businesses, with James Watt having retired from active duties many years ago and Matthew Boulton increasingly incapacitated by painful illnesses.

Leaving by the engine-yard gate, the visitors cross a bridge to a row of houses known as Brook Row, running along one side of the Hockley Brook. The opposite side was laid out as allotments a few years previously, but the terrace of houses was constructed in the earliest years of the Manufactory to house the upper ranks of the Soho workforce. A sharp right turn leads into the steep road that ascends to the principal building, passing more housing on the right. In the middle of this terrace, under a gable, is the former main entrance to the Manufactory, blocked now by the rolling mill. Opposite, on the bank of the mill pool, is a curious low, round building – the main latrine. The visitors are now back at their starting point on the terrace at the front of the principal building, where their carriages are waiting.

The Mint

After a night spent as guests at Soho House, the visitors set off to see the Mint where John Southern, the drawing-office manager and an engineer of considerable achievement, will show them round. It is a short, pleasant walk downhill and the Mint suddenly appears through the trees as a natural part of the garden. Indeed, the Mint began as Boulton's miniature farm, where in 1788 he secretly set up experiments to apply a steam engine to the production of coins. Emerging onto an open grassy slope, the visitors pass between the Mint buildings on the left and the latchet works, an incomplete crescent-shaped building on the right, and turn sharp left into a long narrow yard (Fig. 83). To the left is a two-storey building constructed in 1791 for the Mint in anticipation of obtaining the contract for the national coinage. Unfortunately the contract failed to materialise when expected and Boulton was left with surplus accommodation, which was taken over by the latchet company a few years later, but their occupation in turn has now ceased.

The visitors now enter a taller building on the opposite side towards the end of the yard. This is the old engine house constructed in 1788. The sixteen-inch engine inside is no longer used in the Mint, but instead pumps water from Soho Pool to Boulton's gardens, rooted in the thirsty heathland soils. A fire broke out five years ago destroying the roof, but all is now repaired and in working order. The next room is reached directly from the engine house, and is broad and a generous single storey in height. Within are the remains of a massive circular timber frame, which once held eight coining presses; this was the first coining room in the

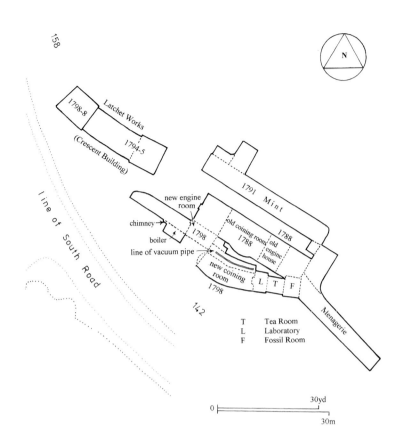

83 Plan of the Mint c. 1805. (George Demidowicz)

world powered by a steam engine. The metal parts such as shafts, cogs and presses have been removed, and the room is in some disarray, despite the significant events that took place within it. New machinery, invented by Southern, has replaced the old. Although the cutting-out of coin blanks is still accomplished in the old way, by the machine seen yesterday, the forces required to impress the design of a coin onto the blanks were found to be so great that the whole frame nearly shook itself to pieces when in operation. When Soho at last won the national coinage contract in 1797, Southern persuaded Boulton to adopt an entirely new method of working the presses, using vacuum power.

Going back into the yard, along the outside of the old coining room and around a tall pedimented building attached to it, facing the latchet works across a broader yard, the visitors circle another low building and suddenly a vista over Soho Pool opens up. Adjacent to a tall chimney and its boiler is a two-storey building, which they enter. This contains another sun-and-planet steam engine with a twenty-three-inch cylinder. Much of the vacuum plant cannot be seen from the visitors' vantage point but Southern is reluctant to show any more.[8] There are many who would wish to learn the secrets of the unprecedented productivity of Soho Mint, and all measures are being taken to avoid the pirating of this new technology.

Leaving the engine room, the group turns left and descends along the double gable end of a broad curving building which respects the contour of the hillside and stands much lower than the main part of the Mint. The curving façade is decorated by a simple but elegant sequence of round classical arches and pilasters.[9] This building is the high point of the visit – the new coining room housing Southern's invention. The eight coining presses stand in a row opposite the windows, both the arrangement of presses and the windows conforming to the gentle curve of the building. Each press is surmounted by what is best described as a vertical trumpet, broadside down, made of cast iron with the slender end reaching to the ceiling where it is surrounded by a star-blaze ceiling rose. Two workers are rapidly supplying coin blanks to the presses using an automatic feed called a 'layer-in', and it takes little time to calculate that coins are being impressed at a rate of about sixty per second. There is no sign of the vacuum tube, and Southern explains that it runs in a trough at the rear of the presses, behind a wall.

Leaving the coining room, the visitors pass through Boulton's laboratory and into the tea room. The tea room, and the laboratory and fossil room which flank it, were constructed some time before the Mint as garden buildings, and they share the pleasing wide aspect over Soho Pool and towards the town of Birmingham. The coining room was in fact constructed in place of earlier menagerie or aviary pens. Some time ago a replacement menagerie in the form of a long, elegant pavilion was constructed, attached to the fossil room, but it, too, has now ceased to house any animals or birds.

After refreshments the visitors have only one more building to see. They retrace their steps to the lawn in front of the curving latchet works, whose unusual appearance has already been noticed. It was clearly intended to take a symmetrical crescent form, but the full plan has not been executed and it is foreshortened. The left-hand wing has been constructed to the requisite two storeys, whilst the right-hand wing is entirely missing.[10] This asymmetry is emphasised by the three-storey central section topped by a pediment.[11] In 1792 Boulton entered into partnership with the Smith brothers, James and Benjamin, to produce the latchet, a patented flexible buckle that could be transferred to different shoes. Boulton threw himself with his usual enthusiasm into the new business, conceiving an elegant crescent to house some of the latchet workshops. The two-storey part of the building was constructed in 1794–5 and the central section in 1798–9. Ambition was, however, not matched by orders and the building was not completed.

The Soho Foundry

After a second night of Boulton's hospitality, the visitors set off for the Soho Foundry. The route from Soho House to the Foundry is circuitous, though the distance as the crow flies is only one mile. The journey is across open countryside so that the Foundry, when first glimpsed, stands isolated amongst fields.

Two terraces of houses stand on the right-hand side of the approach to the main gate, closely resembling Brook Row at the Manufactory. They are occupied, as at the Manufactory, by the upper echelons of the Foundry staff. The Foundry itself is defended like a fortress with a high brick wall. The main gate is flanked on both sides by two-storey buildings attached to the wall, which terminate in two-storey deep bay windows facing each other across the entrance. When construction began ten years ago, in the summer of 1795, this was the first steam-engine manufactory in the world (Fig. 84). The guide today is William Murdock, an amiable, sturdy Scot clearly happier in the workshop than the counting house. He leads his visitors past the Foundry office and counting house on the left, to the attached carpenters' shops beyond, in which wooden patterns of varied intricacy and size are made for the metal castings. The carpenters use great skill in translating a drawing accurately into its first three-dimensional form. The largest patterns are impressed directly into the sand of the Foundry casting floor.

On returning to the entrance, Murdock points out the line of smiths' shops behind the porter's lodge, but he steers the visitors in the direction of the main building in the Foundry enclosure. To the right, they first pass a two-storey magazine which follows the line of the road. This holds every imaginable material that will be needed in the making of steam engines. A gap separates the store from a larger and more irregular-shaped building, into which the visitors are led. The first two workshops they encounter form the turning shop, where nozzles (valves) and piston rods are made using lathes and drills. These are powered by an overhead shaft, driven directly by a beam engine, which projects slightly into the piston rod shop. The engine has only recently been changed from the original 14H (horsepower) to 20H by installing a larger cylinder. Murdock has his own office on the first floor, as the machinery close at hand is some of the most important in the Foundry. Five years earlier he played a vital role in reorganising and improving an ailing boring mill. The smooth, accurate internal boring of a cylinder is crucial to the efficiency of its later operation. The first boring machinery vibrated alarmingly; Murdock solved the problem by removing a clockwork of connecting toothed cogs and replacing it with 'worm drives', or revolving screws. His other innovation was to introduce two of his newly-invented small engines (3H each, the same type of compact beamless engines that were seen at the Manufactory two days earlier) to take the strain off the beam engine and power some of the lathes and boring machines independently. The cutting head of the boring engine is now progressing at a snail's pace into a newly-cast engine cylinder, slicing off thin slivers of metal to produce a perfectly circular interior.

The boring mill has a broad central door in its gable facing the Foundry, through which the engine cylinders are transported directly across a yard on a level from the casting pit. The visitors leave the mill and are taken round the corner of the building to look quickly into the engine house at the rear, where three small

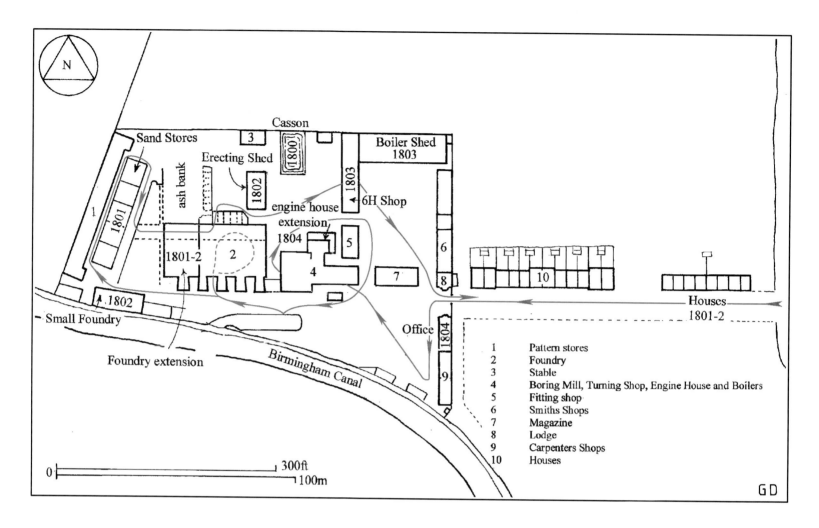

84 Plan of the Foundry c. 1805, showing the tour route. (George Demidowicz)

engines stand close together. One is working the boring machine just seen and another, of 4H, powers an underground shaft leading into the Foundry where another boring machine was installed in 1802. The visitors must now circle the whole of the turning shop with its many lathes and a separate fitting shop to reach the Foundry at the main upper-ground level. To its left is the canal basin, connecting directly to the Birmingham Canal. This is where heavy materials are brought in and completed engines transported to their destination. The main Foundry building is broad and massive with a huge sweep of roof entrapping five great chimneys on the canalside, which vent air furnaces where iron is melted for casting.[12] Within it is very dark, the only light coming from the windows at either end of the building and from the searing flames of one of the air furnaces, from which molten metal is being poured into a freshly prepared mould in the sand floor. The air is filled with the smell of hot metal and singed sand, a scene of apparent chaos with finished castings and wooden patterns lying scattered across the sandy floor. To the right a crane stands over a vast pit, at the edge of which preparations are being made to construct the next cylinder mould out of bricks and casting loam. On the north side is a series of stoves for drying moulds. Moisture needs to be removed from the moulding sand, otherwise it can explode when liquid metal is poured into it.

The visitors leave the Foundry and head towards the west boundary, along which is a long narrow building where the wooden patterns are stored in meticulous order so that they can be readily found for re-use. As the land slopes to the north, about half-way along the stores there was room to construct a low undercroft of brick arches for more storage space. The ground is now well below the main Foundry floor level. In front of the pattern stores are the backs of five large brick enclosures holding different types of sand and loam. On the opposite side of the sand stores and parallel to their open front is a high brick retaining wall. Above this, and set back, stands the broad west gable of the Foundry which the visitors have just left. In the middle of the retaining wall a tunnel can be seen heading in the direction of this same building. The tunnel was constructed to maintain a level link between the pattern stores and the bottom of the casting pit. From the earliest days ash from the furnaces was dumped on the west side of the Foundry, creating a thick deposit of material. A recent extension of the Foundry westwards onto this ash incorporated a tunnel back to the original Foundry. From thenceforth ash was to be dumped on the north side of the new extension.

The visitors are given lanterns and led for about 100 yards along the brick vault, before turning left at an arched opening that leads back into the open air.[13] To the left is the great tongue of ash, gradually creeping forward and retained partly by a series of small brick vaults used for storage. Behind is the great north wall of the Foundry against which is another row of small vaults, which support the chimneys to the drying stoves within the building. Near here is Murdock's laboratory, where only seven years ago and in great secrecy he conducted experiments in gas lighting. He leads his visitors through a narrow gap between the Foundry and its small arches on the right and a workshop to the left. They find themselves back near the engine house and boring mill, already visited. It is now clear that the main Foundry floor is one storey higher than where they are standing. Cylinders are cast at this level in the pit lying within the wide doorway visible in the east gable of the Foundry.

The visitors turn to the left and cross a broad yard to a long narrow workshop. The shop was built two years earlier to erect Murdock's little engines, for which there are many orders, and is itself powered by a 6H engine of the same type. Gas burners can be seen running along the length of the 6H shop, and Murdock stops at one to demonstrate the new source of light that has illuminated the Foundry. The flame is bright and yellow but emits an unpleasant distinctive smell, which would inhibit its use in a fashionable drawing room. More work is needed to purify the gas, but already there are factory owners who wish to install gas light plant in their own premises to save on the expense of oil lamps. The flame is doused by a turn of the tap below the burner. It was only very recently that gas lighting was installed in the turning and boring mill and its engine house and in the counting house, so the Soho Foundry is the first factory in the realm to be lit by gas.

The tour is at an end; the visitors are both exhausted and exhilarated by what they have witnessed over three days. Many marvels of the new age have been shown and explained to them, but all has blurred into one overpowering image of restless industry and creativity. Amidst all the inventiveness and ingenuity applied to the harnessing of invisible forces and the development of mechanical devices, traditional skills of fashioning beautiful objects by hand have not been abandoned; they are equally valued and respected. Art and natural science dance a complicated minuet at Soho, and our visitors are in no doubt that the rare and extended excursion they have been privileged to complete was to an extraordinary place, giving a glimpse of the future.

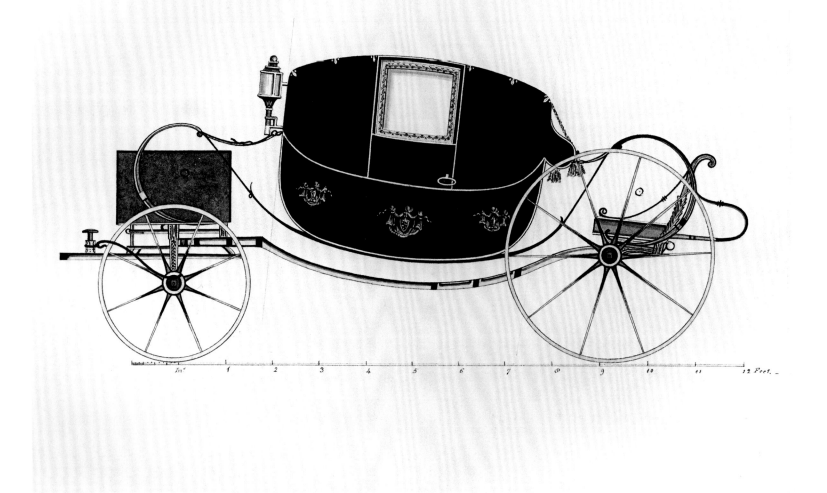

85 Design for a carriage, from Matthew Boulton's notebook 14 (1780) headed 'Thoughts on Carriages – M. Boulton'. Boulton travelled a good deal and often wrote of the discomforts of long journeys. This notebook contains a number of drawings of improvements to carriages. (Birmingham Archives & Heritage, MS 3782/12/108/22)

13: How do we know what we know? The Archives of Soho

Fiona Tait

Archives tend to survive because they are required for legal or financial reasons, and the papers of Matthew Boulton illustrate that. However, this brief description can only hint at the glorious wealth of eighteenth- and nineteenth-century society that comes to life in the correspondence and accounts.

The records of Matthew Boulton and family comprise a substantial archive of business and personal papers, which are owned by the Birmingham Assay Office Charitable Trust and are held on permanent loan by Birmingham Archives & Heritage. The surviving records reflect the variety of businesses in which Matthew Boulton was involved, his wide circle of friends and business acquaintances, and they provide detailed records of his house and personal pursuits.

The papers have recently been catalogued as part of the 'Archives of Soho' project (1998–2003), funded by the Heritage Lottery Fund and Birmingham City Council. This project also saw the production of new catalogues of other related collections, the records of the Boulton & Watt steam-engine business and the papers of James Watt and family.

Planning for this project started after the City Council and private donors raised the money to purchase the papers of James Watt in 1994. Other collections had arrived at various times over the previous eight decades, starting in 1912 when the records of the Boulton & Watt business were given to the city by the engineer George Tangye; the Muirhead papers were given to the Reference Library in the 1930s by the Muirhead branch of the Watt family. The Matthew Boulton Papers, which had been kept by the Boulton family, were loaned to the Birmingham Assay Office in 1920, and this loan was converted to a gift in 1926. When the new Birmingham Central Library opened in 1973, the papers were moved from the Assay Office to the new building where storage and public access could be more easily managed. Ownership of the papers was transferred to an independent charitable trust – the Matthew Boulton Trust – and in March 1974 the papers were deposited in the City Archives. The purchase of the Watt papers in 1994 thus brought together the surviving correspondence and papers of two key figures of the eighteenth and nineteenth centuries with their business records, and made them all publicly available for the first time.

These collections are available in Birmingham Archives & Heritage for anyone to consult free of charge. Between them, the papers are the most frequently used collection in the Library, and are consulted by scholars from all over the world. The catalogues identify each item by a unique number. The collection references for the three groups of records are as follows. MS 3782: the papers of Matthew Boulton and family; MS 3219: the papers of James Watt and family; and MS 3147: the papers of the Boulton & Watt steam-engine business. The Birmingham Central Library holds extensive catalogues of all three collections and there is not room here to cover them all in detail, but the contents of the Matthew Boulton Papers (MS 3782), which occupy some 107 metres of shelving, are summarised below.

The business aspects of the Boulton & Watt steam engine business records are well known. The records of Boulton's and Watt's households and estates are also very substantial and give a detailed picture of the life and times of the families and employees, from groceries to garden plans, house furnishings, housekeepers, travel, tree-planting, medicine and money.

The amount of correspondence in all the collections is vast, and covers all areas of life and work. There are over two hundred letter books containing many hundreds of copies of outgoing letters and about 120,000 incoming letters in the three collections, as well as a huge number of other papers including notebooks and drawings. The invention of the letter-copying press by James Watt in 1779, and its continuous use for both business and personal letters, has resulted in the unusual survival of both incoming and outgoing correspondence in the same place.

Boulton's correspondents include doctors, scientists, inventors, bankers, businessmen and women, employees, friends and family members – people such as James Adam, Aimé Argand, Sir Joseph Banks, Thomas Beddoes, Sir Stephen Cottrell (Secretary to the Board of Trade), Dr Erasmus Darwin, Thomas Day, Charles Dumergue (the Royal dentist), Benjamin Franklin, Samuel Garbett, Sir William Hamilton, Sir William Herschel, Jonathan Hornblower, James Keir, Antoine-Laurent Lavoisier; Lord Macartney (about his embassy to China), Charlotte and William Matthews (Boulton's bankers), Elizabeth Montagu, Dr Joseph Priestley, Rudolf Raspe (geologist and author), John Rennie, Dr John Roebuck, Dr William Small, Josiah Wedgwood, John Whitehurst, John Wilkinson (iron founder), Dr William Withering, Count Woronzow (the Russian ambassador), the Wyatt family of architects, and many more.

With regard to his business records, on 2 May 1790, Matthew Boulton wrote to Henry Stieglitz: 'Every different trade I am concern'd in hath its own Books and Bookeeper' [sic].[1] While not all the 'Books' have survived, enough remain to illustrate the range and variety of Boulton's enterprises.

Boulton & Fothergill, 1762–1782 [MS 3782/1]

Matthew Boulton and John Fothergill were in partnership from 1762 to 1782. The firm's business comprised various manufactures, which operated more or less as separate departments. The articles produced included buttons, buckles, sword hilts, 'toys', silver and plated ware, ormolu, clocks and 'mechanical paintings'. The surviving records include accounting records, letter books and pattern books. Much material relating to the firm of Boulton & Fothergill can also be found in Matthew Boulton's personal papers [MS 3782/12], and the firm's surviving pattern books are found in the 'Additions' part of the catalogue [MS 3782/21]. The earliest letter book contains copies of a few letters sent by Matthew Boulton's father from the original Snow Hill premises. After the Soho Manufactory was built, initially goods were made at both places, but in about 1766 it was decided to move all production to Soho and keep just a warehouse in Birmingham. There were two separate sets of records, comprising in each case letter books and accounting records, one set of which was known as the 'Soho Books', the other set as the 'Birmingham Books'. The former are chiefly concerned with manufacturing operations, while the ledger contains accounts for suppliers, employees and materials, as well as Matthew Boulton's tenants; the main account books for the Birmingham warehouse do not survive, but there are letter books and cash books.

The firm of M. Boulton, 1782–1816 [MS 3782/2]

When the partnership of Boulton & Fothergill was dissolved on 22 June 1782 following Fothergill's death, an inventory was made and a new firm in the name of M. Boulton was begun, with a new set of books. Not all the businesses formerly carried on by the old partnership were taken up by this concern; the main button manufacture, for example, was continued by the partnership of Boulton & Scale (see below).

The surviving records of this firm include letter books, order books, ledgers and cash books for the Soho Manufactory and the Birmingham Warehouse, but do not form a complete sequence. One of the ledgers includes accounts apparently relating to a business carried on at Soho by John Wyatt. The inventory of 1782 is a wonderful description of the tools and stock in all rooms of the premises.[2] After Boulton's death in 1809, these businesses were continued in the name of his son, Matthew Robinson Boulton.

The Soho Mint, 1791–1850 [MS 3782/3]

The Soho Mint business was run successively by Matthew Boulton, 1791–1809, his son Matthew Robinson Boulton, 1809–41, and *his* son Matthew Piers Watt Boulton, 1841–50. It manufactured coins, medals and mint machinery. There are also records of the Soho Rolling Mill, 1825–41. A large quantity of correspondence relating to the business of the mint can also be found among the personal papers of Matthew Boulton [MS 3782/12] and Matthew Robinson Boulton [MS 3782/13]. Initially the mint was solely engaged in the manufacture of coins, medals and tokens, but later it supplied machinery for mints in various parts of the world, though most of this machinery was ordered from other firms, particularly Boulton, Watt & Co. Related records, including drawings of mints and mint machinery, can therefore be found in the Boulton & Watt Collection [MS 3147].

A comparatively large number of the Mint's records have survived. They include accounting records, operational records for coinages and for mint machinery, and correspondence. There are records for the Brazilian, Calcutta, Danish, Mexican, Royal, Russian and Sierra Leone Mints, etcetera.

The London Banking Agency, 1802–1819 [MS 3782/4]

From about 1768, Matthew Boulton and the firms at Soho used William Matthews as their London banker and agent. When Matthews died in 1792, the business was continued by his widow, Charlotte, and on her death in 1802 the banking house was continued by the new firm of M. & R. Boulton, J. & G. Watt, & Company. The main business of the Banking Agency was to accept bills and run accounts for Matthew Boulton and his firms, and for members of the Boulton and Watt families. Two-thirds of this concern were held by Matthew Boulton and his son, the remaining third by James Watt junior and Gregory Watt in equal shares. The new firm appointed Mrs Matthews' two clerks, John Woodward and John Mosley, to act as agents for them in London, but the accounts were to be checked at Soho by the agent William D. Brown.

None of the firm's books exist, neither the originals nor the duplicates. The records that do survive are entirely those made or retained by Brown. These fall into two groups: letters from the agents, mainly containing transcripts of the London House's books; and account books, compiled by Brown himself from information sent to him from London.

Other firms, 1802–1831 [MS 3782/5]

Matthew Boulton had other firms at Soho, for which only one or two records survive. These were as follows:

Boulton & Smith, the Latchet Company, 1793–1819 [MS 3782/5/1]

This partnership, of Matthew Boulton and the brothers Benjamin and James Smith, was formed in 1793, changed its name to Boulton & Smith in 1801 and had closed by 1819. Boulton was introduced to Smith by Samuel Garbett and thought it worthwhile to invest some money in the enterprise of buckle fastenings. In 1801 an inventory of stock was taken. The books were kept by William D. Brown, but only one cash book, 1802–1819, survives. References in this book indicate that it was formerly accompanied by a ledger and journal for the same period. Matthew Boulton's correspondence with Richard Chippindall [MS 3782/12/59] includes copies of a few letters written by James Watt junior, who acted as an agent for the firm whilst abroad, in 1794, together with Chippindall's replies. There are also papers relating to the partnership with Smith, the patent and design, and the latchet works at Soho Manufactory in Matthew Boulton's personal papers [MS 3782/12/106].

Soho Rolling Mill, 1805–1818 [MS 3782/5/2–4]
There are a few records that appear to relate to the Rolling Mill at Soho before the Rolling Mill accounts were incorporated with those of the Soho Mint. They comprise a ledger and a journal, each covering the period 1805–1818 and a bundle of account balances and charges, 1802–1815. Further papers relating to the rolling mill can be found in the records of the Soho Mint [MS 3782/3].

M. Boulton & Plate Company, 1815–1823 [MS 3782/5/5–6]
There is little information about this firm and only two volumes of its records have survived. These are a letter book, 1815–19, containing press-copies of letters issued on the firm's behalf by William D. Brown, and a burnishing book, 1815–23, containing weekly accounts of work done by burnishers, who were all female, and whose names are given. There are also two files of papers and memoranda relating to M. Boulton & Plate Co. among the Miscellaneous Cashiers' and Agents' Records [MS 3782/10].

M.R. Boulton (Button Company), 1825–1831 [MS 3782/5/7]
A letter book, 1825–31, from this company survives, relating to the button trade at Soho, dating from the period when John Bown as acted as agent for this firm.

Boulton & Scale, 1782–1799 [MS 3782/21/10–12]
The firm of Boulton & Scale, manufacturers of buttons, steel chains and sword hilts existed from 1782 to 1799, the original partners being Matthew Boulton and John Scale. Scale had been a partner in the main button trade at Soho from 1777, and when Boulton & Fothergill was dissolved, he and Boulton formed a new partnership to continue that trade. The memorandum of agreement between the partners required the firm of Boulton & Scale to have their own set of account books, and provision was made for these to be kept by James Pearson. Unfortunately, none of the account books survive; the only records that do are a pattern book and an incomplete letter book, 1784–5.

Household books and documents, 1752–1863 [MS 3782/6]
These are the accounts of receipts and expenditure connected with the day-to-day running of Soho House, its gardens, and

86 Design for the entrance hall floor cloth at Soho House. Floorcloths were forerunners of linoleum, made of painted canvas, designed to be easily cleaned. Ink on paper, from a letter from Smith, Baber & Downing of London, 1799. (Birmingham Archives & Heritage, MS 3782/12/44/50)

other estates in Warwickshire and Staffordshire owned by the Boulton family. The various series of records include: 'Household Books and Accounts, 1766–1840', 'Estate Records, 1800–1848', 'Accounts of Matthew Robinson Boulton, 1791–1809', 'House and Land Agency Records, 1822–1848' and 'Miscellaneous Accounts, 1752–1827'.

The first person known to have been appointed by Matthew Boulton to keep his household books was John Scale, whose duties had begun by 15 July 1768, the date of the first entry in the earliest surviving cash book. It is clear from cross-references in the records that similar books were kept at an earlier period, but these do not survive. The keeper of the 'Household Books' also acted as a general house and land agent, collecting rents and supervising workers on the estate, and usually held a position of importance within the Soho Manufactory. John Scale was bookkeeper until 11 August 1787, handing over to John Roberts until 17 February 1791, when Matthew Robinson Boulton took over and kept them until 29 April 1796. For a few months following, the system lapses, but towards the end of the year William Cheshire completely reorganised them and kept them until 1811.

The surviving records are substantial in quantity and very detailed. There are ledgers, cash books, tradesmen's bills, expenditure accounts, books for servants' wages and liveries, separate

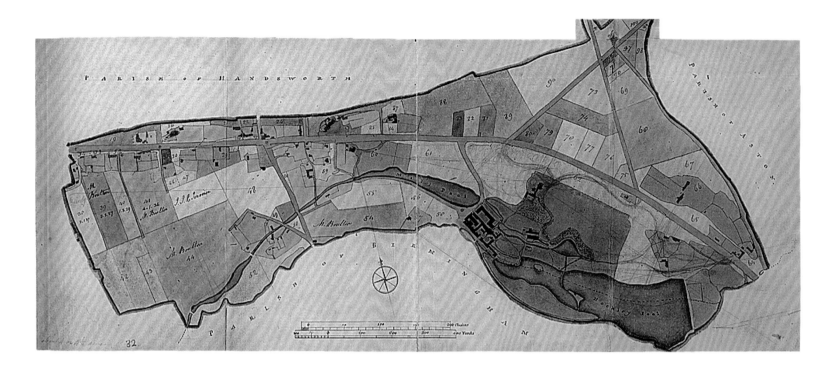

87 Plan of the Soho estate, from Matthew Boulton's notebook no. 27, 1793–1799. This was the period during which he acquired the freehold of the estate. (Birmingham Archives & Heritage, MS 3782/12/108/68)

bills of the architect Benjamin Wyatt for work at Soho House, saddlery and stables accounts, land and house agents' papers, correspondence, rent rolls, tax bills and accounts for the Soho and Yoxall estates. Many of the bills have decorative headings and they can be used to vividly reflect commerce in Birmingham and elsewhere.

Housekeeper's books and accounts, 1787–1826 [MS 3782/7]
Prior to 1787, the housekeeper of Soho House periodically submitted accounts of her expenses to the household agent, who reimbursed her. Some of these accounts are found among the 'Household Bills'. From 1787 onwards, the housekeeper kept separate records of her own, principally a cash book, usually called 'the housekeeper's book.' Ann (Nancy) Mynd (Matthew Boulton's niece), who acted as housekeeper from 1787 to 1794, kept one cash book for that whole period and a ledger. Afterwards a separate cash book was kept for each year, but no ledger.
The cash books are straightforward records of the receipts and payments made by the housekeeper in the course of her duties. The most common payments are for provisions, kitchen and table utensils, washing, servants' clothing, and various expenses of the butler and coachman. The records include bills, and special books for bread, butter, malt and hops, meat, groceries and washing. They form a good illustration of daily living expenses at Soho House.

Private books and documents, 1799, 1802–1855 [MS 3782/8]
These accounting records, kept by the agent and cashier in the Mint Office at Soho, chiefly relate to the private account of Matthew Boulton, and later of Matthew Robinson Boulton. The agents' correspondence relates to private, household and the manufacturing business accounts. The records consist of ledgers, journals, vouchers and correspondence.

Great Tew books and documents, 1815–1844 [MS 3782/9]
Matthew Robinson Boulton purchased the estate of Tew Park in Great Tew, Oxfordshire, in 1815. It comprised, besides the mansion, the lodge, and the adjacent park and plantations, a number of cottages and farms. Most of the farms were let to tenants, but a few were retained 'in hand,' and these are usually referred to in the records simply as 'the Farm'. The records of the estate were kept by the cashier in the Mint Office at Soho. The records that exist include a letter book, 1816–44, cashier's vouchers and correspondence, 1815–21, bailiff and agent's accounts, 1815–21, and a memorandum book of articles sent to Tew, 1816–23. The farm accounts are very detailed and include information on the purchases, births and deaths of livestock, sales of corn, details of the work done by the farm labourers and the amounts of their wages.

Correspondence and Papers of Matthew Boulton, 1760–1809 [MS 3782/12/1–113]

One of the largest sections of the catalogue contains the private correspondence and papers of Matthew Boulton. Several of Matthew Boulton's files of correspondence and papers were continued by his son, and are therefore found among the papers of Matthew Robinson Boulton. Nine of Matthew Boulton's letter-books of outgoing correspondence survive, covering 1766 to 1792, though it is clear that there were more as the existing volumes do not form a complete sequence. No books survive for the period 1773 to 1780; at other times two or more books were kept concurrently.

There are various other volumes, including four relating to steam engines, two of which contain tables recording the performance of Cornish engines, and two have directions and diagrams for assembling engines. Five volumes, including copy letters, from 1787 to 1790, relate to Boulton's dispute with Jean-Pierre Droz, the Swiss die-sinker, over payments for the improvement of dies for the coining press.

The earliest record of the arrangement of Matthew Boulton's papers is contained in a small volume entitled 'Catalogue of Papers, M.B.'s Counting Room, 1760–1790' [MS 3782/13/158]. This catalogue was compiled by Boulton's clerk William D. Brown not long after he arrived at Soho in early November 1791, and was annotated by Matthew Robinson Boulton in the early 1800s, when a substantial re-arrangement of the papers was made. The catalogue has a list of files of correspondence and papers, and an alphabetical index to the list. Three distinct categories are given: 'General Correspondence Files', 'Special Correspondents Files' and 'Special Subjects Files'. This system, which was continued for the files created after this date, was also later adopted (about 1806) by Matthew Robinson Boulton for his own papers.

The first substantial arrangement of the 'Papers of Matthew Boulton and Family' as a whole was undertaken about 1820. Inventories compiled when their transfer to the Birmingham Assay Office took place in 1921 suggest the order remained the same. Some changes were made to the papers while at the Assay Office. In particular, Boulton's general incoming correspondence was rearranged into alphabetical order. The original chronological order as recorded by Brown has now been restored, and the letters arranged in annual files, 1780–1809, with the exception of the two earliest files which cover 1758–79. There are thirty 'Special Correspondents' files, including the very important correspondence from James Watt to Boulton, 1768–91; letters from his son Matthew Robinson Boulton, 1783–1807, from Sir Joseph Banks, Thomas Day, Samuel Garbett, William and Charlotte Matthews, Dr John Roebuck, John Scale, John Vivian; Josiah Wedgwood, John Whitehurst, John Wilkinson, the Wyatt family and so on. The 'Special Subjects' include Albion Mill, the Assay Office, Birmingham Chamber of Manufacturers, the Copper Trade, the Cornish Metal Co., Cornish Mines, Irish Propositions about taxes on imports of iron, lead etc., Lord Macartney's Embassy to China, Mrs Montagu's Glass and the Soho Volunteer Association. The personal papers also include Matthew Boulton's diaries and notebooks. These include notes on business, steam engines, coinage, travel, experiments, schemes, and are a fund of fascinating jottings, drawings and comments.

Correspondence and Papers of Matthew Robinson Boulton, 1787–1842 [MS 3782/13/1–164]

As mentioned above, Matthew Robinson Boulton's papers are arranged in a similar manner to those of his father. The letter book sequence includes copy letters from 1827–40, but not all volumes listed in the original catalogue have survived. The 'General Correspondence' is arranged in four series (1787–1818, 1819–36, 1837–39, 1840–42) and in alphabetical bundles within each series.

The 'Special Correspondents' files include letters from his father Matthew Boulton, correspondence from James Watt, James Watt junior, from his sister Anne Boulton, from Soho agents and employees such as John Woodward, James Lawson and William Cheshire. The 'Special Subjects' files include canals, gas lighting, papers concerning the dispute with George Lander, undertaker, over the cost of Matthew Boulton's funeral, much on 'land speculations' (estates for sale investigated before the purchase of Great Tew was decided on), coinages, the various mints, Miss Anne Boulton and her house, Thornhill, Soho House and garden, bills for his son Hugh William Boulton at Christ Church, Oxford, and so on. There are also his diaries and notebooks, and various catalogues to the Boulton papers.

Papers of Miss Anne Boulton, 1793–1829 [MS 3782/14]

The papers of Anne Boulton, Matthew Boulton's daughter, are mostly financial. The majority of the records begin in 1810, after her father's death. These reflect the financial settlement made for her by her father, and her management of her wealth and property. The papers consist of bills, household accounts, account books for travel and personal expenditure, which includes subscriptions, presents, medical expenses, servants' wages, amusements, house linen and books, clothes and journeys. The earliest surviving book is marked 'To be burnt.' Though this instruction was obviously not carried out in this instance, it may have been in others. There is a small amount of financial correspondence. Papers concerning Thornhill House and garden in Handsworth, which became her home from 1819, can be found with Matthew Robinson Boulton's papers [MS 3782/13/142].

Papers of Mrs Mary Anne Boulton, 1814–1827 [MS 3782/15]

Mary Anne Wilkinson, niece of John 'Iron-Mad' Wilkinson, married Matthew Robinson Boulton in 1817. These papers are, as with Anne Boulton's, mostly financial. They consist of cash books and bills for purchases made by Mary Anne Boulton,

88 One of Matthew Boulton's numerous notebook lists ('Ale and Barrells in my Celler').
(Birmingham Archives & Heritage, MS 3782/12/107/28)

1814–27. The most common items bought are clothes for herself and her children.

Other personal records, 1759–1845 [MS 3782/16]
This section includes the personal records of other members of the Boulton family, and of John Scale, Matthew Boulton's business partner. The following records survive:

Correspondence of Mrs Ann Boulton, 1759–81 [MS 3782/16/1–2]
There are about one hundred letters from Matthew Boulton to his second wife, Ann, and a few from other correspondents. Those from Boulton reveal a very affectionate and sociable nature.

Bills of John Scale, 1766–91 [MS 3782/16/3]
There is a series of personal and household bills of John Scale, Matthew Boulton's business partner, for items such as his son's school fees, brandy, coal, clothing, groceries, subscriptions, travel, etcetera.

Bills of M.R. Boulton's sons, 1840–45 [MS 3782/16/4–5]
The records here consist of bills for Matthew Piers Watt Boulton, from his time at Trinity College, Cambridge, 1840–44, and bills, visiting cards, invitations, etcetera, collected by Hugh William Boulton on a visit to Rome in 1845.

Legal Documents, 1790–1818 [MS 3782/17]
Most of the deeds listed in Matthew Robinson Boulton's catalogues as 'Legal records' are no longer in the collection. However, the 'Rough Catalogue', c. 1820, also records eight 'patents in red boxes' and these do survive. They are Matthew Boulton's coining apparatus letters patent, five licences permitting him to make coins, and Henshall's and Grove's patents for corkscrews and gun locks. There is also a lease by Matthew Robinson Boulton of land to Thomas Allen. Henshall's corkscrews were produced by Boulton at Soho Manufactory. He was presumably planning to do the same with Grove's gun lock, but it is not clear whether he did.

Records of the 1st Battalion, Loyal Birmingham Volunteers, 1794–1805 [MS 3782/18]

The Loyal Birmingham Volunteers were formed as an infantry regiment in 1803 as part of general concern to defend the country against foreign (French) invasion. Matthew Robinson Boulton was an officer from 1803. All his papers relating to the Battalion were kept in a large red box referred to as the 'Birmingham Volunteers' box. The bundles of papers that exist match the list of contents inside the lid of the original box (tactics, orders, government regulations, etcetera), along with some papers not mentioned on the list. Matthew Robinson Boulton also collected various papers relating to the formation of previous battalions, for reference. There are eight files of papers, two notebooks and some printed booklets.

Records of the Committee at Verdun for the Relief of British Prisoners in France, 1808–1813 [MS 3782/19]

The Committee for the Relief of British Prisoners in France was established about 1803 at the depot for prisoners of war at Verdun to direct the distribution of charitable aid to prisoners either in depots or on their march to them, the money for aid being contributed by prisoners' families and friends or collected by public subscription in Britain. There are two volumes of letters and accounts surviving. These are not directly related to any other records in the Matthew Boulton collection, although there are some references to the subscription fund in the papers of both Matthew Boulton and James Watt. The volumes probably came into the possession of Matthew Robinson Boulton through the Reverend William Gorden, one of the committee's principal correspondents who was vicar of the parish of Duns Tew in Oxfordshire, next to Matthew Robinson Boulton's estate. In 1803 he was detained in France and sent as a prisoner of war to Verdun, where he remained for eleven years.

Additions, 1737–twentieth century [MS 3782/21]

Various documents were added to the Matthew Boulton Papers after they came into the custody of the Birmingham Assay Office in 1921, and these have been listed as additions to the other records. The most important of these include a portion of a Boulton & Fothergill Letter Book from 1773, pattern books formerly belonging to Boulton & Fothergill and successor firms, and a letter book and pattern book from the firm of Boulton & Scale (described here in the section on the records of other firms). The letters in the 1773 volume mainly concern the dispatch of orders. The volume was donated to the Reference Library in 1924.

The pattern books containing drawings of wares made by Boulton & Fothergill, M. Boulton (firm) and M. Boulton & Plate Co., were acquired by Elkington & Co., silversmiths and electroplaters, of Birmingham, perhaps in the 1850s when the Soho Manufactory was being vacated and parts of it demolished, or from the steam-engine firm James Watt & Co. at the sale of the assets of Soho Foundry in 1895. The number of books that Elkington & Co. acquired, and exactly which Soho firms they were from, is not known, as the drawings were cut up and rearranged. The pattern books were later acquired by the Library. A further pattern book, preserved in its original format, was purchased by Birmingham City Archives in 2005 (Accession 2005/207).

Other 'Additions' to the Matthew Boulton collection include 'illustrative' material added by the Assay Office, engravings, photographs, newscuttings, sale catalogues and other printed material.

*

In 1796, at a 'rearing feast' to mark the building of the Soho Foundry, Matthew Boulton made a speech to the assembled workers, which seems to sum up his approach to life and to business. He told them:

> I now come as the Father of Soho to <u>Consecrate</u> this place as one of its Branches, I also come to give it a Name & my Benediction. I will therefore proceed to purify the walls of it by the Sprinkling of Wine and in the name of Vulcan & all the Gods & Goddesses of Fire & Water, I pronounce the name of it <u>Soho Foundry</u> – May that name endure for ever & for ever & let all the people say amen amen…
>
> May this Establishment be ever prosperous, may it give Birth to many usefull Arts and Inventions May it prove benificial to Mankind & yield comfort & happiness to all who may be employd [in] it
>
> As the Smith cannot do without his Striker so neither can the Master do without his Workmen, Let each perform his part well & do their Duty in that state to which it has pleased God to call them & this they will find to be the true ground of Equality.[3]

Two hundred years ago the importance of the collection from which that draft speech comes was already being recognised. In 1808 James Weale junior, an author working on a history of the iron trades, asked for access to Matthew Boulton's papers, for, as he said, 'I have no doubt that manuscripts & Printed documents & memoirs are to be found, <u>exclusively</u> in his Collection, which may prove of essential importance to the success of the Work in question'.[4] The recipient of this letter, Boulton's clerk William Cheshire, promised to see what he could do. Ever since then, historians researching many aspects of industry, technology, the development of scientific ideas, and eighteenth-century life in general, have found the Archives of Soho a rich source of firsthand observation and information.

BY THE SKILFUL
EXERTION OF A MIND TUR-
NED TO PHILOSOPHY & ME-
CHANICS, THE APPLICATION OF
A TASTE CORRECT & REFINED, &
AN ARDENT SPIRIT OF ENTERPRISE,
HE IMPROVED, EMBELLISHED, & EXTEN-
DED THE ARTS & MANUFACTURES OF
HIS COUNTRY; LEAVING HIS ESTABLISH-
MENT OF SOHO A NOBLE MONUMENT OF
HIS GENIUS, INDUSTRY, & SUCCESS.
THE CHARACTER HIS TALENTS HAD RAI-
SED, HIS VIRTUES ADORNED & EXALTED.
ACTIVE TO DISCOVER MERIT, & PROMPT
TO RELIEVE DISTRESS. HIS ENCOU-
RAGEMENT WAS LIBERAL, HIS BENE-
VOLENCE UNWEARIED. HONOURED
& ADMIRED AT HOME & ABROAD.
HE CLOSED A LIFE EMINENTLY
USEFUL, THE 17th AUGUST
1809 AGED 81.
ESTEEMED, LOVED, & LAMENTED.

Selling what all the world desires

THE MATTHEW BOULTON BICENTENARY EXHIBITION
CATALOGUE

ABBREVIATIONS

BAH Birmingham Archives and Heritage (at Birmingham Central Library). The Papers of Matthew Boulton and family (MS 3782) are held by BAH on permanent loan from the Birmingham Assay Office Charitable Trust. BAH also houses the Boulton & Watt Papers (MS 3147) and the James Watt Papers (MS 3219).
BAO Birmingham Assay Office
BMAG Birmingham Museum & Art Gallery

All measurements are approximate.
Dimensions of paintings, etcetera, are shown as height × width.

THE STORY BEGINS

The Birmingham into which Matthew Boulton was born in September 1728 was a rapidly growing town with a thriving 'toy' industry. This industry was not concerned with children's playthings; rather, it produced small decorative accessories for personal use and adornment, such as buttons, buckles, watch-chains, trinkets and snuffboxes, in various metals and other materials, such as tortoiseshell and glass.

Matthew Boulton was the third child of one such buckle and toy maker, Matthew Boulton senior, who had come from Lichfield to settle in Birmingham; he and his wife, Christiana, were married at St Martin's Parish Church in 1723. Matthew Boulton the younger was born in Birmingham on 3 September 1728[1] and baptized on 18 September in the church of St Philip (later Birmingham Cathedral). He was educated locally and in about 1745 joined the family business in Snow Hill in the town centre. The business specialised in producing buckles which were exported to Europe.

In order to expand the business further, in 1756 Matthew Boulton senior rented Sarehole Mill in Hall Green to provide additional rolling mill capacity to produce sheet metal. The business was still based in Snow Hill, but Boulton senior lived in the country near Sarehole for the last few years of his life, until his death in 1759. By the time his father died, the younger Boulton had been married and widowed. He had also had nine years' business experience and had been entrusted with much of the management. His entrepreneurial journey to develop the small local toy-making enterprise into a much more ambitious entity, and to establish himself as one of Birmingham's, and Britain's, leading industrialists, was about to begin.

Laura Cox

1 Strictly, 3 September old-style (Julian) calendar. When the Gregorian calendar, already in use in much of Europe, was adopted in Britain in 1752, in order to synchronise with those other countries, a first-year adjustment was made whereby what would have been 3 September was changed to 14 September. From that time on, Matthew Boulton celebrated his birthday on 14 September.

To Mathew Boulton E

This N.E. View of SOHO MANUFACTORY, *is inscribed by his obliged Servt.*
S. Shaw.

1   Lemuel Francis Abbott, Portrait of Matthew Boulton, c. 1798–1801
    oil on canvas, 73.66 × 61.6 cm.
    BMAG, 1908 P 20 (see frontispiece)

Abbott had painted a number of Boulton's friends and associates, including William Herschel, Valentine Green, John Wilkinson and Lord Nelson. Abbott was certified insane in 1798 and died in December 1802, so his portrait of Boulton is likely to be one of his last and may have been done earlier than the 1801 date to which it has been generally attributed.

The portrait shows Boulton holding one of his green notebooks with white metal clasps, still to be found in the Archives of Soho. This example is inscribed 'Mint', reflecting his preoccupation with his Mint and its products in his later years. Abbott's portrait was described by Boulton's biographer H.W. Dickinson as the 'most virile' of Boulton's portraits. It shows a relaxed and confident man in a dark jacket, yellow waistcoat and pink scarf. The eyes and mouth carry a hint of lively good humour, less sombre than the portraits by Sir William Beechey (see no. 248) and Carl von Breda (see no. 373). VL

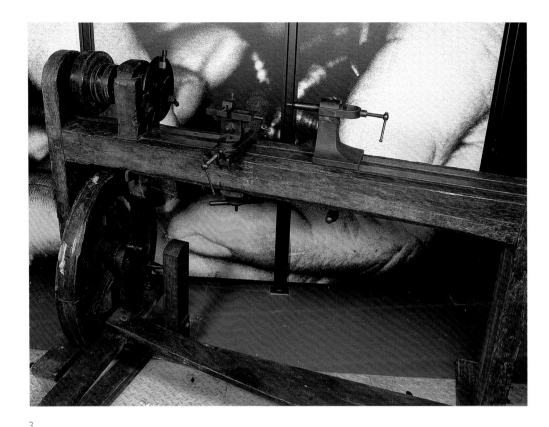

3

2   Joseph Wright of Derby, An Iron Forge, 1772
    oil on canvas, 157 × 145.8 cm.
    Tate Britain, T06670

Joseph Wright was the best-known painter of British scientific and industrial life during the second half of the eighteenth century. In this painting, he depicts a small family-run foundry typical of the kind that would have been found across the Midlands during the period. Wright employs his familiar technique of strongly contrasting light and darkness to add drama to the scene. The detailed depiction of the water-powered drum and tilt hammer is in notable contrast to his slightly earlier *Blacksmith's Forge*, which features a blacksmith manually shaping metal with a hammer. Although water-power was already beginning to be replaced by steam-power by this date, the emphasis placed on the 'modern' machinery in this painting is a celebration of the technological progress of the age. CR

3   Treadle lathe, c. 1762
    wood and iron with leather and lead.
    BMAG, 1951 S 00119.00001, from Thinktank

This lathe belonged to Matthew Boulton and may have been made by him. It was brought from Snow Hill to Soho in 1762, and from 1797 was used to sharpen tools for the Soho Mint, after Boulton won a government contract for coinage. It was subsequently used to train boy apprentices. JK

4–7 Reproduction of a metalworking workshop containing eighteenth- and nineteenth-century items including a blacksmith's anvil, bellows and weights, top and bottom swage, and hand tools.
    BMAG, 1988 F 1733; 2004.0856.1–3; 1995 F 195.13–14; 1995 F 195.1

An anvil, a pair of bellows and a selection of hand tools formed the basic toolkit for Birmingham metalworkers. These items were a common feature in the numerous forges and workshops which were part of the Birmingham landscape in the eighteenth century. SC

THE BIRMINGHAM TOY TRADE

Birmingham has a long tradition of metalworking. The proximity of good supplies of coal and iron ore, together with the increase in demand for luxury goods, contributed to the growth and development of metal industries during the eighteenth century. In particular, Birmingham specialised in producing small decorative metal wares known as toys. This term included a multitude of goods made in steel, brass, iron, gold and silver, as well as other materials.

The two most important branches of the toy trades were the buckle and button trades, both catering for a fashion-conscious market. By 1760 it was estimated that in Birmingham around eight thousand people were employed in making buckles, and that the business generated £300,000 worth of trade. And during the second half of the century the production of buttons helped to secure Birmingham's international reputation as the world's leading button-making centre.

The toy industry grew in importance throughout the eighteenth century. By 1759 it was estimated that twenty thousand people worked in these trades in Birmingham and its vicinity. Although some principal manufacturers, such as John Taylor and Matthew Boulton, employed large numbers of people, most worked from small workshops where all members of the family were involved. These small workshops highlighted another important aspect of Birmingham's toy trades: the subdivision of labour. Skilled workers concentrated on producing individual parts of an object and often did not see the completed article. A visitor to Birmingham in the 1750s noted that it took seventy people to make a button, with each person working on a different process.

The enormous variety of metalworking skills employed in the toy trades helped to make Birmingham craftsmen supremely adaptable. When, for example, in the 1790s, the fashion for wearing shoe buckles waned in favour of 'shoe-strings' (shoelaces), many buckle makers transferred their skills to producing buttons and jewellery. This laid the foundation of the jewellery trade, which by the mid-nineteenth century had become one of the most important industries in Birmingham. A number of other industries also emerged as offshoots of the toy trades. Plating, die-sinking, papier-mâché and japanning were all crafts that grew in importance during the eighteenth century.

By the end of the eighteenth century an increasing amount of Birmingham's trade was being conducted overseas. Birmingham manufacturers sent agents to many countries to open new markets, including America and Russia, and sales from exported goods accounted for a large proportion of the total earnings from trade. Birmingham goods had now achieved international renown. As the MP and philosopher Edmund Burke remarked, Birmingham had become 'The Toyshop of Europe'.

Sylvia Crawley

8–12 Five buckles, c. 1700–50
brass, steel and iron, heights 35–72 mm.
BMAG, 1934 F 41; 1934 F 34; 1934 F 39.1; 1934 F 30.1; 2000 F 43

These buckles were made in the Birmingham area in the first half of the eighteenth century. One carries the mark of William Green, a buckle maker from Wolverhampton; the rest are unmarked. sc

13 Watch chain, c. 1765
gilt metal, length c. 15 cm.
BMAG, 2003.0007.26.2

This gilt-metal watch chain was made in Birmingham. The fashion for wearing gilt metal grew during the second half of the eighteenth century. Before Boulton, John Taylor was Birmingham's largest producer of gilt-metal buttons, and many other items, such as watch chains, were produced in workshops across the town. sc

14 Purse rings, 1700–1800
steel.
BMAG, 1906 F 19.1–4

These metal rings made from chiselled steel are typical of the type of small metal 'toy' produced in Birmingham during the eighteenth and nineteenth centuries. Rings of this sort were used for many personal items, including stocking purses which became popular in the Victorian period. sc

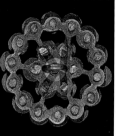

15 front and back

15–16 Buttons, c. 1750
cut steel, pewter, diameters 15–18 mm.
BMAG, 1964 F 2609 and 1971 F 135

From 1700 onwards Birmingham was producing buttons in a variety of different metals and finishes. Highly fashionable cut-steel buttons, like the one here, demonstrate one method of industrial recycling as the tiny steel studs which decorate them were at first made from old horseshoe nails, cut in facets and polished so that they sparkled in the candlelight. Birmingham also produced more functional buttons and supplied many for the armed forces. The pewter button (16), made after 1751, was made for the 17th (Leicestershire) Regiment of Foot. sc

8

THE BIRMINGHAM OF MATTHEW BOULTON'S BOYHOOD

The first known plan of Birmingham was produced by William Westley, a local builder, in 1731 (see no. 17). It provides us with a great deal of information, including an estimate that the population was 15,032, as well as revealing key features of the town's character and appearance. Birmingham was a small manufacturing and market town, which had grown up along the main thoroughfares running to and from its mediaeval core around the Bull Ring. Throughout the late seventeenth and early eighteenth century, developments in the north of the town saw the focus shifting from the Bull Ring. High Town, situated up on the sandstone ridge in the northern part of the town, became Birmingham's prestigious residential area reflecting the new wealth of the town's manufacturers. Builders and architects embraced classical architecture in the construction of prominent buildings such as St Philip's Church, and residential developments such as the Square and Temple Row. Low Town, to the south, consisted mainly of older vernacular timber-framed buildings in narrow streets. It was also the location of many of the town's forges and workshops.

Overall Birmingham was still quite green. On Westley's plan, fields and orchards can be seen in the central areas around New Street and south of St Philip's Church, and most houses in the central and northern parts of the town have gardens. Genteel leisure pursuits are reflected in the town's bowling green, while St Philip's was surrounded by pleasant walks suitable for fashionable promenading.

Westley's plan pictures prominent buildings both old and new, including the town's two churches, St Martin's in the Bull Ring and the recent St Philip's. Birmingham's mediaeval heritage is reflected in two moated sites, the Parsonage Moat and Birmingham Moat ('The Ancient Seat of the Lord Birmingham'). Other vignettes illustrate the steel houses of Steel-house Lane and Cole's Hill Street; the Old Cross and Welch Cross; the free school on New Street (King Edward's); and the Baptist's and Presbyterian meeting houses.

In 1751 Samuel Bradford produced a second plan of Birmingham (no.18). By his estimate Birmingham's population had grown to 23,688. More land has been developed in the north and northeast, and the cherry orchard near St Philip's has now disappeared, though Birmingham still has gardens, bowling greens and tree-lined streets, and, as a result of some long-term leases, fields and orchards can still be seen at the end of New Street. In addition to the public buildings illustrated on Westley's plan, Birmingham has now acquired a

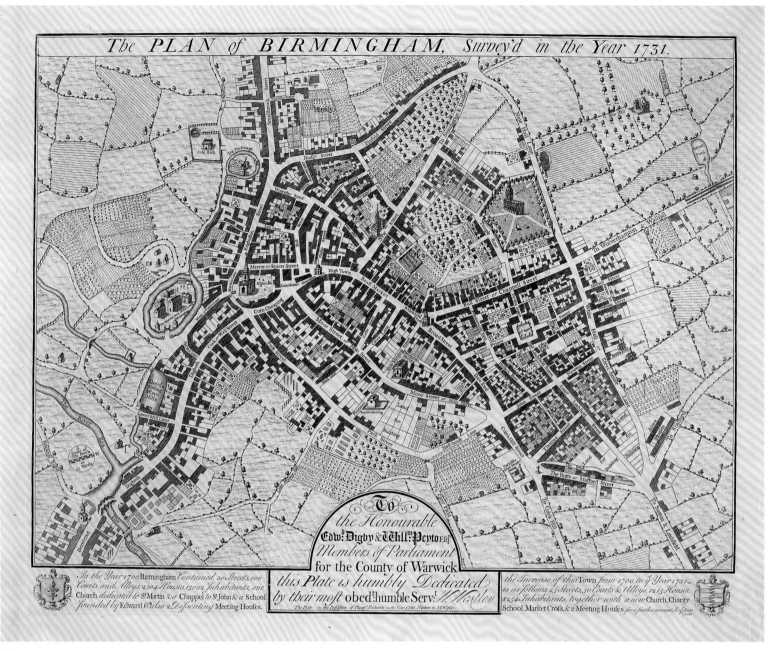

17

charity school, a workhouse with infirmary, a playhouse, several new meeting houses and a new chapel.

In 1755 a visitor described Birmingham as 'London in miniature'. He described the town as being in two halves, the lower 'filled with workshops and Ware-houses of the manufacturers and consists chiefly of old buildings', while the upper part of the town is compared to St James's and 'contains a number of new, regular streets and a handsome Square'.

Jo-Ann Curtis

17 William Westley, *Plan of Birmingham*, 1731
engraving, 48.6 × 69.1 cm.
BMAG, 1997 V 81

18 Samuel Bradford, *Plan of Birmingham*, 1751
engraving, 20.0 × 25.7 cm.
BMAG, 1998 V 4

William Westley's plan of Birmingham is especially important because it presents the town in its entirety for the first time. Birmingham was then a small town with 105 streets, 150 courts and alleys, and 3,719 houses. Note: on Westley's plan north is at the right-hand side and west is at the top.

Samuel Bradford's plan shows how Birmingham has grown in the intervening twenty years.

JC

19 William Westley, *Prospect of Birmingham from the East*, 1732
engraving, 40.5 × 86.7 cm (framed).
BMAG, 1912 P 5

20 Samuel and Nathaniel Buck, *Prospect of Birmingham from the East*, 1753
watercolour, 41 × 84.5 cm.
BMAG, 1934 P 398

To have a prospect of a town produced in the eighteenth century was a source of civic pride for the inhabitants and a way of promoting it to the outside world as a centre of civility and politeness. Westley's prospect reflects his plan of the town produced in 1731 (no. 17), illustrating prominent buildings, often in exaggerated proportions. The

Buck Brothers' later prospect is a much more naturalistic view of the town, although the spires of St Martin's and St Philip's churches still dominate the horizon. JC

INCOMERS AND NATIVES

Birmingham in the eighteenth century saw many able, enterprising and famous men, some of them natives and some incomers. John Baskerville came to the town from Worcestershire in the 1720s. He traded in papier-mâché and japanned goods, but is remembered as a printer and the inventor of the Baskerville typeface. The first book he published was an edition of Virgil's poetry in 1757: the subscribers included Matthew Boulton and Benjamin Franklin. Baskerville's greatest typographical achievement is his Bible, published in 1763, a copy of which Boulton bought (see no. 95).

The young Samuel Johnson arrived in Birmingham from his native Lichfield in 1732, lodging for three years alongside his old schoolfriend Edmund Hector, the apothecary-surgeon who would treat Boulton's family years later (see no. 126). In 1740 Thomas Aris came from London to set up as a printer and bookseller. On 16 November the following year he launched *Aris's Birmingham Gazette*, Birmingham's first weekly (later daily) newspaper. Another incomer was William Hutton; born in Derby, he visited Birmingham for the first time in 1741 and recorded 'I had been among dreamers, but now I saw men awake. Their very step shewed alacrity. Every man seemed to know what he was about. The town was large, and full of inhabitants, and those inhabitants full of industry.' He returned nine years later and became a prosperous bookseller.

Benjamin Franklin came to Birmingham in 1758 and returned several times while in Britain as the agent and representative of his native colony, Pennsylvania. A printer by trade and a scientist by inclination, Franklin struck up a friendship with John Baskerville, who introduced him to Matthew Boulton, who also became a good friend (see nos. 37 and 60). Boulton in turn introduced his long-standing collaborator and friend Samuel Garbett to Franklin. Garbett was a refiner of gold and silver and a manufacturer of vitriol (concentrated sulphuric acid). Like Boulton, he was a supporter of the General Hospital and a Guardian of the Assay Office. He was also one of the founding partners in the Carron Company, the iron foundry at Carron in Scotland. Two other Birmingham-born men deserve a mention. Although little is known in detail of his business, John Taylor was the town's leading toy maker before Boulton's rise to prominence. The other man was Sampson Lloyd II, the scion of a Welsh family which had come to Birmingham in 1698 and set up as ironmasters who specialised in supplying bar and rod iron to the nailers of the region. In 1765 Taylor and Lloyd, and their respective sons, set up a bank at 7 Dale End, Birmingham. It was ultimately to grow into Lloyds TSB.

The Methodist preacher John Wesley, who had been to Birmingham twice before (once being pelted with stones and dirt by a mob), called again in 1782, this time taking in the Soho Manufactory, about which he wrote 'I walked through Mr. Bolton's [sic] curious works. He has carried every thing which he takes in hand to a high degree of perfection, and employs in the house about five hundred men, women, and children. His gardens, running along the side of a hill, are delightful indeed; having a large piece of water at the bottom, in which are two well-wooded islands. If faith and love dwell here, then there may be happiness too. Otherwise all these beautiful things are as unsatisfactory as straws and feathers.'

David Symons/Shena Mason

22

21 Enamelled snuffbox, c. 1750–55
    enamel on copper and gilt metal,
    length 80 mm.
    BMAG, 1957 M 99

Snuff (powdered tobacco taken by sniffing) was popular in the eighteenth and nineteenth centuries, and great numbers of decorative little boxes were produced to contain it. This enamelled snuffbox was made in Birmingham around 1750–55. French designers working in Birmingham used the newly invented process of transfer printing to create a variety of classical and romantic scenes. John Taylor was a principal manufacturer of such enamel snuffboxes, many of which were exported to continental Europe and America. SC

22 Snuffbox, early to mid-eighteenth century
    tortoiseshell
    BMAG, 2005.1431

This snuffbox was owned by William Hutton. Hutton settled in Birmingham in 1750 as a bookseller and stationer, and later wrote the *History of Birmingham* (1782), the first published history of his adopted town. SC

BIRMINGHAM'S OTHER INDUSTRIES

Apart from the specialised toy trades for which Birmingham had become famous, and which were to launch Matthew Boulton on his manufacturing career, the town was also a major centre for the production of brassware, guns, edge-tools and iron wares. Birmingham's success in these industries was largely due to the growth of the canal system from the 1760s. Until then, poor roads and the heavy costs involved in transporting goods had favoured the lightweight toy trades at the expense of larger, heavier products. However, by 1772 Birmingham was linked to the major ports of Bristol, Liverpool and Hull, which brought down the cost of transporting raw materials into the town and made it easier and cheaper for Birmingham-made goods to reach markets further afield.

Although some brass had been manufactured in Birmingham since 1740, most was obtained from Cheadle and Bristol. It was expensive to transport and subject to frequent price rises, but in 1781 Birmingham brassware manufacturers established their own brasshouse alongside the canal. The subsequent expansion in the industry and the proliferation of specialist trades producing industrial brassware, coffin furniture, cabinet brassware, and items such as candlesticks, made

Birmingham the main brass-manufacturing centre in Britain.

Gun-making was another complex process involving a variety of specialist skills. In the eighteenth century each part of a gun was handmade separately in small workshops and then assembled and dispatched. The wars with France and America from the 1770s onwards kept the trade busy. Working under contract to the Board of Ordnance and trading organisations such as the East India Company, Birmingham's gun trade flourished, and by 1815 the town had become England's largest gun-making centre. Domestic and industrial hardware, from cauldrons to scythes, were also made in the town and sold to both national and international markets.

It was, however, the people of Birmingham who were chiefly responsible for its industrial success. Birmingham was not an incorporated borough subject to the restrictive trade practices associated with trade guilds found elsewhere. This freedom attracted many Quakers and other Dissenters who brought a fresh, entrepreneurial spirit to the town. This, together with the ingenuity and flexibility of a skilled and adaptable workforce, put Birmingham at the forefront of industry, a position it was to develop further during the nineteenth century.

Sylvia Crawley

24

25

23

23 Handmade nails, 1700–1800
    iron.
    BMAG, 1979 F 587

Until the 1780s all nails were made by hand. The industry was mainly located in the Black Country and Bromsgrove areas, though many nails, like these examples, were made in Birmingham. The work was carried out mainly by women and children who could produce around 250 nails an hour. SC

24–25 Two candlesticks, 1715–50
    brass, heights 17.5–17.9 cm.
    BMAG, 1969 M 153–154

Birmingham was the centre of the brass industry in the eighteenth century, with over sixty makers specialising in brass candlesticks. Although few makers marked their goods at this time, the restrained designs used point to the influence of Huguenot and other Protestant craftsmen who were able to take advantage of Birmingham's freedom from restrictive trades' guilds. SC

26–29 Pewter tableware made in Birmingham, 1700–40
    tankard, height 19.05 cm; salt, height 50.8 mm; wine quart baluster, height 17.78 cm; and plate, diameter 52.07 cm.
    BMAG, 1956 M 59. 36, 40, 125 and 310

Pewter was originally an alloy of tin with lead, copper and other metals (modern pewter alloys do not contain lead). In the seventeenth century most well-to-do households in England owned a selection of pewterware. However, by the mid-eighteenth century competition from the ceramics trade forced many Birmingham manufacturers to turn their skills towards producing tinplate, brass and copper goods. The pewter plate (see overleaf) was made by John Duncomb, who worked in Birmingham between 1702 and 1720. He was a significant manufacturer, producing around twenty tons of pewterware a year for customers as far afield as Chester, Shrewsbury and Grantham. SC

26

29

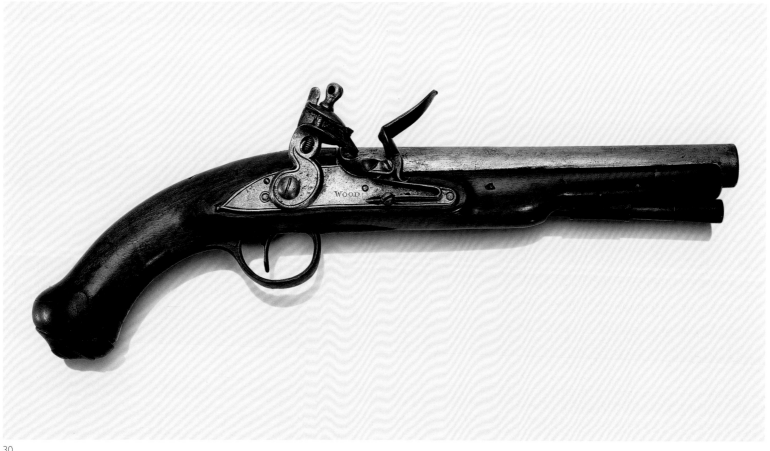

30

30 Flintlock pistol with Ketland proof marks, probably made in Birmingham, c. 1780
wood and steel, length 43 cm.
BMAG, 1990 S 03181.00214

This is a typical Birmingham-produced flintlock pistol. As a means of quality control, it was tested at the private Ketland Proof House in the days before the Birmingham Gun Barrel Proof House was established in 1813.   JA

31–32 Two shoe buckles, c. 1750
steel, height 75 mm.
BMAG, 1934 F 30 and 1934 F 42

These shoe buckles were made in Birmingham. In the second half of the eighteenth century buckles began to increase in size and the fashion-conscious man often owned several pairs which he could wear with different shoes. Towards the end of the century the fashion waned as 'shoe-strings' (shoelaces) came into favour, and many buckle makers had to turn to making other products. Many of them became jewellers, the foundation of Birmingham's major nineteenth-century jewellery industry.   SC

33 Long-case clock mechanism, made by John
    Greaves, Birmingham, c. 1720
    brass and steel, height 2.65 m.
    BMAG, 1953 S 00348

Birmingham craftsmen were skilled in producing precision instruments, including clocks and watches. John Greaves, who had a workshop in the High Street from 1715 until his death in 1771, was part of a well-known family of clockmakers working in the town during the eighteenth century. SC

34 Door lock, c. 1700–40
    brass, steel and iron, height 15.2 cm.
    BMAG, 1985 M 3

This rim lock, probably made in Birmingham, was used on a door at Aston Hall, Birmingham. Although the main part of the lock-making industry was based in Wolverhampton and Willenhall, Birmingham produced many fine brass-cased locks. SC

35–36 Two watchcocks, c. 1720–70
       brass, length 34 mm.
       BMAG, 1930 M 469–70

These gilt brass watchcocks demonstrate the skill and intricacy of the Birmingham brass workers and engravers, and reveal a striking contrast to the larger, utilitarian brassware normally associated with the town. Watchcocks were made to cover and protect the delicate mechanism used in watches, and because they were hidden inside the watch case they were rarely seen. SC

33

34

35

36

37

38

THE EIGHTEENTH-CENTURY INTELLECTUAL CLIMATE

Matthew Boulton was a product of, and a contributor to, what we now call the 'Age of Enlightenment' – the intellectual and philosophical revolution of the eighteenth century that ushered in the modern world. He was typical of the new breed of 'natural philosopher' who emerged during the period. His interests ranged exuberantly over astronomy, geology, meteorology, chemistry, electricity, medicine, the fine arts, classics, music and the new fashion for landscape gardening.

Across Europe and the New World, the period witnessed a surge of interest in science and empirical reasoning. The thirst for knowledge was felt as keenly in a provincial market town like Birmingham as it was in London, Paris and beyond. Such was the interest in the new sciences that people from all walks of life flocked to attend public lectures and demonstrations, and in many towns and cities new 'philosophical societies' were formed.

Boulton was a founder-member of one of the most influential of these new scientific groups – the Lunar Society. Their meetings (at full moon) were delightfully described by Erasmus Darwin as 'a little philosophical laughing'. It was a select band – there were never more than fourteen members (see pp. 7–13 and 133–5) – yet it comprised some of the outstanding minds of the day, and their shared conversations, letters and experiments made major contributions to scientific understanding. Ten of them were Fellows of the Royal Society, and four were also Fellows of the Linnaean Society. Between them they were in correspondence with many of the leading thinkers across Europe and beyond. Boulton himself counted as friends such people as Benjamin Franklin, Sir Joseph Banks and Sir William Herschel. Possibly the honour that would have pleased him most came in 1785, when he was elected a Fellow of the Royal Society. This was a prestigious accolade for a man who had experienced relatively little formal education, and it was the ultimate expression of the regard in which he was held by the men of science whose companionship he had sought and valued so much.

But Enlightenment theories were not just restricted to scientific advances; they were part of a wider intellectual movement which questioned the nature of eighteenth-century government and in particular the traditional hierarchies of Church and State. Many members of the Lunar Society were nonconformists and freethinkers who believed that the same rational principles as were used in scientific investigation could and should be applied to make the world a better place through social and political change. At their most extreme, such views were to provide the intellectual justification for the American and French Revolutions. In Britain, however, the association of reform with revolutionary excess served only to preserve the political and social status quo. The events in France in particular led to a 'patriotic' backlash against radicals, and civil unrest in Birmingham (see pp. 201–2).

Chris Rice/Shena Mason/Laura Cox

39

37 Joseph Siffred Duplessis, *Portrait of Benjamin Franklin*, 1778
 oil on canvas, 71.1 × 57.2 cm.
 National Portrait Gallery, NPG 327

Benjamin Franklin was one of the Founding Fathers of the United States of America. He was one of the five delegates chosen to write the Declaration of Independence in 1776 and was respected internationally as a printer, inventor, scientist, diplomat and social reformer. He spent some years in England and became a friend and frequent correspondent of Matthew Boulton, with whom he discussed subjects as diverse as electricity and his invention of 'musical glasses', an instrument that enjoyed some popularity. Franklin took his friend John Baskerville's 'Baskerville' typeface back to America, where it was adopted by the new federal government.   BF

38 James Millar (attributed), *An Allegory of Wisdom and Science*, 1798
 oil on canvas, 125.7 × 100.9 cm.
 Courtesy of Wolverhampton Art Gallery, 1740/50 – Birmingham 1805

This painting shows Wisdom (Sophia) as a young woman with winged brows surrounded by the emblems of the liberal arts and sciences. Children gather at her knee for her benign instruction. Art is symbolised by the sculpted bust and palette while the orrery, bell jar and retort represent the world of Science. The balloon celebrates mankind's recent conquest of the air. In the background is a statue of Britannia and the picture may have been a design for a frontispiece for an edition of *Encyclopaedia Britannica*. The encyclopaedias were seen as a vital instrument of the Enlightenment, dispelling ignorance and heralding the new Age of Reason.   BF

39 Joseph Wright, *An Experiment on a Bird in the Air Pump*, 1768
 oil on canvas, 183 × 244 cm.
 National Gallery, NG725

Joseph Wright of Derby was interested in depicting the development of the Industrial Revolution and the scientific advances of the Enlightenment. The painting shows a natural philosopher demonstrating the new technology of the air pump to a scientifically curious audience. The people observing this experiment into the nature of air and its ability to support life exhibit a variety of reactions. Wright was a close friend of two members of the Lunar Society, John Whitehurst and Erasmus Darwin. The full moon illustrated in the window to the right of the painting is a possible reference to the Lunar Society.   LC

MATTHEW BOULTON'S INTERESTS

A browse through Matthew Boulton's notebooks gives us some insight into his interests, with sketches and jottings on matters as diverse as double-glazing and bed-quilts. His earliest surviving notebook, begun in 1751, contains detailed notes on, among other things, 'Ferenheath's [Fahrenheit's] thermometer', the different kinds of microscopic organisms found in different species of ground meal, the variation in human pulse rates between young and old, the difference between the solar and the sidereal day, and electricity and chemistry.

Some notebooks contain lists of scientific instruments or books owned or wanted, but others show that Boulton also took an interest in the arts and humanities. In 1757 he bought the works of Pope, Shakespeare and Locke, and bound volumes of journals including *The Spectator*. At other times he mentions volumes of Italian drawings, and novels of the day, including Laurence Sterne's *Tristram Shandy* (1759–69) and Fanny Burney's *Evelina* (1778). He also shared his daughter Anne's keen enjoyment of music.

Science in its many forms remained Boulton's main preoccupation, however, and in 1771 he jotted down his ideas for the kind of private museum he would like at Soho House:

> A round building for my Study, Library, Museum or Hobby Horsery to hold 6 handsome Book Cases with drawers in the lower parts to hold things which relate to subjects of the books wch are in upper parts e:gr: a Book Case containing Chymical Books should have drawers under wch contain Metals Minerals & Fossells … & under the Space between ye upper parts of ye Cases should be fixed such instruments as Baromotor, Thermomotor, Pyromotor, Quadrants, all sorts of Optical, Mathematical, Mechanical, Pnumatical & Philosophical instruments also Clocks of Sundry kinds both Geographical & Syderial, Lunar & Solar System & one good regulator of time. A table in ye middle of ye room & a skylight in ye middle of the doomical roof wch roof may be covered either with Sail Cloth or brown paper. Out of this round room should open a private door into a passage in which passage should open doors into sundry convent: rooms such as Cold & Warm Bath, a Labritory a dressing & powdering room & an observatory for a transit instrumt &c

Although the museum with the 'doomical roof' never quite took shape in the way he envisaged, Soho House was certainly equipped with a range of scientific instruments.

Shena Mason

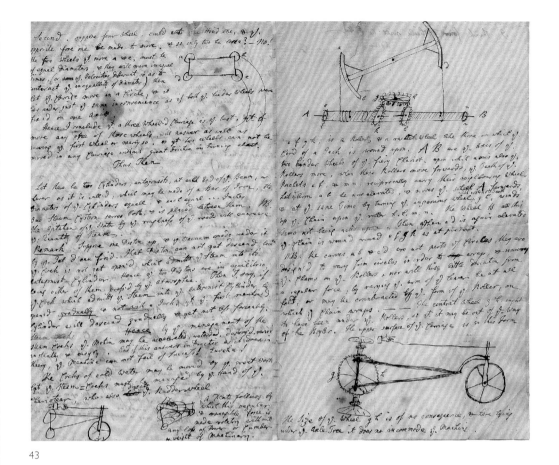

43

40 Notice advertising 'Mr. Nicholson's Philosophical and Chemical Lectures', 1799
BAH, MS 3782/12/99/19

This printed notice advertises three series of scientific lectures given by the chemist William Nicholson at his house, 10 Soho Square, London. The series is promoted as 'a Course of entertaining and popular Instruction' with an appeal for a general as well as a specialist audience. Scientific lectures became popular in the eighteenth century and travelling lecturers toured the country. VO

41 John Rawstorne, *Proposed plan and elevation for alterations to Soho House*, October 1788
ink and wash on paper, 50.8 × 71.8 cm.
BAH, MS 1682/2/1 (see p. 12)

Various plans were drawn up for alterations to Soho House by different architects in the late 1780s. Most plans, like this design, were not built. It proposes a huge new front on the existing house (shown in grey). The rooms to be added reflect Boulton's scientific interests – rooms designated for wet and dry chemistry, astronomy, and a greenhouse for botany. VL

42 Matthew Boulton's notebook No. 8, 1772
paper with brown leather cover.
BAH, MS 3782/12/108/7 (see p. 9)

Matthew Boulton's pencil-written list (on the left-hand page) shows scientific equipment which he already owns, while on the facing page he has made a list of items on order or required. Boulton's interest in lenses comes across clearly from these lists, which include a telescope, a microscope, a 'cammora' (probably a camera obscura), and a query about whether he could use his 'solar microscope' as a magic lantern. SM

43 Letter from Dr Erasmus Darwin to Matthew Boulton, 1760s
BAH, MS 3782/13/53/13

Darwin tells Boulton that 'As I was riding Home yesterday, I consider'd the Scheme of yᵉ fiery chariot, and yᵉ longer I contemplated this favourite idea, yᵉ [more] practicable it appear'd to me.' He then goes on to discuss, and illustrate with a number of sketches, how a vehicle powered by steam might work:

> These things are required. First a Rotary Motion. 2 easyly altering its Direction to any other Direction. 3 To be accelerated, retarded, destroy'd, reviv'd instantly & easily. 4 The Bulk

of Weight & Expence of the machine as small as possible in proportion to its use.

The letter demonstrates how Boulton and his circle were interested in all the ways that the power of steam might be put to use. DS

44 William Pether, print after Joseph Wright's *A Philosopher Giving a Lecture on the Orrery*, 1768
print on paper, 48.4 × 58.2 cm.
Derby Museum & Art Gallery, DBYMU 1954–219/9 (see p. 13)

This print after Joseph Wright's masterpiece depicts a small audience listening to a philosopher's lecture on planetary motion. The philosopher is using an orrery to demonstrate the workings of the solar system. LC

45 Receipted account from George Donisthorpe of Birmingham to Matthew Boulton, 9 January 1773
BAH, MS 3782/6/6/1047

Donisthorpe was a local clock and instrument maker who regularly attended to the Boulton family's watches and clocks, and also made some scientific instruments for Boulton. This account, which covers a long period, shows that on 12 December 1765 he supplied 'a new Electrical Orrery' to Matthew Boulton for the sum of £1 8s. 0d. The cost suggests a relatively uncomplicated instrument and it may have been similar to an electrical orrery made by George Adams of Fleet Street, London, c. 1765. Adams was the scientific instrument maker to George III. His electrical orrery, which is in the Science Museum, London, has brass rods with brass balls representing just the sun, earth and moon (known as a tellurian). The orrery relied on the power of static electricity generated by an electrical machine to make it revolve to indicate the movement of these bodies. SM

46 Orrery, made by Robert Bate of London, 1818
brass and ivory, 34.2 × 36.5 × 51.5 cm.
The Galton Collection, University College London

One of the earliest orreries was made by John Rawley, a London instrument maker, for Charles Boyle, 4th Earl of Orrery, and his name came to be used for the machine itself. The orrery shown here was bought in 1818 by Samuel Galton junior, one of the Lunar Society members. The maker,

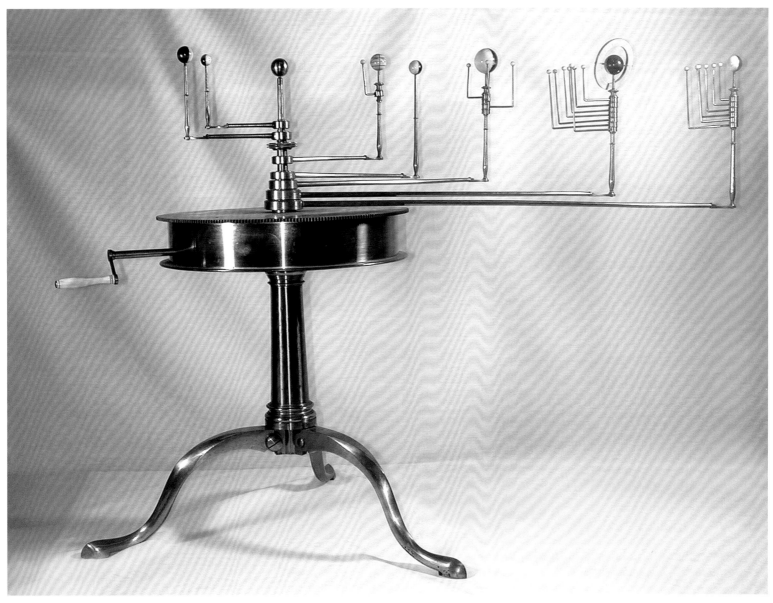

46

47

Robert Bate, was born in Stourbridge in 1782 and became an instrument maker in London. Although manually operated, this orrery is almost certainly a more complex, accurate and expensive model than the 'electrical orrery' that Matthew Boulton possessed (see no. 45). SM

47 John Phillp, Measured drawing of a Gothic building, probably the observatory at Soho, 1796
 pen and ink on paper, 98 × 98 mm.
 BMAG, 2003.0031.58

This drawing probably shows Boulton's observatory in the grounds at Soho and is similar in design to Copenhagen Observatory. By the time Phillp drew the observatory in 1796 it had been disused for around twenty years and he may not have understood its original use, which is perhaps why he showed it with smoke emerging from the roof. VL

48 Receipt for eight copies of William Pryce's Mineralogia Cornubiensis, 23 May 1778
 BAH, MS 3782/6/192/45

Pryce, an apothecary and surgeon from Redruth in Cornwall, was part-owner of the nearby Dolcoath copper mine. His book, a treatise on Cornish minerals, mines and mining, was clearly important to Boulton, who had significant business interests in Cornish copper mining. He spent twelve guineas buying copies for himself and various friends, including Erasmus Darwin and James Watt. DS

49 Bill for the supply of fossils, 25 April 1783
 BAH, MS 3782/6/192/101

This bill is from Thomas Pearson of Matlock, Derbyshire, who is supplying Boulton with seventy-two fossils at a cost of 3½d. each, plus 1s. for the 'packin box', making a total of 22s. Boulton became very interested in geology and built up a large collection of minerals and fossils. DS

50–53 Samples from Matthew Boulton's mineral collection:
 Quartz (BMAG, 1993 G 03.403); hematite (BMAG, 1993 G 03.871); malachite (BMAG, 1993 G 03.956); fossil-rich limestone (BMAG, 2008.1826)

Boulton was a keen collector of geological specimens and fossils, and had two special cabinets made to house his collection. These can still be seen at Soho House. Many of the samples he collected were of rocks and minerals that might be useful in his manufactures, like metal ores or decorative stones such as blue john. DS

54 Faujas de St Fond, Descriptions des Experiences Aerostatiques de MM Montgolfier, 1783–4
 plus an associated ink drawing of a balloon, with annotations and sketches
 book, 20 × 13.0 cm; drawing, 39.7 × 29.7 cm.
 BMAG, 1997 F 764.1–3 (see p. 11)

The possibility of taking to the air proved enticing in the early 1780s. Sir Joseph Banks wrote to Boulton in 1783 with questions about the construction of 'wings', which he had heard a Birmingham man was building to enable him to fly. Boulton made enquiries but was not able to gain any conclusive information. However, the country did become gripped with balloon fever. Matthew Boulton witnessed a balloon launch in Oxford in December 1783 and became greatly interested in this newly emerging science. A business contact in Paris sent him this book detailing the first unmanned flight by the Montgolfier Brothers in Paris in November 1783. The following year Boulton conducted his own balloon experiment into the nature of thunder, with inconclusive results. Boulton, like the Montgolfiers, used hot air for his balloon, but others favoured hydrogen. The ink drawing shows the highly dangerous work involved with ballooning. The men are creating hydrogen by adding sulphuric acid to a barrel of iron filings. SC

50–53

THE LUNAR SOCIETY

The list of people regarded as belonging to the Lunar Society is somewhat elastic. In addition to Matthew Boulton, the following seem to have been the chief 'Lunaticks':

Dr Erasmus Darwin, FRS, FLS (1731–1802) was born near Nottingham, and set up practice in Lichfield in 1756. He moved to Derby in 1784, but stayed in touch with his Lunar Society friends. Physician, philosopher, inventor and poet, he was one of the founder-members of the Lunar Society, with Matthew Boulton and Dr William Small. He published books on medicine, botany and education. His ideas on evolution contributed to the theories put forward by his grandson Charles Darwin.

Thomas Day (1748–1789) was born in London. Independently wealthy, he studied law, but never practised. A writer and social reformer, he was best-known for his children's book *The History of Sandford and Merton*, in which he emphasised educational ideals and the importance of hard work. He campaigned against the slave trade and gave financial support to some of the Lunar Society's activities, and took a liberal attitude to both the American and French Revolutions.

Richard Lovell Edgeworth (1744–1817) was born in Bath into a family of Irish Protestant gentry. He was interested in science, mechanics and exact measurement, and invented a 'perambulator', a device for measuring the size of a plot of land, for which he was awarded a medal by the Society of Arts in 1767. His other achievements included designs for carriages and his collaboration with his daughter Maria on a book about education for girls.

Samuel Galton junior, FRS, FLS (1753–1832) was born in Birmingham. Although Quakers, his family were prominent arms manufacturers, and this led to him being disowned by the Society of Friends. He was interested in chemistry, natural history and the study of colour, and proved that the seven colours of the spectrum are made up from three primary colours.

James Keir, FRS (1735–1820) was born in Stirlingshire. He studied medicine at Edinburgh, where a fellow-student was Erasmus Darwin. After graduating, Keir joined the army, serving from 1756 to 1768. Thereafter he settled in the Birmingham area and became a manufacturer of chemicals, glass and metals. He owned a glassworks at Stourbridge and a chemical works at Tipton.

William Murdock (1754–1839) was born in Ayrshire. He came to Birmingham to work for Boulton & Watt, and spent from 1779 to 1798 working for the company in Cornwall, erecting steam engines. He made a number of important contributions to the development of the engines, as well as building Britain's first steam locomotive capable of running on roads (see no. 64). He is best-known today as the inventor of gas lighting, using gas that he produced by heating coal in a retort. His house at Redruth in Cornwall was the first domestic building in the world to be lit by gas.

Dr Joseph Priestley, FRS (1733–1804) was born in Batley, Yorkshire. A Unitarian minister, he was elected to the New Meeting in Birmingham in 1780. He published many books including *Experiments and Observations on Different Kinds of Air* in 1774. His experiments played a vital role in the development of modern chemistry, with his most celebrated contribution being the discovery of oxygen (his 'dephlogisticated air').

Dr William Small (1734–1775) was a Scot. From 1758 to 1764 he was Professor of Natural Philosophy and Mathematics at William and Mary College in Williamsburg, Virginia. During this time one of Small's students was Thomas Jefferson, the future third President of the United States of America. Small's interests included chemistry, steam engines and the science of measurement.

James Watt, FRS (1736–1819) was born in Greenock. Interested in all scientific subjects, he worked as a mathematical instrument maker at Glasgow University, and as a canal surveyor. He spent a year in London in the 1750s gaining further training. Given a model of a Newcomen steam engine to repair, he thought he could improve on it and began experimenting. His development work was initially supported by Dr John Roebuck, but when Roebuck ran into financial difficulties Matthew Boulton stepped in and invited Watt to come to Birmingham, where he arrived in 1774.

Josiah Wedgwood, FRS (1730–1795) was the greatest of the Staffordshire potters. He worked closely with Boulton and others on the production of small ceramic ornaments to be mounted onto metal snuffboxes, buttons and the like. Responsible for many advances in the production of mass-produced pottery, his scientific interests reflected his industrial needs: mineralogy, thermodynamics and chemistry. His Etruria works became as famous as Boulton's Soho. He was a strong opponent of slavery.

John Whitehurst, FRS (1713–1788) was born in Congleton, Cheshire, the son of a clockmaker. He settled in Derby in 1736, where he followed the same trade. He became an expert maker of scientific instruments, especially those concerned with all kinds of measurement. A keen geologist, Whitehurst was the first to tell Boulton about the blue john stone that he was to use in such quantity in his ormolu production. Whitehurst also made many of the movements for Boulton's ormolu clocks. In 1775 he became Keeper of Stamps and Weights at the Royal Mint.

Dr William Withering, FRS, FLS (1741–1799) was born in Wellington, Shropshire. He studied medicine in Edinburgh and came to Birmingham in 1776 to replace the late Dr Small, serving as physician at the General Hospital from 1779 to 1792. Keenly interested in botany, especially the medicinal uses of plants, one of his biggest achievements was the discovery of the effectiveness of the drug digitalis in the treatment of heart disease. His book *An Account of the Foxglove* was published in 1785 and is now known as one of the most important resources in medical history.

There are two further members of the Lunar Society about whom relatively little is known: Dr Robert Augustus Johnson, FRS (1745–1799), and Dr Jonathan Stokes, FLS (1755–1831). Johnson is known to have been interested in chemistry. Stokes was particularly interested in plant classification. Although both are occasionally referred to in other Lunar Society members' correspondence, they do not seem to have played a very active role within the group.

Laura Cox/David Symons

55 John Russell, *The Face of the Moon*, c. 1795
pastel on board, 59 × 44.7 cm.
BMAG, 1957 P 23

John Russell was a leading portrait painter, but he was also interested in the scientific developments of his day. In particular he was a keen astronomer and made this amazing drawing using a telescope. Appropriately, it is now normally displayed at Soho House, scene of so many meetings of the Lunar Society. Russell's pastel portrait of Boulton's friend Sir Joseph Banks, shown holding a copy of *The Face of the Moon*, is illustrated on p. 5.
DS

56 Matthew Boulton's diary for April 1781
BAH, MS 3782/12/107/12

The entry for Monday 9 April reads 'Lunar Society at Dr. Priestly's'. Unlike his notebooks, Boulton's diaries are often blank for long stretches and what entries there are can be very brief, like this one. Often his diaries simply record details of expenses he has incurred. This entry is one of his rare direct references to the Lunar Society. DS

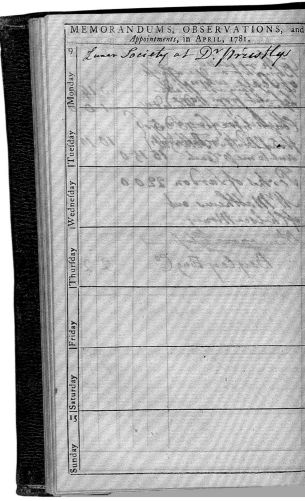

56

57

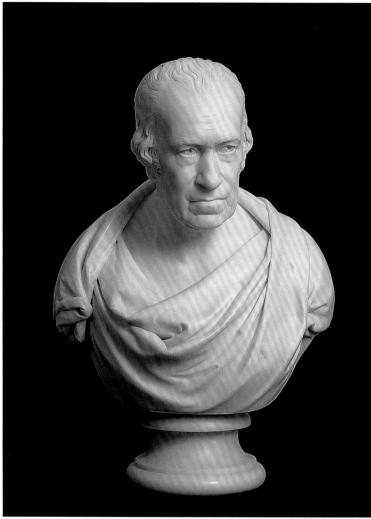

58

57 John Flaxman, *Portrait Bust of Matthew Boulton*, c. 1815
   Carrara marble, height 73.66 cm.
   BMAG, 1995 P 9.

This portrait bust of Matthew Boulton was commissioned by his son Matthew Robinson Boulton after his father's death in 1809. The portrait is very similar to that on Boulton's tomb at St Mary's Church in Handsworth. John Flaxman, who had trained at Soho, was a leading exponent of the neo-classical style in this period. However, his portrait of Boulton is rendered in a realistic style and depicts the sitter in contemporary rather than Roman dress.   LC

58 Sir Francis Chantrey, *Portrait Bust of James Watt*, 1818
   Carrara marble, height 66.04 cm.
   BMAG, 1995 P 10.

This bust was commissioned by Watt's son, James Watt junior, not long before Watt's death in 1819. The portrait is closely related to the monument to Watt in St Mary's Church in Handsworth. Watt is shown with his shoulders draped in a toga. The bust, inscribed 'James Watt aged 83 Chantrey sculptor 1818', is thought to commemorate Watt's life achievements and his contribution to manufacturing.   LC

59 Letter from James Keir to Matthew Boulton, 2 October 1772
   BAH, MS 3782/12/65/3

This letter from James Keir to Boulton refers to glass equipment he is making for Boulton's experiments:

Your orders shall be executed as speedily as possible, especially those for your own experiments.... I should also be sorry to check by delay your present hobby horsicality for chemistry.   LC

60 Letter from Benjamin Franklin to Matthew Boulton, introducing Dr William Small, 22 May 1765
BAH, MS 3782/13/53/26

In 1764 Small returned to Britain from Virginia. Armed with a letter of introduction to Matthew Boulton from Benjamin Franklin, he came to Birmingham in 1765. Franklin writes, 'I beg leave to introduce my friend Doctor William Small to your acquaintance….I would not take this freedom if I was not sure it would be agreeable to you…' LC

61 Note from James Keir to Samuel Galton junior and James Watt junior explaining the process for drawing the winner of the Lunar Society library, 1813
BAH, MS 3219/4/51/57

By 1813 most of the original members of the Lunar Society were dead, and those who remained seldom met. They decided to wind up the Society. The end of the Lunar Society was marked by a draw between James Watt junior, Samuel Galton junior and James Keir for the Lunar Society's collection of scientific books, which seems to have been kept in Boulton's library at Soho House. Galton junior won the draw on 8 August 1813 with his 'lucky slip', subsequently sending a cart to collect the books. Their present whereabouts is unknown. The note from Keir explains the process for drawing the winner of the library:

> There are three sealed papers each containing an enclosed ticket A, B or C, one of which contains the ticket declaring the person who draws that paper to be entitled to the whole of the books of the scientific library.

The note is further annotated by Keir: 'Mr Galton was the fortunate drawer and that all the books belonged to him.' LC

62 Vase and cover, Wedgwood & Bentley (mark impressed on base), 1771–80
black basalt stoneware, height 34.3 cm.
BMAG, 1885 M 2758

Wedgwood introduced his new black stoneware, known as basalt or black basaltes, in 1768. Dense and high-firing, it was particularly appropriate for the elegant neo-classical vases for which Wedgwood was famous. Here, the body has been lathe-turned, and decorated in relief with moulded drapery swags and an applied oval medallion of the Three Graces. ME

60

63 Angle barometer, by John Whitehurst, c. 1775
mahogany, silvered brass and mercury,
104 × 11 cm.
BMAG, 1997 M 22.

John Whitehurst established a clock-making business in Derby in 1736, but also specialised in the production of complex and sophisticated scientific instruments. He maintained a long collaboration with Boulton (who had a Whitehurst barometer at Soho House) and provided movements for many of the Soho ormolu clocks. ME

63

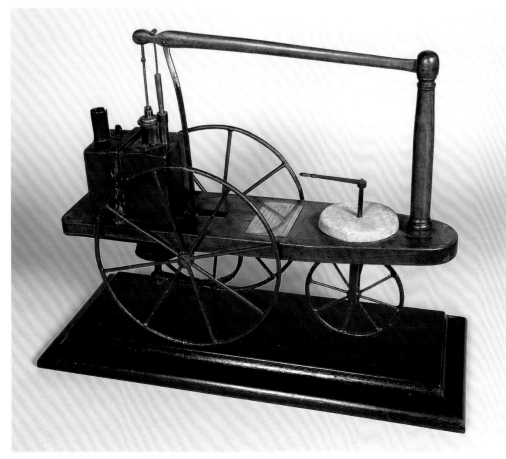

64 Locomotive (model steam engine) by William Murdock, c. 1784
various metals, timber and stone; length 55.5 cm.
BMAG, 1951 S 00088.00008

William Murdock felt there was a market for steam-powered vehicles and built this model about 1784. It was tested on a road in Cornwall, making it the first steam-powered vehicle ever to run on a public road in Britain. Matthew Boulton considered the idea too ambitious and persuaded Murdock to abandon the project. It was some twenty years before steam vehicles showed much success. JA

HEADING FOR SOHO

The period 1759–65 marked a leap forward in Matthew Boulton's remarkable industrial career. It was clear that the workshop in Snow Hill was not big enough, so in 1761 Boulton, who had remarried the previous year, purchased the lease on thirteen acres of land at a site called Soho, two miles north of Birmingham in Handsworth (then over the county boundary in Staffordshire), and began to erect the great Soho Manufactory at a cost of some £10,000. It was an impressive sight, with workshops, showrooms, offices and stores, and accommodation for some of the workforce. It was a major industrial attraction and an icon of the manufacturing empire that he created.

In 1762 Boulton went into partnership with John Fothergill. Fothergill's expertise was in trading. He had travelled widely on the Continent as an agent for other manufacturers and had good connections and a wealth of marketing experience. The partnership was of great benefit to Boulton as it allowed him to expand the business to become the largest in Birmingham, and to export toys and decorative wares across Europe. He continued to produce buttons and buckles, but increased the range to include other metal goods such as cut-steel jewellery, watch chains, toothpick cases and sword hilts. Birmingham-produced goods had such a bad reputation that the term 'Brummagem goods' had become synonymous with gaudiness and inferiority. Boulton did not want his goods to be associated with this derogatory term and saw himself as being in competition with London makers. While Boulton concentrated on improving the Manufactory and extending the range of products, Fothergill was responsible for their promotion and acquiring orders. Between 1762 and 1775 the Soho Manufactory established a strong reputation for excellent quality and artistic design and craftsmanship. During their partnership, which lasted until Fothergill's death in 1782, Boulton gained a greater commercial knowledge and acquired many contacts that would prove valuable in his future business.

Laura Cox

65

65 Unknown artist, *Portrait of Matthew Boulton*, n.d.
    oil on canvas, 57.8 × 49.5 cm.
    National Portrait Gallery, NPG1532
This unsigned painting is probably a nineteenth-century copy. The clothes suggest the original was painted around 1770, and it is possible that it is a copy of a portrait of Boulton by Birmingham artist James Millar, which is known to have been exhibited at the Royal Academy in 1784. The present location of Millar's portrait, which would answer this question, is unknown. VL

THE SOHO MANUFACTORY

The Soho Manufactory was founded by Matthew Boulton on Handsworth Heath in 1761. A water mill, only a few years old, was taken over, quickly converted and the site around it expanded over the next five or six years into an extensive button, buckle, toy and plated-ware manufactory. The dominant 'principal building,' designed by James Wyatt, was added in 1765–7. Apart from being intended to impress, the 'principal building' contained the new plated-ware manufactory, warehouses, the main office or counting house and accommodation for senior staff. Soho was to be the largest factory in the Birmingham area for many decades.

The Manufactory continued to expand until the early nineteenth century, but mostly in relation to the new steam-engine and (from 1780) letter-copying businesses. In the first two decades of the Boulton & Watt partnership they did not provide complete working engines, but supplied licences to build them. The early buildings of the Boulton & Watt engine works were, therefore, on a small scale, making only specialist parts that local foundries could not undertake, since most of the parts had to be manufactured as near to the intended site of the engine as possible. Following the construction of the Soho Foundry in 1795–6 (see pp. 105–7), a major expansion of the Manufactory engine works took place between 1802 and 1804 under the direction of Matthew Robinson Boulton.

After Matthew Boulton's death the momentum of the original Soho businesses was lost, leading to the sale in 1810 of the button company by Matthew Robinson Boulton. In 1833 the plated company was sold to Robinson, Edkins and Aston, who traded as the Soho Plate Company until 1852, when the premises were taken over by A. & J. Toy, bedstead and chandelier manufacturers. Lastly, the demolition of the remainder of the Manufactory was overseen by Matthew Boulton's grandson, Matthew Piers Watt Boulton. This took place between 1859 and 1863 with the 'principal building' being the last tragic loss. Had the buildings survived to the present day they would undoubtedly have become a nationally and internationally significant site of industrial heritage.

George Demidowicz

67

66 Plan of the Soho Manufactory, June 1788
 ink on paper, 33.3 × 43.5 cm.
 BAH, MS 3147/5/1447

Detailed plans showing the whole of the Manufactory with internal layouts are not known. This sketch plan of external dimensions identifies some of the buildings, including the rolling mill, substantially rebuilt in 1785. The workshops were constructed around a series of courts, the largest, the 'great court', lying at the rear of the 'principal building'. The engine works at this time consisted of only two workshops at the top left-hand corner of the plan. GD

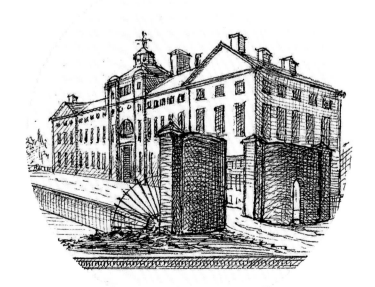

70

71

67 John Phillp, *View across Birmingham Heath*, December 1796
   watercolour, 17.0 × 24.6 cm.
   BMAG, 2003.0031.28

This view makes Soho's rural setting apparent. Soho House is partly concealed by trees on the top of the hill near the centre of the image. The Manufactory sits among the trees to the left of the house. On the right-hand side an engine cylinder is being transported on a wagon. VL

68 View of Soho Manufactory, from Stebbing Shaw's *History and Antiquities of the County of Stafford*, engraved by Francis Eginton junior, 1801
   etching and engraving with aquatint and some hand (?) colouring, 30.2 × 48.9 cm (paper) / 27.7 × 38.2 cm (plate mark).
   BMAG, 2003.0031.89 (see p. 119)

This view was produced for inclusion in Stebbing Shaw's *History and Antiquities of the County of Stafford* (published in two volumes, 1798–1801) along with a view of Soho House and Pool (see p. 14). The inclusion of the family coat of arms under the image and the plate of the house emphasise Boulton the gentleman, the Sheriff of Staffordshire, rather than Boulton the manufacturer. VL

69 The Soho Manufactory near Birmingham, from *The Monthly Magazine*, 1797
   line engraving, 19.0 × 26.6 cm.
   BMAG, 1940 v 788 (see p. 28)

Like most of the published images of the Manufactory, this view focuses on the principal building with the buildings of the rolling mill to the right and the latchet works to the left. VL

70 John Phillp, *View of Soho Manufactory*, c. 1793–1800
   pen and ink on paper, diameter 80 mm.
   BMAG, 2003.0031.105

This appears to be a sketch design for a medal showing the principal building with the terrace and the 'canal', one of the waterways linking the mill and Soho pools. No medal of this design was ever produced. VL

71 John Phillp, *View of Soho Manufactory taken on the Spot*, 1796
   pen and ink, 42.0 × 61.8 cm.
   BMAG, 2003.0031.41

This view is taken from the mill pool with the side of the principal building in the centre. The arched entrance to its right, with the clock above, was used by the workers. Soho House is hidden in the trees on the hill to the left. VL

72

73

72 F. Calvert (illustrator) and T. Radclyffe (engraver), *Soho from the Nineveh Road*, from *Picturesque Views, and Descriptions of Cities, Towns, Castles, Mansions and Other Objects of Interesting Features in Staffordshire and Shropshire*, 1830
steel engraving.
BMAG, 1933 V 321.13
This is the first published view of the Manufactory from the back, showing the rear of the principal building with workshops and housing in front of it. It is unlikely that the surroundings were as densely wooded as this image suggests; the setting has been depicted to fit in with the picturesque approach of the publication it was produced for. VL

73   John Phillp, *View across Soho Pool*, 1796
       watercolour, 17.2 × 24.8 cm.
       BMAG, 2003.0031.26

Phillp's watercolour looks across Soho Pool towards the engine works and the bridge across Hockley Brook. Fruit was grown on the south-facing side of the brick wall in front of the buildings.   VL

74   Boulton & Scale pattern book, c. 1775
       ink on paper with touches of watercolour,
       60.5 × 81 cm (open).
       BAH, MS 3782/21/11

Extensive pattern books were kept at the Soho Manufactory, showing the ranges of jewellery, silverware and ormolu. These finely-executed ink drawings of sword hilts, buckles, watch-chains and buttons all have pattern numbers, and some carry additional information. Many of the drawings show clearly the mass of tiny, facetted steel studs with which the goods were ornamented. After Fothergill's death in 1782 Boulton took his bookkeeper and manager John Scale into partnership in the button and jewellery business, which became Boulton & Scale. This Boulton & Scale pattern book has been dated c. 1775 from dates on some of the drawings; this date puts it some years before Fothergill's death but it was perhaps an ongoing collection of patterns which was started in the 1770s. After the Soho Manufactory was closed down, the pattern books were bought by the major nineteenth- and twentieth-century silver and plated ware manufacturers Elkington & Company, who cut them up and pasted the drawings into books of their own. Some of the context and information contained in the original Soho books has therefore been lost, but they remain valuable evidence of Soho's production.   SM

75   'A List of the Articles Manufactur'd at Soho', 1771
       BAH, MS 3782/12/108/6

This notebook contains a list of the items made at the Soho Manufactory in 1771 with their pattern numbers (where relevant), and also a note of the prices charged. We learn that a plated tankard with lid cost 42s., while a plated half-pint saucepan was 6s. (However, you could buy a set of six saucepans for just 21s.) Filigree-plated buttons with painted plates had been reduced from 28s. to 24s. 6d. a box.   DS

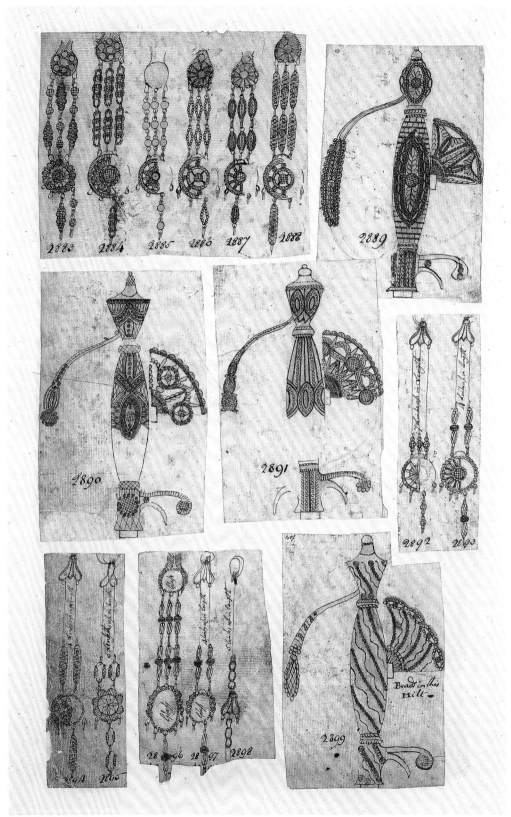

74

76

76 Shoe buckle, Boulton & Smith, c. 1795–1806
    steel and leather, height 60 mm.
    BMAG, 1934 F 45

This steel shoe buckle with its simple border and black leather insert was made in Birmingham to a patent design of Boulton & Smith, a company formed around 1792 by Matthew Boulton with the brothers James and Benjamin Smith. It dates from the final period of the buckle industry when changes in fashion from buckles to 'shoe-strings' (shoelaces) led to a severe decrease in demand for buckles and many makers turned their skills to button-making and jewellery. SC

77 Purse, Boulton & Fothergill, c. 1780
    steel, height 90 mm.
    BMAG, 1953 F 545

This cut-steel purse was made by Matthew Boulton at the Soho Manufactory around 1780. It is made from steel 'chain mail' links and would have had a cotton or velvet lining. Items such as this were part of a wide range of steel toys made at Soho at this time. SC

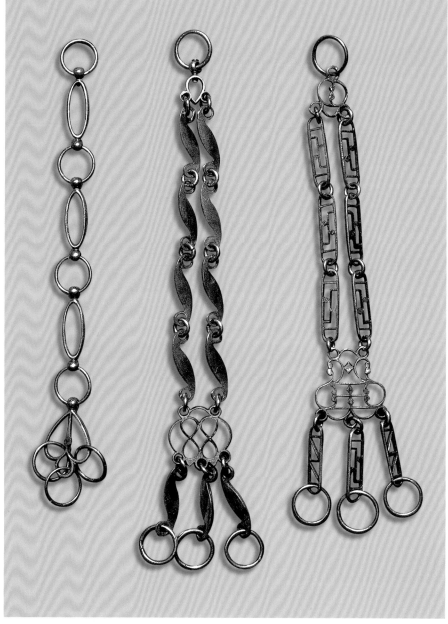

78, 78a, 79

78, 78a, 79 Fob watch chains c. 1780
    cut steel, length c. 150 mm.
    BMAG, 2003.0007.23.1, 2, 3

Watch chains like these were among the various kinds of inexpensive metal items that made up Boulton's successful 'toy' business. The profits from the toy business helped to finance the production of Boulton's finer wares in Sheffield plate, silver and ormolu. However, unlike silver and Sheffield plate, few of the toys produced at Soho carried any identifying marks so it is often difficult to attribute them to Boulton with absolute certainty. These cut-steel watch chains were acquired from the Watt family in 2003 and are therefore likely to have been manufactured at Soho. Similar watch chains are also shown in the Soho pattern books. CR

80 Watch key, c. 1750–1800
    steel, 35 × 12 mm.
    BMAG, 1934 F 228.1

Delicate, high-quality cut steel had been made in Woodstock, Oxfordshire, from the sixteenth century. By the 1750s much of the industry had become based around Birmingham and Wolverhampton. This watch key shaped like a flintlock pistol, with a ring attachment, is a novelty version of a practical everyday item. SC

81–84 Chatelaines, Boulton & Fothergill, c. 1765
    gilt-metal, length c. 13.5 cm.
    BMAG, 2003.0007.15–21

Chatelaines were decorative clasps worn on women's belts from which were hung useful items like scissors, thimbles, keys, etcetera. These particular chatelaines are from a group of seven acquired from the Watt family in 2003, and were almost certainly made at the Soho Manufactory. They were made in openwork versions (illustrated at 81–4) or in 'solid' versions like the examples shown below, right. They are thought to have been taken as samples by James Watt himself when he first visited Soho in May 1768, as they are still sewn on to their original card, with tissue and blue paper wrapping. CR

81–84

85 Sword attributed to the Soho Manufactory, c. 1790
steel, length 95.4 cm.
BMAG, 1991 D 34.77

This gentleman's fashion accessory, typical of the period, has a hilt decorated with cut-steel beads, probably produced at Matthew Boulton's Soho Manufactory. The steel was of such high carbon content that it did not rust easily and could be polished to a sparkling finish. JA

86 Small sword, c. 1775
hilt by Boulton & Fothergill
cut steel and copper wire, length 60.5 cm.
Victoria & Albert Museum, 141–1889

This is a child's version of the type of small sword worn by adults on formal occasions in the late eighteenth century. The hilt is made of cut and polished steel and was made at the Soho Manufactory. The blade was made in Solingen, Germany, which was famous for sword manufacture. Similar hilts are shown in the Soho pattern books. Boulton was able to produce cut-steel products at Soho more cheaply than his competitors in Wolverhampton and Woodstock (the other main centres of production) by utilising water and later steam to power the polishing wheels in the workshops. CR

87

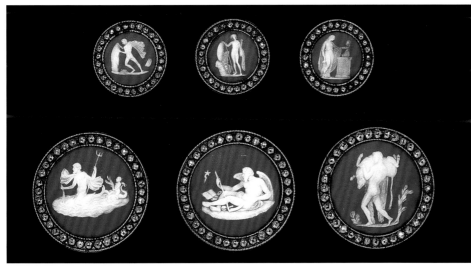

88

85 detail

87 Belt buckle by Josiah Wedgwood and Matthew Boulton, c. 1810
jasper and back-painted glass mounted on cut steel, diameter 86 mm.
Victoria & Albert Museum, 414:1294–1885
Perfected by Wedgwood in 1775, jasperware was a hard, fine-grained stoneware which was used for a whole variety of items including vases, cups and saucers, teapots and decorative cameos. When used for buttons and jewellery, jasperware was often mounted in cut steel. As well as performing the practical function of encasing and protecting the ceramic, the sparkling cut-steel facets complemented the blue and white medallions.

This jasper and cut-steel buckle probably formed part of a woman's belt, which would have been worn high under the bust as was the fashion after 1810. The jasper reliefs are of the signs of the Zodiac. These are based on a relief that Wedgwood bought from the London plaster shop of Mrs Lander in 1774. CR

88 Coat and waistcoat buttons by Josiah Wedgwood and Matthew Boulton, c. 1785–1800
jasper mounted on cut steel, diameters 32 mm (coat), 18 mm (waistcoat).
Victoria & Albert Museum, 276 to N-1866
Although Wedgwood sold small numbers of steel-mounted jasper medallions at his London showroom, the majority were mounted and sold by other manufacturers like Matthew Boulton, to whom these examples are attributed. Wedgwood wrote to Boulton in 1786:

> I have left a few sets of my cameo buttons to be mounted, and shall be glad to increase our connection in this way, as well as selling you cameos for your trade, as in having them mounted by you for mine, both in gilt metal and steel. CR

89 Memorandum dated 28 August, 1769
BAH, MS 3782/12/23/143)
The memorandum gives details of an order for buckles and buttons for a Mr Fountain of Marylebone, London, and for vases, chains and plated buttons for a Mr Yates. The wax-seal impressions show the decoration required on the livery buttons. FT

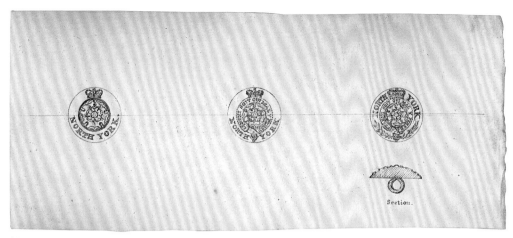

91

90 Notebook belonging to Matthew Boulton, 1790
    leather, paper, gold leaf, 12.5 × 18.5 cm (closed).
    BAH, MS 3782/12/108/56

One of Matthew Boulton's ninety-six notebooks, this has 'Gilt Buttons 1790' written on the cover, and contains notes, drawings and observations about button design, gilding and manufacturing. It also contains a small piece of gold leaf. The notebooks are a rich source of information about Matthew Boulton's many and varied interests, his shopping habits and other matters.   FT

91 John Phillp, Copy press drawing of possible designs for buttons for the North Yorkshire Local Militia Regiment, c. 1808
    8.2 × 18.9 cm.
    BMAG, 2003.0031.142

The North Yorkshire Local Militia regiment was based at Richmond in Yorkshire and existed 1808–16. The middle design was the one finally chosen. Military buttons remained an important market, even after the fashion for showy buttons on everyday clothes declined.   DS

A FAMILY MAN

Although Matthew Boulton's varied business interests took him away from home frequently, it is clear from his correspondence that his home and family were tremendously important to him. There were no surviving children from his first marriage to Mary Robinson, who died in 1759. Within months of her death Boulton was writing love-letters to her younger sister, Ann Robinson, addressing some of them to 'An Angel at Mrs Robinson's in Boar Street' (the Robinson family's Lichfield address). He and Ann married in June 1760, and there are many loving and affectionate letters from Boulton to his second wife in the Matthew Boulton Papers.

Matthew and Ann Boulton began their married life at the Boulton family home at Snow Hill in Birmingham, close to the family business. Boulton's widowed mother and his two sisters, Mary and Catherine, also formed part of the household, until Mary married Zaccheus Walker, who became chief clerk at Soho, and Catherine married a local toy maker, Thomas Mynd.

When Boulton acquired the site for his new Manufactory at Soho in 1761, Soho House was already there but it was not until 1766 that Boulton and his second wife moved in. Their two children, Anne Boulton and Matthew Robinson Boulton ('Matt') were born at Soho House. Boulton was an affectionate father, constantly concerned with his children's upbringing, health and education. He referred to Anne as 'the Fair Maid of the Mill' and later 'the greatest blessing and comfort of my Old Age'; Matt was his father's 'General of Soho'. Anne shared her father's love of flowers and music, and played the harpsichord and pianoforte. Matt would later go on, with James Watt junior, to run the business after their fathers retired.

Some other members of the Boulton family also worked at Soho. Boulton's younger brother, John, worked there for a time before moving to Bedfordshire. Boulton's nephews, 'Zack' Walker junior and George Mynd, were also employed. Zack became a continental traveller for the firm and Matthew Boulton thought well of him. He spent some time in Paris (where he narrowly escaped the guillotine thanks to the intervention of Robespierre, with whom he had become friendly), and in America. By 1803 he was in St Petersburg for Boulton & Watt, attending to the construction of the St Petersburg Mint. George Mynd on the other hand proved troublesome to his uncle, who eventually dismissed him; George's sister Nancy became Miss Anne Boulton's companion and took charge of running Soho House for a while until her marriage.

Shena Mason

92 Poem and prayer written by Matthew Boulton to his first wife, Mary, August 1759
BAH, MS 3782/12/112)

Matthew Boulton and his first wife, Mary Robinson, were married in 1749. Mary died in August 1759. This poem is endorsed on the back:

> Upon seeing the Corps of my dear Wife Mary many Excellent Qualitys of Hers arose to my Mind which I could not then forbare acknowledging Extemporary with my pen & depositing it in her Coffin, of which this is a Copy.

The poem refers to Mary 'bearing many children', but records have been found of only three daughters from the marriage, all of whom died before their mother. SM

93 Bill from J. Whiston & B. White, publishers, Fleet Street, London
for supplying 180 copies of *The case of marriage between near kindred particularly considered with respect to the doctrine of scripture, the law of nature and the laws of England*, by John Fry (1756), 22 April 1760
BAH, MS 3782/6/190/5

Boulton distributed this booklet, which he referred to as 'Fry on Marriage', to support his case for marrying his late wife's sister, Ann Robinson, in 1760. The marriage, which took place at St Mary's Church at Rotherhithe, London, only ten months after the death of Boulton's first wife, was controversial though not, at the time, illegal. Boulton negotiated a substantial discount with the publishers for the bulk order. SM

94 Jean-Etienne Liotard, *Miniature Portrait of Matthew Robinson Boulton*, 1773
pastel on paper, 27.5 × 20.0 cm.
BMAG, 1987 P 3 (see p. 17)

Liotard was one of the greatest masters of pastel drawing. This work was completed during his visit to England in 1772–4, when he was commissioned to draw the three-year-old son of Matthew Boulton. He had also painted the children of the Prince of Wales during a previous visit to England in 1753. LC

92

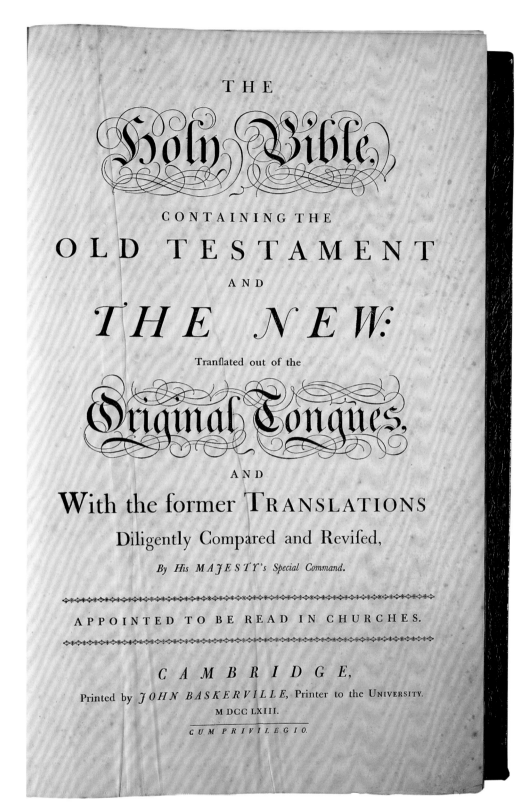

95

95 Boulton family Bible, printed by John Baskerville, 1763
paper with morocco binding, height 50 cm.
BAO

Boulton's friend John Baskerville began printing Bibles in 1763 while he was the university printer at Cambridge. This Bible was printed the same year, and contains notes recording the births of Matthew Boulton, Matthew Robinson Boulton and Matthew Robinson Boulton's children. SB

96 John Phillp, Painting of a spaniel's head, 1796
watercolour, 20.4 × 24.7 cm.
BMAG, 2003.0031.85

This watercolour of a spaniel's head by John Phillp depicts Dash, the favourite dog of Matthew Boulton, who is often mentioned in letters by Boulton and by some of his friends. LC

97 Peter Rouw, *Portrait of John Phillp*, 1807
wax relief.
On loan from a private collection

This miniature wax portrait illustrates John Phillp. It has been suggested, though not conclusively established, that he was Matthew Boulton's illegitimate son, who was born in Falmouth, Cornwall. A talented artist, he came to train and work at Soho in 1793. He went on to produce designs for silver, medals and tokens, and his sketches and watercolours are the main source of information about the Soho estate in Boulton's lifetime. LC

98–99 John Phillp, Metalware designs and sketches, 1807 and c. 1795–1810
pen and ink; ink with crayon, 22.5 × 37.5 cm; 22.0 × 17.3 cm.
BMAG, 2003.0031.125 and 2003.0031.113

Phillp produced many drawings and designs for various departments at Soho, including the Mint and the metalwares businesses. These could be sketch designs, working drawings used by those making the products, or drawings for catalogues to show potential customers. Those shown here are a design for the centre of a candelabrum with one arm shown in detail and a sketch for a lion's head sword hilt. The candelabrum was both designed and drawn by Phillp (indicated by 'invt. delt.'). VL

A FAMILY MAN

96

97

101

100 Plan of the Soho estate, 1834
engraving hand-tinted in watercolour,
11.4 × 11.5 cm.
BMAG, 1970 V 1082 (see p. 19)

This insert from Arrowsmith's 1834 map shows Soho and the surrounding area when it was owned by Matthew Robinson Boulton, although the layout of the park remained much as it had been towards the end of Boulton's life. The kitchen gardens, which provided food for the house, can be seen opposite Mr Evett's house. VL

101–102 John Phillp, *Views of Soho House from Birmingham Heath*, 1796 and June 1799
watercolour; pen and ink with wash,
17.3 × 24.8 cm; 33.0 × 44.6 cm.
BMAG, 2003.0031.29 and 2003.0031.36

These views across Soho Pool show Soho House before and after the remodelling work carried out by James and Samuel Wyatt in 1796. The later view shows a smoking chimney at the Manufactory as well as making it clear that further landscaping work has been carried out in the park. It also highlights the importance of the pool for recreation; people can be seen boating, fishing and walking round it. VL

102

104

103 John Phillp, *View over the Soho Manufactory*, June 1796
    watercolour, 33.0 × 47.3 cm.
    BMAG, 2003.0031.32 (see p. 27)
This view looks from an outbuilding and water trough, across the gardens, with a feature tree and garden urn on a grassy knoll, to the mill pool and beyond. The top of the principal building of the Manufactory is visible above the water trough. VL

104 John Phillp, *Lawn and Park at Soho House*, 1801
    pen and ink, 75 × 106 mm.
    BMAG, 2003.0031.12
This view shows the boundary between the formal lawn with its flower beds and the parkland which was grazed by animals. The chain-link fence prevented the animals straying. VL

105 John Phillp, *Soho House Stables*, 1799
    pen and ink, 90 × 136 mm.
    BMAG, 2003.0031.24
The new stable block was designed by William Hollins in 1797 and built by Benjamin Wyatt of Sutton Coldfield, a cousin of the architects James and Samuel Wyatt. Hollins was a stonemason, sculptor and architect who taught Phillp architectural drawing and had worked on the alterations to Soho House. VL

106–107 John Phillp, *Views of the Hermitage*, 1795 and 1799
    pen and ink and watercolour, 113 × 77 mm; 100 × 152 mm.
    BMAG, 2003.0031.11 and 2003.0031.23
Boulton built a number of small buildings within his grounds. These pictures show the hermitage which he also referred to as 'a building adapted for contemplation'. It was built 1778–9, probably with bark-faced exterior walls and ling (heather) thatch on the roof. The interior view shows a portrait surrounded by attributes connected with gardening and farming, and a plaque inscribed 'A faithful record of the virtues of….' This has been left incomplete, which suggests that Phillp may have drawn from imagination rather than showing something to be found in the building. VL

108–109 John Phillp, *Temple of Flora*, and *Soho Pool and Garden Buildings*, c. 1795–1800
    pen and ink, 98 × 140 mm; 99 × 150 mm.
    BMAG, 2003.003116 and 2003.0031.22 (see also p. 18)
The Temple of Flora was built in 1775–7 on the shore of the smallest of the pools at Soho, the Shell Pool (named from the large carved stone shell which was installed at its edge). The Temple

105

106

107

110

was probably inspired by the Temple of Bacchus at Painshill in Surrey which Boulton had visited in 1772, and where he had made several pages of notes and sketches.

The second view shows Soho Pool in the foreground with the cascade (or waterfall) flowing down from the Shell Pool and the Temple of Flora in the distance. The building at the edge of Soho Pool may be the 'cascade building' which included a library room. It was built in 1776 and pulled down in 1801. VL

110–111 John Phillp, *Soho Pool with the Boat and Boathouse, and Soho Pool with the Boathouse and a Side View of the Manufactory*, 1796
pen and ink, 41.4 × 60.5 cm; pencil, ink and wash, 32.8 × 46.2 cm.
BMAG, 2003.0031.39 and 2003.0031.35

The pools at Soho provided opportunities for sailing, which were to prove too exciting for some. In August 1793 Patty and Mary Fothergill (daughters of the late John Fothergill) and Nancy Mynd thought they would like to have a sail in the boat, but 'unfortunately Mary was in such a terrible fright when we put up the Sails we were oblig'd to make to the shore, as fast as possible, and as we could not get quite close Mr H carry'd Mary out of the Boat in his arms.' The more adventurous Patty and Miss Mynd stayed out in the boat until driven in by heavy rain and were late for tea. VL

111

112

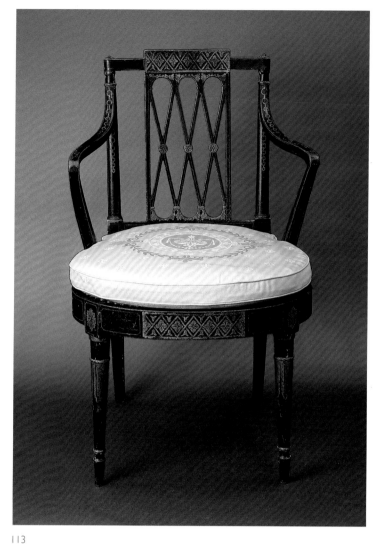

113

112 Klismos chair, by James Newton, 1805
 mahogany with cane seat, height 89 cm.
 Victoria and Albert Museum, w.2:1, 2–1988
It is thought that Matthew Boulton ordered this elegant chair for Soho House around 1805. It is a very distinctive example of a 'klismos' chair, 'klismos' being the Greek word for throne. The seat of the chair was usually fitted with a leather cushion pad called a 'squab'. The chair is one of a pair; the other chair is on display in Matthew Boulton's study at Soho House.   LC

113 Japanned armchair, by James Newton, 1798
 japanned beechwood, gilt, with cane, height 87 cm, depth 45.5 cm, width 51.7 cm.
 BMAG, 1990 M 100.1
This japanned armchair was one of a set of chairs made for the drawing room at Soho House by the London cabinet maker James Newton in 1798. Japanned wood was often used to make elegant furniture as a way of imitating imported Oriental decorative objects.   LC

114 Boulton family tea vase, 1775–6
 silver, 48.0 × 20.3 cm.
 BMAG, 1988 M 204
This silver Boulton & Fothergill tea vase was made at the Soho Manufactory. It is engraved with the Boulton family coat of arms and was for the personal use of Matthew Boulton at Soho House.
 LC

114 facing page, and detail above

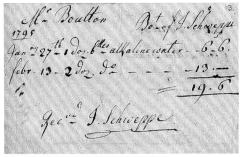

120

117 Bill from Thomas Carless druggist, Birmingham, to Matthew Boulton, for 25¾ lbs (c. 11.68 kilogrammes) of sugar, 10 March 1769
BAH, MS 3782/6/191/4

Sugar consumption seems to have been high in the Boulton household, judging from bills. Annual sugar consumption in England and Wales rose from 430,000 hundredweight in 1700 to over three million hundredweight in 1800, though it was an expensive luxury during most of that period and its damaging effects on teeth were not understood. SM

118 Bill from Joseph Cooke to Matthew Boulton, for oysters, 9 October 1769
BAH, MS 3782/6/191/24

Oysters were a popular food in the eighteenth century and barrels of them were regularly sent to Soho House from London, Colchester and elsewhere. This bill, for oysters supplied in 1767 and 1768, also gives some indication of the length of time suppliers often had to wait for payment. SM

119 Bill from Edward Ruston of Birmingham to Matthew Boulton, for green tea and 'bolivado', 15 August 1763
BAH, MS 3782/6/190/130

The Boultons regularly bought supplies of various kinds of tea from suppliers in both Birmingham and London. Enquiries have failed to reveal what 'bolivado' was, but two ounces of it cost 8½ d. SM

120 Bills from J. Schweppe of London, to Matthew Boulton, for a dozen bottles of 'alkaline water', 13 February and 23 March 1796
BAH, MS 3782/6/195/13

Jacob Schweppe took up Joseph Priestley's idea of impregnating water with gas and began marketing it. Boulton bought regular supplies of it for Soho House, to the extent that in 1802 Schweppe asked him whether he could find him a Birmingham agent for it. SM

115 Bill from Davenport & Farrant, tailors, London, to Matthew Boulton, for a suit, 2 June 1790
BAH, MS 3782/6/194/40c

Boulton liked to be fashionably dressed and there are many bills from tailors among his papers. This one records the purchase of an olive-coloured coat, a striped silk waistcoat trimmed with a figured border, and a pair of lavender-coloured silk breeches. This ensemble cost him £9 17s. 2d., about three months' pay for the average working man. DS

116 Bill from Mary Greaves, dressmaker, Birmingham, to Matthew Boulton, for dressmaking for Mrs Boulton, 6 October 1769
BAH, MS 3782/6/191/23

Like her husband, Mrs Ann Boulton liked to dress fashionably. This bill is for various items of dressmaking work that Mary Greaves has carried out, both making up and repairing clothes for Mrs Boulton. Boulton often bought lengths of silk and lace and jewellery for his wife when he was in London, and kept her informed on what was in fashion in the capital. DS

122

121 List of wines consumed at Soho House in 1804
BAH, MS 3782/13/37/26

This list, covering a total of 976 bottles of various wines and spirits, shows a total cost for the year of £278 6s. 6d. Matthew Boulton and his son kept a well-stocked cellar at Soho House. Port and Madeira were evidently particular favourites, and Matthew Boulton is known to have drunk diluted Madeira when he was unwell. The servants also received allowances of ale or small beer. SM

122 Unknown artist, Portrait of Dr William Small, c. 1765–75
pencil sketch on paper, c. 30 × 20 cm.
BAO

William Small arrived at Soho in 1765 with a letter of introduction from Benjamin Franklin, whom he had met while he was teaching at William and Mary College in Virginia between 1758 and 1764. Small did much to encourage James Watt's researches in steam-engine development before Watt came to Birmingham, and was also a naturally inventive man himself. His death in February 1775 was a great blow to all the Lunar Society members and affected Boulton particularly deeply. SB

MISS BOULTON'S HEALTH

Matthew Boulton's daughter Anne had orthopaedic problems from infancy. It is possible that she was born with a club foot. Whatever the exact nature of the problem, Miss Boulton does not, generally speaking, appear to have been an invalid, although some books describe her as such. The disability did not put Anne off travelling. Throughout her life she made regular trips to take the waters at various spas, including Bath, Buxton, Scarborough and Malvern; she went to Southampton, Ramsgate and Brighton for sea bathing, and to London for shopping. In 1794 Anne did become more seriously disabled, due to two falls. These resulted in a knee injury which caused acute pain and made walking difficult. Various cures were tried, ranging from 'electricity' to hot and cold bathing, bleeding, 'blistering plasters' (seen at the time as drawing impurities from the system) and leeches. At one time it was said Miss Boulton would lose either her leg or her life. Fortunately nothing so drastic happened, though it was over six years before there was much improvement. Miss Boulton died after suffering what was probably a stroke at the age of 61.

Shena Mason

123 Letter from Erasmus Darwin to Matthew Boulton, 9 June 1769
BAH, MS 3782/13/53/40

In this letter Dr Darwin discusses eighteen-month-old Anne Boulton's orthopaedic condition, giving his opinion (with sketches) of the likely problem and suggestions for treatment. Mr and Mrs Boulton later consulted the London physician John Hunter about their daughter's condition. Throughout her life Anne Boulton wore a shoe with a thick built-up sole on her left foot, to compensate for an apparent inequality in the length of her legs. SM

124 Thomas Rowlandson, Transplanting of Teeth, 1787
cartoon, hand-coloured engraving on paper, 33 × 45 cm.
© The Trustees of the British Museum, BM 1892, 0714.449

From the mid-eighteenth century, live tooth transplants began to be performed. At around five guineas, the treatment was only for the well-to-do; it was also painful and often led to serious infections. Thomas Rowlandson's cartoon sums up the process. A ragged boy and girl are leaving the dentist's surgery, having been paid a small sum to have healthy teeth removed. On the sofa, a chimney sweep is just undergoing the same process, watched with distaste by a fashionably dressed older woman clutching a smelling-bottle; the sweep's tooth will be inserted into her gum in place of her own extracted rotten tooth. Meanwhile the young lady in the chair submits reluctantly to the dentist's attentions and will perhaps receive one of the children's teeth. In the background a dandy examines his mouth in the mirror. The tooth donors were usually poor, and ironically it was their poverty that ensured that their diets would have included far less tooth-damaging sugar than those of the wealthy recipients of their teeth.

In the year that Rowlandson drew this cartoon, Boulton's nineteen-year-old daughter, Anne, had such a transplant, because, as her father observed, 'the pulling out of foreteeth is a matter of great consequence to a young woman'. The operation was carried out by Charles Dumergue, a Frenchman living in London, who apart from being a Boulton family friend was also the Queen's dentist. SM

125 Letter from Matthew Boulton to Matthew Robinson Boulton, Versailles, 29 June 1787
BAH, MS 3782/13/36/8

This letter, from Boulton to his son, who was continuing his education in France, describes Anne Boulton's tooth transplant operation:

Your Sister hath had a tooth drawn & at the same time another was drawn out of a young Girls head about 14 & after cleaning it & washing it in Spirits of Wine & Campher it was planted in your Sisters jaw where it seems to fix firm & I have no doubt but it will prove a good tooth & usefull. SM

126 Bill for medical services, 1792
BAH, MS 3782/6/194/38

This bill for £7 11s. 6d. is payable to the estate of the late Edmund Hector, a Birmingham surgeon. It covers the period from January 1790 to January 1792. In that time Hector had visited Soho House thirty-one times (at 5s. a visit) and had supplied 'draughts', 'mixtures' and pills to Boulton's family and servants. The entry for 16 January 1791 records a charge of 10s. 6d. 'For the Cure of Miss Boultons Toe'. DS

127 Bill from L. M. Stretch, Twickenham, to Matthew Boulton, for Matthew Robinson Boulton's schooling from December 1782
BAH, MS 3782/6/192/96a

Matthew Robinson Boulton (born 1770) went away to school in September 1778 at the same time as his sister. Matt was initially a pupil at a school for young gentlemen run by a Mr L. M. Stretch at Twickenham. Later Matt moved to a school at Stoke-by-Nayland in Suffolk, before going to France and Germany to finish his education. SM

124

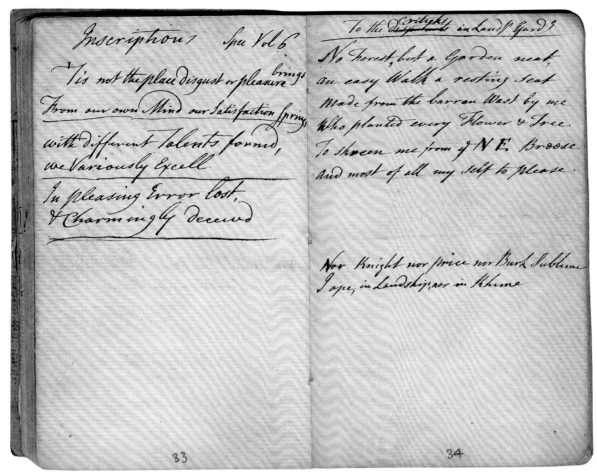

129

128 Bill from Mrs Elizabeth Wilkes to Matthew Boulton, for Anne Boulton's schooling, 1 May 1781

BAH, MS 3782/6/192/85

Matthew Boulton's daughter, Anne (born 1768), went away to school in 1778, first to a London school, Camden House, where a fellow pupil was her cousin, Nancy Mynd. In 1780 Anne moved to Mrs Elizabeth Wilkes's school for young ladies at Richmond, Surrey, an establishment recommended to her parents by Elizabeth Montagu. Anne was at Mrs Wilkes's school for about two years and was visited by Mrs Montagu there. Half a year's tuition and board cost her father £71 19s. 0d. Matthew Boulton regularly visited both his children at their schools and reported back to their mother on their health and progress. SM

129 A poem by Matthew Boulton about his garden at Soho house, from a bound notebook, 1795

BAH, MS 3782/12/108/70

Matthew Boulton's many notebooks contain widely varied material. This one ranges from sketches of house façades to a note on the cost of shipping India corn to Liverpool. Among it all is this poem, in his own idiosyncratic spelling:

132

To the Crityks in Land(ski)p Gard(enin)g

No Forest, but a Garden neat,
An easy Walk a resting seat
Made from the barren Wast by me
Who planted every Flower & Tree
To skreen me from ye NE Breese
And most of all my self to please.
Nor Knight nor Price nor Burk sublime
I ape, in Landskip, nor in Rhime.

Boulton was greatly interested in the subject of landscape gardening and is known to have bought books on the subject by Uvedale Price, Sir William Chambers, William Gilpin and others, and was clearly aware of the work of Richard Payne Knight and Edmund Burke. DS

130 Bill from Brunton and Forbes, nurserymen, to Matthew Boulton, for the supply of garden plants for Soho House, 9 April 1783
BAH, MS 3782/6/192/100

This bill is for supplying 18 pinks or carnations and no fewer than 304 herbaceous plants, all duly listed under their Latin names, at a cost of £3 1s. 3d. (a month's wages for many workers). Boulton put a great deal of effort into developing the gardens at Soho House and he clearly derived much pleasure from them, which his daughter, who was taught botany by William Withering, shared. DS

131 E. Gray Saunders, Pair of sphinxes for the Soho House gardens, 1795–6
stone, height 1 × 1.22 m length × 61 cm width.
BMAG, 2001 P 37.1–2.

132 John Phillp, Measured drawing of one of the sphinxes, 1796
pencil on paper, 60 × 41.1 cm.
BMAG, 2003.0031.44

133 Bill from William Hollins for erecting the sphinxes, covering 25 September 1795–19 March 1796
BAH, MS 3782/6/195/15

These stone sphinxes were originally situated either side of the 'Sphinx Walk' in the landscaped park of the Soho estate. The beautifully designed gardens and parkland have long since been completely built over, and the sphinxes are the only surviving statuary from Boulton's estate. Boulton paid the sculptor E. Gray Saunders a total of £30 for the sphinxes, with packing and transportation by canal costing a further £17. William Hollins was paid £8 14s. 0d. for erecting the sphinxes on stone plinths in the gardens. James Watt's wife, Ann, was disparaging about them, writing to her son, Gregory, 'Mr Boulton is going on in spending money… [he] has placed two Gigantic synphaxes near the house…. I believe he is gone crazy.'

John Phillp made a measured drawing of one of the sphinxes in 1796. It was this drawing that enabled the sphinxes to be positively identified as Boulton's. LC/SM

131

EXPANDING THE MARKET

With his Manufactory complete and reputation for quality secured, Boulton embarked on a quest to introduce new, upmarket, artistic lines. While his toy trade continued and even expanded, he started to diversify into other areas of metal-working, such as sterling silverware, ormolu, and Sheffield plate, and soon became one of the largest producers of such goods in the country. Josiah Wedgwood declared him 'the Most compleat Manufacturer in Metals in England'.

Despite the productivity of the Manufactory, one major obstacle to the sterling-silver trade was the law that required sterling silver to be assayed and hallmarked at an assay office. As there was no assay office in Birmingham, Soho silver goods had to be sent to Chester or London for hallmarking, until after a two-year campaign, led by Boulton himself, Birmingham Assay Office opened in 1773. But meanwhile he had already begun producing Sheffield plate. This increased the market for 'silverware' by manufacturing nearly identical goods to the sterling-silver range at greatly reduced prices.

Another branch of metalworking that Boulton set about developing at Soho was the manufacture of ormolu (gilded) wares, which he hoped would raise standards of taste and design. He began to produce fashionable decorative objects in ormolu, such as candelabra, perfume burners and clocks, aiming to rival the ormolu of French manufacturers.

By 1770 the Soho Manufactory was said to be employing up to one thousand workers and its goods were widely exported. Boulton delighted in entertaining visitors at Soho and notable tourists from around the world flocked to visit his Manufactory. The visits became so popular that Boulton had extra facilities built in the grounds of the Manufactory where people could be served refreshments after their factory tour.

Laura Cox

134 J. S. C. Schaak, *Portrait of Matthew Boulton*, 1770
oil on canvas, 73.7 × 61 cm (unframed).
BMAG, 1987 F 330

This polished portrait shows a young, prosperous and confident Boulton, at the height of his powers. Schaak painted two portraits for which he charged £10 10s. It is not known who the other portrait represented; it may have been Mrs Boulton, or it may have been John Fothergill (see p. 3). VL

135 Inventory of the Soho Manufactory engine works, September 1784
BAH, MS 3147/9/8

Matthew Boulton maintained a room within the Manufactory engine works during its early days. This was probably situated in the workshops that divided the main upper and lower levels of the Manufactory. None of Boulton's voluminous paperwork is mentioned, but it is interesting to note the presence of one of the letter-copying machines that were produced in the adjacent workshops of James Watt & Co. GD

136 Sample of copper ore from Poldory Mine, Gwennap, Cornwall
137 sample of tin ore (cassiterite) from St Agnes, Cornwall
both from Matthew Boulton's mineral collection
BMAG, 1993 G 03.1 and G 710

Matthew Boulton amassed a considerable mineral and fossil collection, and had a particular interest in examples from localities involved in his business. He used large amounts of copper in the manufacture of his plated objects and, from 1786 onwards, in the Soho Mint. Tin was also an important raw material for him. He became interested in Cornish tin and copper mines from 1777 when Boulton & Watt steam engines were increasingly used to drain water from the mines. Poldory, where this copper ore sample came from, was one of the mines equipped with a Boulton & Watt steam engine. Boulton helped to organise the Cornish Mining Company in 1785, and was also involved in setting up various refining processes for copper, such as smelting works and rolling mills. ST

138 Two pieces of blue john stone, from Matthew Boulton's mineral collection
BMAG, 1993 G 03.934 and 1147

Blue john is a rare variety of the common mineral fluorspar which is characterised by its bluish-purple and yellowish-white banding. It is found only in caverns near Castleton in Derbyshire. The evidence suggests that blue john first began to be used as an ornamental stone in the early eighteenth century, but it really only became popular during the 1760s. Boulton was one of the first to exploit its potential, using it especially in his ormolu vessels. PM

139 Bill for the supply of blue john stone, 2 March 1769
BAH, MS 3782/6/191/2

Boulton used so much blue john stone in his ormolu products that in 1768 he actually investigated the possibility of leasing or buying the mines. This bill records him paying £81 1s. 6d. for 14 tons and ¾ cwt (14,263 kilogrammes) of the best quality stone. The seller, John Platt, probably owned the Treak Cliff mines at Castleton, Derbyshire. DS

SILVER AT SOHO

Matthew Boulton's ambition for his silver products is clear from the earliest known examples. Unlike the simple and often crude domestic objects produced by many provincial silversmiths in the eighteenth century, the Chester-marked candlesticks (no. 146) and coffee pot (no. 207) are of exceptional design and in the height of fashion. Looking to international and London models, employing leading architect-designers such as William Chambers, James Stuart, Robert Mylne and the young James Wyatt, and developing a network of influence which included politicians, aristocrats and ambassadors, Boulton clearly set out to establish the Soho Manufactory as a major centre of silversmithing to rival the leading London firms.

From the opening of Birmingham Assay Office in 1773, for the next four years the Soho Manufactory produced an outstanding range of subtle, refined and beautifully conceived neo-classical plate, indicated by the richness of designs in the Soho Pattern Books. However, from the later 1770s production began to decline. Boulton was taken up with his development of the steam engine with James Watt and the pioneering work of the Soho Mint.

After the death of John Fothergill in 1782, when the Matthew Boulton Plate Company was formed, silver manufacture faced a series of difficulties which it never really overcame. Silver design and production became increasingly pedestrian as it commanded less of Boulton's time and energies, competed for an elusive market, and lacked adequate finance to invest in skilled labour, new machinery or design. In terms of financial return, silver was quite overshadowed by the success of the Sheffield plate enterprise, which reached markets both in the middle classes and the London trade. London competitors, such as the Bateman and Hennell families, dominated the middle market, particularly following their adoption of steam technology in the 1790s.

Some finer work was produced, the epergne (no. 157), for example, but even so, later Soho silver design had none of the panache, ambition or quality that had distinguished it in the early years, though production continued until Matthew Robinson Boulton sold the Matthew Boulton Plate Company in 1833.

Martin Ellis

140–142 Three Masonic candlesticks, Boulton & Fothergill, 1768

base and column: Sheffield plate; capital: silver; plinth: plate; candle holder: silver; height 92.71 cm.

Photograph by permission of the Royal Alpha Lodge No. 16

The Duke of Cumberland commissioned these 'three Great Solomonean Candlesticks' for the Royal Lodge in 1767. Royal Lodge supplied the designs, which necessitated sinking a special set of dies. The production costs of £141 were enormous, and nine years later Boulton had still not been paid! His penchant for accepting seemingly prestigious commissions frequently proved unprofitable. The candlesticks are still in regular use.

GC

143, detail

140–142

143

143 Six dinner and six soup plates
(BMAG, 2000 M 2); six dinner and six soup plates
two sauce tureens (BAO 335 and BAO 1442)
all Boulton & Fothergill, hallmarked at Birmingham, 1776–7
silver, diameter of a dinner plate, 24.4 cm.
These pieces form part of a large dinner service (described on pp. 43–4), which was purchased by Mrs Elizabeth Montagu, whose coat of arms appears on some of the pieces. Other items from this service are in other parts of the world, including two soup tureens at the Speed Art Museum at Louisville, Kentucky, USA (see p. 43).

A long-time correspondent and friend of Matthew Boulton, Mrs Montagu was a celebrated intellectual of the period, often referred to as 'the Queen of the Bluestockings'. She was the widow of Edward Montagu, MP, and an important patron of Boulton. ME

144 Pattern Book 1, Boulton & Fothergill, 1762–90
45.2 × 61 cm (open).
BAH, MS 3782/21/2
This is the earliest of the Boulton & Fothergill Pattern Books which provide a remarkable insight into the range of goods produced by the firm, from jewellery, buckles, buttons and sword hilts, to silver and Sheffield plate and ormolu ware. These Pattern Books also contain useful drawings of stamped or pierced borders and other decorative elements which help in the identification of pieces. SM

144

146

147

146 Pair of candlesticks, Boulton & Fothergill, 1769
sterling silver, hallmarked at Chester, height 32 cm.
BMAG, 2002 M 364.1–2.

Two of a set of four candlesticks, rare examples of the few remaining items of silver by Boulton & Fothergill marked in Chester. Known as 'lyon-faced' candlesticks, they are modelled on French design of the 1760s and demonstrate Boulton's enthusiastic embracing of the most sophisticated fashions. Lyon-faced candlesticks and candelabra were also produced in ormolu during the early 1770s. The remaining two of this set are in the Grosvenor Museum, Chester. Further examples are at the Speed Art Museum, Louisville, Kentucky (see p. 42). ME

147 Dish ring, Boulton & Fothergill, 1773
sterling silver and fruitwood, hallmarked at Birmingham, length 41.4 cm.
BAO, 833

A dish ring would have been used to keep a dish of food hot at the dining table. This one has a spirit burner in the centre which is swivel-mounted so the ring can be used either way up. It is hallmarked for Boulton & Fothergill, Birmingham, 1773, the year the Birmingham Assay Office was established as a result of a campaign led by Boulton. SB

145 Pair of rococo candlesticks, Boulton & Fothergill, 1768
sterling silver with Sheffield plate nozzles, hallmarked at Chester, height 29.6 cm.
BAO, 1140 (see p. 52)

These exuberantly rococo candlesticks, with their swirling bases and richly ornamented stems, are among some of Boulton & Fothergill's early silver and at the time they were made would have been very much 'London style'. Like all silversmiths, Boulton & Fothergill had to send their silverware to an assay office for it to be tested and hall-marked. Before the Birmingham Assay Office was established in 1773, the nearest Assay Office to Birmingham was Chester. SB

148

148 Two candlesticks from a set of four, Boulton & Fothergill, 1774
sterling silver, hallmarked at Birmingham, height 30.0 cm.
BAO, 906

These candlesticks, all four of which are illustrated above, are in a restrained and elegant neo-classical style, based on a design by the architect James Wyatt. They appear in Pattern Book 1. By using components from several designs in different combinations, it was possible to produce a range of variations relatively simply. Boulton & Fothergill used Wyatt's patterns more than those of any other designer, and candlesticks in this style became a popular line for the Soho Manufactory. SB

149 Flagon, chalice and cover, and standing paten from St Bartholomew's Chapel, Birmingham, Boulton & Fothergill, 1774–5
sterling silver, hallmarked at Birmingham, height of flagon 54.3 cm.
On loan from the Diocese and Lord Bishop of Birmingham, 1937 L 1.1, .3 and .5

This is one of two sets of altar plate from St Bartholomew's Chapel. The Chapel was built in 1749 in Masshouse Lane, to the northeast of the town centre, as a chapel of ease for St Martin's. It was closed in 1937 and subsequently demolished. Traces of the churchyard still remain.

The chalice and paten are both engraved with the sacred monogram, IHS (derived from the first three letters of the name 'Jesus' in the Greek alphabet), while the flagon bears a cast figure of Faith in an oval medallion, surrounded by a ribbon-tied wreath of vines and wheat (representing the bread and wine of the Christian Eucharist).

The bold neo-classicism of this set, particularly of the flagon, with its vase-form body and high loop handle, lend an authority and confidence to the design which is often lacking in later eighteenth-century ecclesiastical plate.

All three items are engraved with the inscription: 'The Gift of Mrs Mary Carles of Birmingham to St. Bartholomew's Chapel.' ME

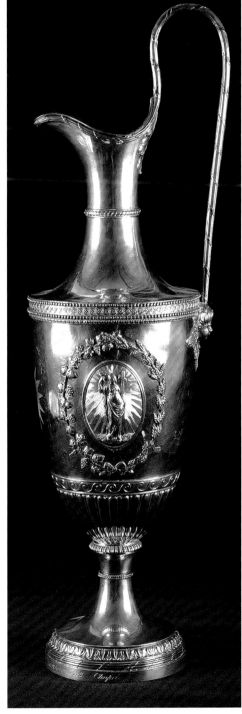

149 (Flagon 1937 L 1.1)

150–151

152

150–151 Waiter and wine jug, Boulton & Fothergill, 1774–5
   sterling silver, hallmarked at Birmingham, diameter of waiter 27.2 cm.
   BMAG, 1930 M 1170–1171

These are two of a group of three pieces, representing part of a drinking set, presented anonymously to Birmingham Museum and Art Gallery in 1930, and engraved with the same, as yet unidentified, coat of arms. The third piece, a smaller waiter, is currently on loan to Sheffield Museums Trust (The Millennium Galleries). Several designs for similar jugs are recorded in the Soho Pattern Book No. 1, and a more elaborate version, with chased rosette and Vitruvian scroll banding, is in the Museum of Fine Arts, Boston. ME

152 Pair of candlesticks, Boulton & Fothergill, 1776–7
   silver, height 32 cm.
   BMAG [Civic Plate], AG 12

The design of these candlesticks was inspired by the classical architecture of ancient Greece and Rome. The stem takes the form of a fluted column. This is surmounted by a Corinthian capital. By the 1770s, the dignified neo-classical style was well established throughout Europe. FS

153 Cup and cover, Boulton & Fothergill, 1777
   sterling silver, hallmarked at Birmingham, height 30.1 cm.
   BAO, 539

This cup and cover with entwined snake handles bears the engraved marital arms of Cornelius O'Callaghan, MP, who was created Baron Lismore of Co. Tipperary. The guilloche pattern around the foot appears on a number of pieces of Boulton silverware, for example no. 158. SB

154 Bread basket, Matthew Boulton & Plate Co., 1788
   sterling silver, hallmarked at Birmingham, height 29 cm.
   BAO, 569 (see p. 45)

Made from heavy-gauge silver wire, this basket is a fine example of wirework, with its three-strand handle and decorative hinges. Inside the base is an engraved coat of arms. Boulton considered wirework to be a speciality of the Soho Manufactory, and he made considerable improvements to the machinery used to make wire. A number of designs for such baskets are to be found in the Pattern Books. They were made in both silver and Sheffield plate. Matthew Boulton & Plate Co. was the company that succeeded Boulton & Fothergill after John Fothergill's death in 1782. SB

155 Egg frame and egg cups, Matthew Boulton & Plate Co., 1789–90
silver, height 11 cm.
BMAG, 1929 M 500.1–7

This egg frame contains six egg cups, but unlike many egg frames does not hold egg spoons or a salt cellar. The egg cups have a pierced openwork border below the rim. The frame was constructed using the wirework technique. FS

156 One of a set of four sweetmeat dishes, Matthew Boulton & Plate Co., 1794–5
silver, length 14.5 cm.
BMAG, 1929 M 632.2

The sweetmeat dish is in the form of a scallop shell. It is engraved with the Maxwell family crest and probably formed part of a large dinner service. The dish was designed to hold dried fruits preserved with sugar, such as peaches, pears, melons, nuts and orange peel. FS

153

155

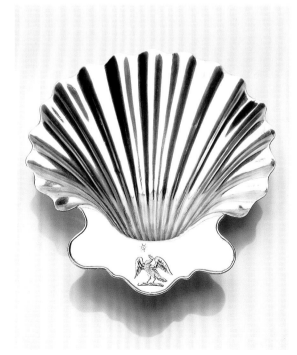
156

157

158

ORMOLU

In the 1770s Matthew Boulton became well-known for his ormolu ornaments. He sold them to the leaders of fashion in England and exported them to many parts of Europe. He was minutely involved in the processes of their design and production, and took the lead in marketing them, in conscious competition with the *bronziers* in Paris, who seemed to have the market to themselves. Like them, he saw the scope for cashing in on the craze for vases by making vase-shaped candlesticks, candelabra, perfume burners, clock cases and timepieces. His approach to the design of his ornaments was eclectic. He culled ideas from many sources, and drew widely from the repertoire of classical ornament, as did Wedgwood and other manufacturers. Usually several versions of each ornament were made, often incorporating marble or blue john from Derbyshire, reflecting Boulton's policy of making different combinations of the same models and his factory-based methods of production.

The best prospect of selling such goods was to sell direct to patrons, and Boulton cultivated his aristocratic patrons assiduously. In March 1770 he received orders from King George III and Queen Charlotte, which gave a great stimulus to his marketing plans. The sophisticated objects that he made for them (nos. 160, 163, 164, 165 and 167) were something completely new in the repertoire of English metalworkers.

In 1770 Boulton & Fothergill staged a sale at James Christie's sale room in London, targeting the fashionable classes. A further and larger sale took place at Christie's in the spring of 1771 (no. 166). This sale contained nearly four hundred objects. A further sale, for which no catalogue has yet been found, took place in 1772 and included some of Boulton's grandest objects. It was a financial failure. Even the two great clocks, the geographical and the sidereal (nos. 169, 170), failed to sell.

The ormolu business continued through the 1770s, but it never fully recovered from the blow of the failure of the 1772 sale. By the late 1770s there was excess stock, which a fourth sale at Christie's in 1778 tried but failed to clear. By the time of John Fothergill's death in 1782 the business had effectively died.

The ormolu trade shows how works of art were produced in a factory environment at the height of the 'neo-classical' period. It shows how a Birmingham metal manufacturer could, with determination, attention to detail and fashionable designs, capture the imagination of 'fine folks.' And Boulton's later successes in the production and sale of engines and coinage owed not a little to the lessons learned and the contacts made in the earlier business.

Nicholas Goodison

157 Epergne, Matthew Boulton & Plate Co.,
1805–6
silver parcel-gilt and glass, height 34.7 cm.
BMAG, 1931 M 636.1

Epergnes are a type of centrepiece for a sideboard or a dining table. They originated in the French Court of King Louis XIV during the late seventeenth century. The name derives from the French 'epergner' meaning 'to save'. Epergnes, with their tiers of small dishes, save space on the table. FS

158 Vase candelabra, by Diederich Nicolaus
Anderson (attributed), c. 1765–70
ormolu, height 43.94 cm.
BMAG 2005, 4465.1–2

Anderson was a Danish metalworker in London. Boulton copied a model of a classical tripod perfume burner that Anderson had made to a design by the architect James Stuart. It is highly likely that Boulton derived other mounts from Anderson's pieces, such as the collared spiral stem of this type of vase, the design of which can also be attributed to Stuart. Other examples are in the Royal Collection, the Victoria and Albert Museum, at Spencer House, and at Althorp. This pair of vases came from Hagley Hall. NG

159 Candle vases, Boulton & Fothergill, c. 1770
blue john and ormolu, ebonised wooden base, height 37.85 cm.
BMAG, 1946 M68–69

Several examples of this type of vase survive, some with a central reversible finial/nozzle, some with three branches, some with marble bodies. The vases are one of Boulton's earliest designs and are decorated with mounts that are probably derived from the architect James Stuart (nos. 168, 177). The ormolu rim is pierced for the emission of perfume; there is a silvered copper lining. Several of these vases were in Boulton's ormolu sale at Christie's in 1771. Examples survive in the Royal Collection at Frogmore, at Blenheim Palace, at Harewood House and in the David Collection, Copenhagen. NG

160 Goat's head candle vases, Boulton & Fothergill,
c. 1770
blue john and ormolu, height 20.83 cm.
The Royal Collection © 2008, Her Majesty Queen Elizabeth II, RCIN 6828.1–2

A large number of goat's head vases were made, the earliest in 1769. Vase bodies were most often of blue john, but gilt, green and blue enamelled copper and white marble are also recorded. The finial reverses to become a nozzle. Some vases have

159

160

161

medallions, as here, of Alexander the Great with the horns of Ammon, derived from a classical gem. This pair of vases may have been among the set supplied to King George III and Queen Charlotte in 1770. The design corresponds to a drawing in Boulton & Fothergill's Pattern Book I, p. 171. NG

161 Cleopatra candle vases, Boulton & Fothergill, c. 1770
blue john and ormolu with lacquered glass panels simulating aventurine, height 21.59 cm.
BMAG, 2000 M 15–1–2

Many vases of this pattern survive. The name 'Cleopatra' derives from a cameo or intaglio. Only one pair of vases with her portrait, stamped on a gilt pedestal, is so far known. Others have a portrait of Ceres, derived from a classical gem in the Duke of Marlborough's collection, on their pedestals; others, like these vases, have no medallion. The design corresponds to a drawing in Boulton & Fothergill's Pattern Book I, p. 171, which like many surviving vases has a cap with a finial. NG

162 Candle vases, Boulton & Fothergill, c. 1770–1
white marble and ormolu, height 39.88 cm.
City of Leeds Art Gallery, Temple Newsam

These three-branched candle vases have a central finial that reverses into a nozzle, ram's head mounts, and caryatic mounts on the 'altar' pedestal. The branches were based on a French model and, like most of Boulton & Fothergill's mounts, appear on other ornaments, in particular the 'lion-faced' candlestick (no. 146). There is a gilt lining. This type of vase, probably known as 'Burgoyne's' vase in Boulton & Fothergill's records, was also made in blue john but with different branches. This pair of vases was sold by the Soviet Union from the Stroganoff Collection in 1931, and may have been the pair recorded at Pavlovsk in 1903. NG

163 King's clock, Boulton & Fothergill, 1770–1
blue john and ormolu, height 47.75 cm.
The Royal Collection © 2008, Her Majesty Queen Elizabeth II, RCIN 30028

This clock is part of the *garniture de cheminée* in the Queen's private sitting room in Windsor Castle. The case, with its distinctive spirally grooved feet, decorative mounts and engraved back door, was designed by the royal architect William Chambers and echoes sketches of ornament that he drew in France and Italy. The three-train movement was made by Thomas Wright but was later heavily modified by the royal clockmaker B. L. Vulliamy. Boulton probably received Chambers's design in

162

163 facing page

the spring of 1770. The making of the case, and of the movement, are the subject of detailed correspondence in the archives. The clock was delivered in the spring of 1771. Boulton wanted to put a second version of the clock into the sale at Christie's in 1771 but his agent William Matthews persuaded him that it would not be tactful. Other versions were, however, made. One was in the Christie's sale in 1772. It was unsold and may have been the King's clock in the 1778 sale. Later versions used many of the same mounts but were not identical: one of them, with a full chiming movement by Wright, has a case similar to the geographical clock. Another, which is now in the Courtauld Institute of Art, has glass panels simulating lapis lazuli. NG

164 King's vases, Boulton & Fothergill, 1770–1
blue john and ormolu, with tortoiseshell-veneered base, height 56.13 cm.
The Royal Collection © 2008, Her Majesty Queen Elizabeth II, RCIN 21669.1–2

These candle vases are part of the *garniture de cheminée* in the Queen's private sitting room in Windsor Castle. The design is attributable to William Chambers, who exhibited 'various vases, etc., to be executed in or moulu, by Mr Bolton for their Majesties' at the Royal Academy in 1770. A sketch of the design survives in Boulton & Fothergill's Pattern Book I, p. 19. The branches are held aloft by demi-satyrs. The base is strongly reminiscent of Chambers's Franco-Italian taste. The rim of the vase is pierced for the emission of perfume, and the linings are gilt. Typically, Boulton copied the design for other patrons, and there were two vases with branches supported by 'demy satyrs … after a model that hath been executed for his majesty' in the 1771 sale. There are two more in the Royal Collection, and five well-documented six-branched vases at Saltram (National Trust). Single vases are at Harewood House, Hinton Ampner (National Trust) and in the Art Institute, Chicago. NG

165

165 Sphinx vases, Boulton & Fothergill, 1770–1
blue john and ormolu, coloured glass, tortoiseshell-veneered base, height 31.75 cm.
The Royal Collection © 2008, Her Majesty Queen Elizabeth II, RCIN 21668.1–2

These perfume burners are part of the *garniture de cheminée* in the Queen's private sitting room in Windsor Castle. They were almost certainly two of the vases in the designs exhibited by William Chambers at the Royal Academy in 1770. The lining is gilt copper and the rim is pierced for the emission of perfume. The base is veneered with tortoiseshell, the upper plinth has painted glass panels simulating aventurine. Several of the mounts appear on other models of vases, but the grotesque masks on the body of the vase are unique. Boulton produced many of these vases, with slight variations of mounts and materials. No fewer than ten were in the 1771 sale. Two others survive in the Royal Collection, given to George VI and Queen Mary in 1937 by the Prince Regent and Princess Paul of Yugoslavia. NG

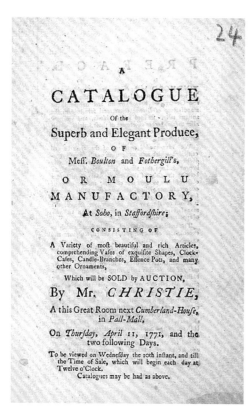

166

166 James Christie, Catalogue of the sale of 'the superb and elegant produce of Messrs Boulton and Fothergill's Or moulu Manufactory, at Soho, in Staffordshire', 11 April 1771 and the two following days
Christie's

This was Boulton & Fothergill's second sale of ormolu ornaments at Christie's sale room in London, the first having taken place in April 1770. Boulton saw the sales as a means of attracting the nobility and gentry to his products, which they did. Two copies of the catalogue survive in Christie's archive. The first has reserve prices, buyers' names and the prices reached. The second appears to have been an accounting record of the prices realised, the buyers and the lots actually sold, showing that a large number of the ornaments were unsold. The sale contained a wide range of products consisting of nearly 400 objects in 265 lots, many of them easily identifiable. Only eight objects were not ormolu. The catalogue gives an idea of the quantities made of some of the models of vases – forty-seven pairs in one case and twenty-eight pairs in another. The sale raised about half of what Boulton hoped for. He was disappointed, but it didn't put him off holding a further sale in 1772. NG

167 Tea urn, Boulton & Fothergill, c. 1770–1
ormolu on wooden base veneered with ebonised fruitwood, height 55.12 cm.
The Royal Collection © 2008, Her Majesty Queen Elizabeth II, RCIN 55429

The design of this tripodic tea urn, along with a similar urn at Syon House, owes a lot to the architect James Stuart and there is archival evidence that he probably designed them in 1769. There is a tinned copper lining and the underside of the lid is also tinned. The tap handles, which are decorated with grotesque masks, are of mother-of-pearl. The heating burner is suspended on chains and has a chased cover with a flame finial when not in use. The three urns in the frieze are lightly engraved with classical scenes, two of them representing Hercules. Boulton was anxious to take a tea urn when he visited Buckingham Palace in 1770, but it was not this one, which was bought by Queen Mary in 1938. NG

168 Vase perfume burner, Boulton & Fothergill, c. 1771
blue john and ormolu, height 27.18 cm.
BMAG, 1949 M 26

The blue john here is of a rich red colour, suggesting that it came from the 'Bull Beef' vein of the mine. The lining is of gilt copper, and the lid is pierced for the emission of perfume. The vase was also made in white marble with a marble pedestal. The design corresponds to a drawing in Boulton & Fothergill's Pattern Book I, p. 171, although the drawing appears to have a round pedestal and an acanthus-leaf pierced lid. NG

167

169 Geographical clock, Boulton & Fothergill, 1771–2
ormolu, tortoiseshell-veneered base, enamelled dial, height 77.47 cm.
Private collection

170 Sidereal clock, Boulton & Fothergill, 1771–2
ormolu, silvered dial, height 104.14 cm.
BMAG, 1987 M 70.1
Pedestal attributed to James Newton, 1797. Painted wood.
BMAG, 1987 M 70.2
(Both illustrated overleaf.)

Boulton commissioned the geographical and sidereal clock movements from his friend John Whitehurst in 1771, wanting to exploit the growing interest in scientific enquiry and the fashion for the antique taste. The geographical clock had a turning terrestrial globe at the top, with the sun moving along the ecliptic. The sidereal clock had a revolving dial showing the movement of the sun against the stars, each studded into the dial. Both clocks had twenty-four-hour dials. The inspiration for the movements was the astronomer James Ferguson. Boulton's letters to Whitehurst spell out in detail what he wanted.

The case of the geographical clock owes significant parts of its design to the King's clock case (see no. 163) designed by the royal architect William Chambers, showing how Boulton often reused casting models. The figures are copied from three of the four figures on a silver-gilt crucifix made by Antonio Gentile da Faenza c. 1582, two of which are illustrated in Chambers's *Treatise on Civil Architecture* under the heading 'Persians and Caryatides'. Boulton bought the plaster casts from the sculptor John Flaxman in December 1770 as 'a group of Hercules and Atlas'. The enamelled dial was supplied by Weston in Smithfield, London, and the globe is signed by Nathaniel Hill who worked in Chancery Lane.

The case of the sidereal clock also owes a debt to Chambers's King's clock case, but the source of Urania, the Muse of Astronomy, who reclines on the top, is not yet known. Below the dial is a bas relief depicting 'Science explaining the Laws of Nature by the globe and the solar system', with the engraved abbreviated line from Virgil's *Georgics*: *'Felix rerum cognoscere causas'* ('Fortunate is he who could understand the causes of things'), referring to the philosopher Lucretius.

Boulton put both clocks into the sale at Christie's in 1772, but they failed to sell. Boulton's disappointment was keen (see p. 61). He thought of sending both to Russia, where he thought Catherine the Great's Court would be more sympathetic to such remarkable pieces. In the event only the sidereal clock went to Russia, from where it returned unsold in 1787.

This is the first time that these two clocks – arguably the finest objects ever made at Soho – have been shown together since the exhibition at Christie's in 1772. NG

168

170 detail

171

171 Candle vases, Boulton & Fothergill, c. 1772
blue john and ormolu, height 16.51 cm.
Private Collection

The red colour of the blue john suggests that it came from the 'Bull Beef' vein of the blue john mine. The finials are reversible, to become candle nozzles, fitting into gilt cylindrical copper linings. The vases correspond to a drawing numbered 859 in Boulton & Fothergill's Pattern Book I, p. 170. Other examples survive with white marble bodies, and some are mounted on circular 'altar' pedestals. NG

172 Wing-figured candle vases, Boulton & Fothergill, c. 1772
white opaque glass and ormolu, white marble base, height 37.59 cm.
BMAG, 2002 M 26.1–2

Boulton supplied vases of this pattern with both glass and blue john bodies. The glass bodies were supplied by his friend James Keir from his glass works in Stourbridge. Two pairs of similar vases, one of them mounted on square pedestals with blue john panels, were supplied to the banker Robert Child in 1772. They survive at Osterley

172

173

(National Trust), along with a pair of blue john vases. The design corresponds to a drawing in Boulton & Fothergill's Pattern Book I, p. 156. There is a pair at Pavlovsk, St Petersburg. NG

173 Ewers, Boulton & Fothergill, c. 1772
blue john and ormolu, height 48.77 cm.
BMAG, 1946 M 70–1

The design, with its satyr masks on the vase and the handle, was conceived for Sir Harbord Harbord in 1772. He wanted a pair of 'ures such as are proper for the gods to drink necter'. The richly veined blue john is made of a number of pieces, as was the case on many of Boulton's larger vases. The design corresponds to a drawing in Boulton & Fothergill's Pattern Book I, p. 83. Another pair, made for the Earl of Sefton in 1772, is now in the Gerstenfeld Collection. NG

175

174 Duchess of Manchester's cabinet, 1774–5
oak, mahogany, satinwood, rosewood and other woods, *pietre dure* panels, ormolu, height 188.47 cm.
Victoria and Albert Museum W.43-1949

One of the most decorative pieces of furniture of its period, this cabinet was made by the London cabinet-makers Mayhew & Ince for the Duchess of Manchester as a means of displaying the eleven intarsia coastal and lake scenes, which were made in Florence by Baccio Capelli in 1709. The architect Robert Adam made drawings, one of which is fairly closely related to the finished design. The capitals of the pilasters are modelled on an engraving of a capital in Adam's *Ruins of the Palace of the Emperor Diocletian at Spalatro in Dalmatia* (1764). Mayhew & Ince commissioned Boulton & Fothergill to make the ormolu mounts in December 1774, and they were finally completed late in 1775. The duchess installed the cabinet in her home at Kimbolton Castle. NG

175 Candle vases, Boulton & Fothergill, c. 1775
white marble and ormolu, height 38.86 cm.
Weston Park Foundation

The construction of this type of vase is illustrated in 'Ormolu Ornaments', Fig. 5, in this volume. The eight gilt brass medallions on the pedestals were copied from ceramic cameos supplied by Wedgwood. They represent on one vase Filial Piety, Neptune, Achilles Victorious and Hygieia, and on the other Pomona, a Heroic Figure (twice) and Aesculapius. The vases were probably acquired in the mid-1770s by Sir Henry Bridgeman, who succeeded to the Weston estate in 1764, or by his wife. The vases and the branches correspond to drawings in Boulton & Fothergill's Pattern Book I, pp. 11, 171. There are single vases at Pavlovsk, St Petersburg, and in the Fondazione Whitaker, Palermo. NG

174 facing page

176

176a

176 Titus clock, Boulton & Fothergill, c. 1775 white marble and ormolu, height 42.16 cm. BMAG, 1966 M 141

The Roman Emperor Titus utters the words '*Diem perdidi*' ('I have lost a day'), which he is supposed to have said when he passed a day without doing good. The words, which convey the idea of time passing all too quickly but also the admonition that doing good is an end in itself, are engraved on the plaque mounted on the vase, which could be interpreted as a funerary urn. The vase's lid is pierced for the emission of perfume. One of several timepieces with an allegorical theme, the Titus clock (strictly a watch) corresponds to a drawing in Boulton & Fothergill's Pattern Book I, p. 169 (176a). The figure is derived from an engraving in A. F. Gori's *Museum Florentinum*, vol. 3 (1734). Several Titus clocks were made, some with bronzed figures. Two were in the 1778 Christie's sale. An earlier wholly gilt version, sold to George III and later re-gilt and supplied with a movement by B. L. Vulliamy, is at Kenwood (English Heritage). Other versions are in the National Gallery of Victoria, Melbourne, at Sans Souci, Potsdam, and at Fasanerie.   NG

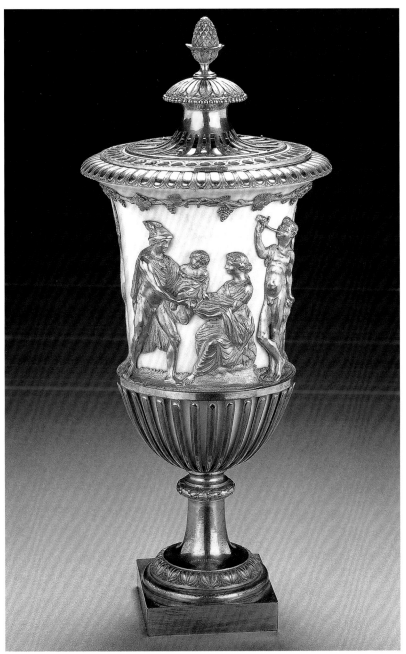

177

178

177 Bacchanalian vase, Boulton & Fothergill, c. 1775–7

white marble and ormolu, height 40.64 cm.
BMAG, 2000 M 13

Perhaps the most elegant of Boulton & Fothergill's vase designs, the classical shape may have been derived from an engraving in Montfauçon's *L'Antiquité expliquée*, which Boulton owned. The frieze, appropriately for a vase shaped like a wine vessel, represents Mercury, accompanied by Bacchic figures, delivering the infant Bacchus to the care of Ino. The theme derives from the Athenian sculptor Salpion's 'Gaeta' vase, now in Naples. Boulton may have copied it from Wedgwood's interpretation of it in a jasper plaque, but a more likely source may be the Roman eighteenth-century bronze-workers Giacomo and Giovanni Zoffoli who produced vases of a similar shape with the same frieze. The lid is pierced for the emission of perfume. The lining is gilt copper. The vase corresponds with a drawing in Boulton & Fothergill's Pattern Book I, p. 171, although the vase in the drawing has no lid. Two Bacchanalian vases were in the 1778 sale at Christie's. There is a single Bacchanalian vase at Syon House. NG

178 Narcissus clock, Boulton & Fothergill, c. 1777

white marble, ormolu, bronzed base, glass beads; height 41.91 cm.
BMAG 1992 M 16

Narcissus gazes at his reflection in the pool. His dog looks on. One of several timepieces with an allegorical theme, the Narcissus clock (strictly a watch) corresponds to a drawing in Boulton & Fothergill's Pattern Book I, p. 77. The medallion on the obelisk represents Hygieia, who appears on other ornaments (see no. 175). The beaded rim of the bezel suggests that this timepiece was made for the export market. Two Narcissus clocks were in the 1778 sale. A marble and wholly gilt version is in the Leeds City Art Gallery, Temple Newsam. NG

SHEFFIELD PLATE

The process of plating by fusion was invented in 1742 by the Sheffield cutler, Thomas Boulsover, but never patented. Sheffield plate was produced by fusing copper with silver on both sides and rolling them together to produce a tri-metallic sheet, which was used to manufacture goods that had the outward appearance of silver, at a much lower price than those of silver all through. By the mid-1750s, items of Sheffield plate, such as snuffboxes and buckles, were included amongst the manufactures of the Birmingham toy trade. But it was Joseph Hancock's idea, in the late 1750s, to utilise this new medium in the production of domestic hollow-ware, that gave rise to the huge expansion of the industry which took place during the 1760s. Hancock's own plated candlesticks were being exported to St Petersburg by 1762. Sheffield's success in promoting this new industry was clearly the spur that persuaded Boulton to embark on production too.

In 1764 commercial production of Sheffield-plate candlesticks began, still at Boulton's Snow Hill site. By the end of the year he was supplying them to London retailers. Production was hampered by the lack of a suitable rolling mill, and only with the move to Soho, and the lease of Holford Mill, could larger items like tea urns be made.

Orders for plated ware abound amongst the Boulton Papers from members of the aristocracy, including the Prince of Wales, while some clients of great wealth, like Lord Rockingham or the Duke of Queensbury, specified a preference for plated articles to those in silver. The Royal Goldsmith, Thomas Heming, sent a service of silver dishes made for the Russian ambassador to Soho to have plated covers fitted. Likewise, Lord Craven and the Duke of Montagu ordered plated lids and covers for sets of silver dishes. Mrs Montagu's plated ice pails were probably made to accompany her silver dinner service. In 1806 Sheffield plate produced a profit of £6,536, compared to £408 from buttons.

Gordon Crosskey

179 Pair of candlesticks, Boulton & Fothergill, c. 1765

Sheffield plate, height 28.5 cm.
BAO, 374 (see p. 42, Fig. 1)

For a long time Boulton & Fothergill ran the production of Sheffield plate in tandem with sterling silver, often utilising the same designs in this cheaper alternative. In the process they perfected their industrial production techniques and developed a wider market. These Sheffield plate candlesticks show the gradual move from the ornate style of the rococo to the simple lines of the neo-classical. SB

180 Pair of candlesticks, Boulton & Fothergill, c. 1765

Sheffield plate, height 31.4 cm.
BAO, 819

These columnar candlesticks have the lower portion fluted in contrast to the plain, highly polished upper portion. The columns terminate in Ionic capitals and are supported on gadrooned bases (see no. 182) which bear the marks 'B&F' with a small crown either side that identify Boulton & Fothergill Sheffield plate. SB

181 Tea urn, Boulton & Fothergill, c. 1770

Sheffield plate on mahogany base, height 46.35 cm, capacity 4.26 litres.
Private collection

This urn, or 'tea kitchen', of neo-classical style, retains handles of rococo form typical of the mid-1760s. The body lifts off the base via a bayonet fitting to reveal a charcoal heater. Such urns were used to refill teapots with hot water, not to hold tea itself. The original cost was about ten guineas.
GC

181

183

184

182

182 Pair of chambersticks, Matthew Boulton & Plate Co., 1805–25
Sheffield plate, height 72 mm.
BMAG, 1931 M 593.1–2
The edges of the drip pan and tray on these chambersticks are decorated with gadrooning. This term describes the application of convex curves or inverted fluting to the edges of a curved surface. The neo-classical style favoured the simplicity of gadrooned decoration. FS

183 Pair of two-light candelabra, Matthew Boulton & Plate Co., 1797–1800
Sheffield plate, height 42 cm.
BMAG, 1931 M 599.1–2
By the time these two-light candelabra were made, between 1797 and 1800, Soho was producing goods that would appeal to a much wider market. Sheffield-plate candelabra like these would have been cheaper to buy than silver. FS

184 Pair of wine coolers, Matthew Boulton & Plate Co., 1805–25
Sheffield plate, height 26 cm.
BMAG, 1931 M 601.1–2
Wine coolers, designed to hold bottles of wine plunged in lumps of ice, were made at Soho. These are decorated with acanthus leaves, a popular classical form of ornamentation based on the thick, prickly, scalloped leaves of the acanthus plant. FS

185

186

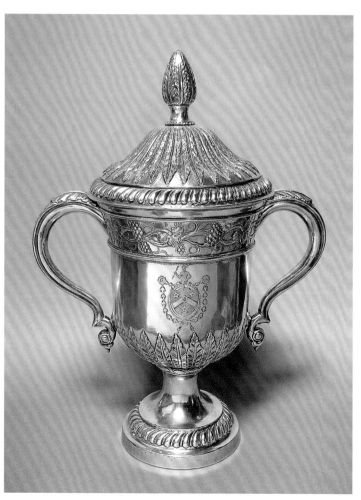

187

188

185 Entrée dish and heater base, Matthew Boulton & Plate Co., 1805
Sheffield plate with silver handles, width 34.92 cm, height 24.13 cm.
Private collection

Made for Matthew Boulton's own use, this dish bears his armorial and crest. The quality is exceptional as all the handles are cast silver. It comprises: cover, two serving dishes, a hot-water compartment, and base unit for holding a hot iron bar. GC

186 Beer tankard, Boulton & Fothergill, late 1760s
Sheffield plate, height 18.73 cm, capacity .85 litre.
Private collection

The lidded tankard with unidentified armorial is marked 'B&F' on the body, with three crowns stamped on the lid. Pieces, other than candlesticks, bearing this early mark are exceptionally rare. The original price was about 40s. GC

187 Two-handled cup and cover, Boulton & Fothergill, c. 1776
Sheffield plate, height 36.83 cm.
Private collection

This cup with hand-raised decoration bears the arms of Sir George Shuckburgh, elected MP for Warwickshire in 1780. The quality of workmanship is exceptional as the cup was entirely hand raised – no dies were used. The cup was possibly made for Shuckburgh as part of a £92 order supplied in 1776. GC

189

188 Coffee pot, Matthew Boulton & Plate Co., c. 1790
Sheffield plate with wooden handle, height 19.37 cm, capacity .85 litre.
Private collection

This coffee pot, of remarkably modern design, is illustrated in the Soho pattern books. Such pots were sometimes supplied with stands or spirit heaters. GC

189 Coaster, vine pattern, Boulton & Fothergill, c. 1772
Sheffield plate pierced work, turned wood base, diameter 127 mm, height 47.62 mm.
Private collection

Plated vine-pattern bottle stands were made at Soho in the early 1770s, costing about 24s. a pair. They were made mechanically by bending stamped and pierced strips of plated metal into circles and soldering the join. GC

190 Coaster, lace pattern, Boulton & Fothergill, 1770
Sheffield plate pierced work, turned wood base, diameter 142.88 mm, height 38.1 mm.
Private collection

This coaster shows piercing at its most delicate. It was made with a fly-press, where sections of metal were cut out using a series of beds and punches. Designs for piercing can often identify manufacturers, as many designs are unique to particular makers. GC

191 Snuffer tray, Boulton & Fothergill, c. 1775
Sheffield plate pierced work with wooden base, width 193.67 mm, height 25.4 mm.
Private collection

This unusual snuffer tray has a pierced gallery, but with an inset wooden base covered in baize. The original price was about 18s. Most eighteenth-century snuffers were steel, and priced about 4s. a pair. GC

190

191

192 Wig powderer box, Matthew Boulton & Plate Co., c. 1785
Sheffield plate, height 15.24 cm, diameter 11.75 cm.
Private collection

This powderer retains its original swansdown powder puff. It was probably part of a lady's *toilette*, those comprehensive suites comprising looking glasses, a variety of boxes, needle cushions, vases, candlesticks, etcetera, which could cost as much as £84. GC

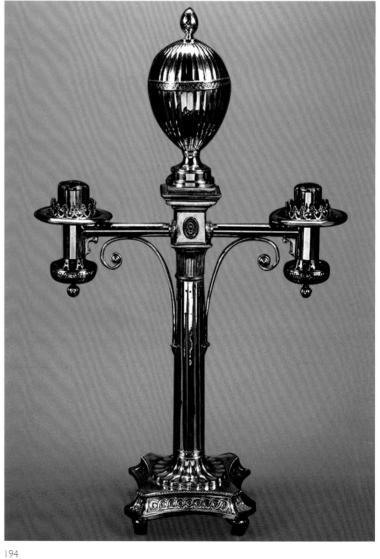

193–194 Two Argand lamps, Matthew Boulton &
   Plate Co., c. 1795 and c. 1809
   Sheffield plate, heights 54.5 and 63.5 cm.
   BMAG, 2009.0001 and 2009.0002

One of Boulton's many business ventures was with the Swiss inventor Ami Argand. Argand developed a revolutionary new oil lamp which was not only much brighter, but did not produce the smoke and smell of other oil lamps of the time. Although the venture was fraught with difficulties and legal entanglements, the lamps finally went into production in the 1780s. Matthew Robinson Boulton took one as a gift in 1785 to his schoolmaster in Suffolk, Reverend Samuel Parlby, writing to his father that the Parlbys admired it but found it too bright, and he had failed to convince them that putting on the lampshade would soften the light. The Soho pattern books contain a number of designs for such lamps (see p. 44), and many were exported to the USA: George Washington and Thomas Jefferson both owned examples. Boulton showed particular enthusiasm for this project, to the extent that a plated-lamp warehouse was opened in Bond Street.   CR

BIRMINGHAM 1759–1809

When Thomas Hanson produced his plan of Birmingham in 1778 (no. 195), its population had grown to 42,550 inhabitants. The increase in population is mirrored in the continued expansion of the town (by 1801 the census would show that Birmingham's population had risen to 73,670). Hanson's plan shows St Philip's Church, which once sat at Birmingham's most northerly point, now surrounded by streets and houses situated at the town's centre. There are newly laid-out streets that are as yet undeveloped. By the later eighteenth century, Birmingham has a new hospital, founded by Dr John Ash on the outskirts of the town, a new theatre in New Street, and the newly-built Hen and Chickens Hotel, also in New Street. Many new chapels and meeting houses have also been built. Although Birmingham is still described as a market town, the influence of manufacturing increasingly shapes it. The diversity of trades and businesses has increased tremendously: Birmingham is well on its way to becoming the 'city of a thousand trades' of the nineteenth century.

The cutting of the Birmingham Canal in 1769 linked Birmingham's industries to the coalfields of the Black Country, Wolverhampton and beyond. The canal, with its docks and wharfs, can clearly be seen on the 1778 plan. The nearby land, still relatively undeveloped except for a few iron foundries, will develop quickly and it will not be long before the canal wharfs are surrounded by warehouses and workshops. By 1793 one hundred boats will be at work on the canal every day, and by 1800 the canals in Birmingham will have reached most parts of the town.

The Bull Ring is still one of the major thoroughfares of the town, although now it is one of Birmingham's most problematic areas due to its jumble of overcrowded buildings and narrow streets. William Hutton describes some of Birmingham's worst streets as scarcely admitting 'light, cleanliness, pleasure, health or use'. In 1769 an Act of Parliament was passed that set up a Board of Street Commissioners tasked with lighting the streets, fixing the street level, regulating the street line and removing buildings that caused severe congestion. Progress is slow; by the end of the century Birmingham has achieved only a fraction of its ambitions. The most significant achievements have been the lighting of streets with oil lamps, and opening up the thoroughfare at the end of New Street.

Jo-Ann Curtis

195 Thomas Hanson, Enlarged reproduction of
   the *Plan of Birmingham*, 1778
   based on a print produced by Birmingham Public Library in 1968.
   BMAG

Hanson's 1778 plan demonstrates Birmingham's continued growth since Samuel Bradford's plan of 1751 (see no. 18). The plan includes several topographical studies of prominent buildings including the churches of St Martin's and St Philip's, the meeting houses and chapels, schools, hospital, hotel and the house of Mr Joseph Green, a wealthy wine merchant whose house on New Street – the Portugal House – was said to be the finest in the town.   JC

196 James Bisset, *A Poetic Survey Around Birmingham*, 1800
   printed book.
   BMAG, Social History Library, L/Birm953

This poem by James Bisset was intended as a guide to Birmingham for strangers to the town. It briefly describes places of use and interest such as Matthew Boulton's Soho Manufactory and the General Hospital, which opened in 1779. Bisset has this to say about Soho:

On yonder gentle slope which shrubs adorn,
Where grew of late rank weeds, gorse, ling and thorn,
Now pendant woods and shady groves are seen,
And Nature there assumes a nobler mien,
Here verdant lawns, cool grots, and peaceful bow'rs
Luxuriant, now are strew'd with sweetest flowers,
Reflected by the lake, which spreads below;
And Nature smiles around – there stands Soho!
Soho! – where GENIUS and the ARTS preside;
EUROPA'S wonder and BRITANNIA'S pride.   DP

JAMES BISSET (1761–1832)

James Bisset was a man of many talents: poet, writer, craftsman, entrepreneur and collector. Born in Perth, Scotland, Bisset moved with his family to Birmingham when he was fifteen. He attended one of the new design schools in the town in order to learn arithmetic and other technical subjects, and became apprenticed to a japanner, becoming a miniature painter by 1785 and a fancy painter by 1797. His abilities and inventiveness soon made him a wealthy and prominent man.

Bisset was an avid collector. His collection was eclectic, but principally consisted of natural history and objects from African and Pacific cultures which were regarded as exotic curiosities. He opened a shop and a museum, a 'Cabinet of Curiosities', on New Street which was run by his wife Dolly.

Bisset was a member of many local clubs and societies. These societies provided a forum for men to air their views and discuss national and local issues. One such group, of which Bisset was a regular member, was Freeth's Circle (see no. 223), also called the Birmingham Jacobin Society, which met at The Leicester Arms. The members came from Birmingham's growing middle class – merchants, manufacturers and entrepreneurs. Like them, Bisset was involved in many schemes designed to improve Birmingham; for example, he served, with Matthew Boulton, on the committee that set up regular night patrols in part of the town in order to prevent robberies.

In 1812 Bisset and his family moved to Leamington Spa. He moved his museum there the following year. However, he continued to visit Birmingham and to promote the town's development.

David Powell

197–198 Two medals produced by James Bisset, c. 1800–1812

white metal, diameters 48 mm and 37.5 mm.
BMAG, 1885N1526.182; 1885N1527.308

These medals were struck by Bisset in order to promote his trade as a miniaturist and a glass painter, and to advertise the many attractions and industries in Birmingham which might be of interest to visitors. These include the Theatre Royal and the principal manufactories. Both medals also advertise Bisset's 'Museum and Grand Picture Gallery'; the smaller even served as a token allowing free admission to the museum, which was located on New Street. DP

199 Collage of illustrated lists of trades, professions and industries of Birmingham, c. 1808
BMAG, 1933F321.7; 1965F221.60; 1995F270–271/281/291–292/299/302; 1996F35/53/61/64–65/67

These illustrated lists are drawn from James Bisset's *Magnificent Directory* and other sources. They demonstrate the tremendous variety of industries and businesses in Birmingham at the start of the nineteenth century. Some list the miscellaneous industries in a particular area or on a specific street, while others are arranged by type of business; Bisset himself can be found among the 'toymakers'. Each list is illustrated with an appropriate Birmingham scene or with items appropriate to the trade covered. For example, the sword cutlers are listed on a banner set against a backdrop of swords, axes and lances, while the gentlemen of the town are listed in front of a backdrop depicting the estates of Richard Ford and Matthew Boulton, at Hockley Abbey and Soho respectively. DP

200 Set of twenty-seven tokens depicting Birmingham buildings, issued by Peter Kempson, late 1790s
copper, diameter 28.5 mm.
BMAG, 1885 N 1527.247/257–262/265–271/273–281/283; 1939 N 168; 2006.1042; 2008.1372

Although they sometimes circulated as halfpennies, Peter Kempson's 'tokens' were really small medals intended for sale to collectors. The tokens share a common obverse design which describes Kempson as a 'Maker of Buttons, Medals, &c, Birmingham'. The reverses feature a range of churches, chapels, public institutions, public schools, commercial buildings (including, of course, the Soho Manufactory) and civic amenities. DS

197

198

201 Sixpence token of the Birmingham Workhouse, 1812
Copper, diameter 50 mm.
BMAG, 1900 N 17

This huge copper token shows the Workhouse as it appeared at the time that Boulton died. The Workhouse stood in (the now vanished) Lichfield Street. The central portion was built first in 1733, and the two wings were added in 1766 and 1779. As finally enlarged it could house six hundred people. DS

MATTHEW BOULTON'S PUBLIC LIFE

For three days in September 1768, music lovers from Birmingham and beyond flocked to the theatre in King Street and to St Philip's Church, to listen to some of the most exciting music of the day. The programme included Handel's *Messiah*, composed less than thirty years earlier and already a popular work (especially with Boulton, who loved Handel's music). The 1768 musical festival was a significant milestone in the cultural life of Birmingham: it was to be the first of a series that lasted well into the twentieth century.

Boulton's friend, Dr John Ash (Cat. 202), had been campaigning since 1765 for the establishment of a hospital. Boulton's name was associated with the campaign from its early days, and the Hospital Committee instigated the musical festivals as fundraisers. The hospital ultimately opened in 1779, and its success was thanks in no small part to the profits raised through music. Boulton also played a large part in the establishment of the General Dispensary, which delivered outpatient treatment. It was typical of Boulton to become involved in this kind of initiative. His later collaboration with Joseph Moore, a button maker and keen promoter of musical events, led to the establishment of a series of private concerts in 1799, alongside the already successful Oratorio Choral Society which Boulton also supported.

Considering Boulton's boundless energy and wide interests, it is no surprise that his name crops up in many different areas of Birmingham's public life. His campaign for the Assay Office is referred to elsewhere in this book. He was one of the founders of the New Street Theatre in 1774, writing to Lord Dartmouth in 1779 that it attracted well-to-do visitors to Birmingham and encouraged them to stay longer and thus spend more money in the town. Towards the end of his life he helped obtain the royal licence that turned the theatre into the Theatre Royal, Birmingham. Boulton served on a committee to organise nightly patrols to reduce crime. He supported the local volunteer (home defence) forces, providing funds for arms. In 1794 he was elected to serve as High Sheriff of Staffordshire. He advocated the establishment of a Botanical Society, and addressed a public meeting about this in 1807 at the advanced age of seventy-nine. He and his Lunar Society associates set up a library, which he continued to support when it later became the Birmingham Medical and Scientific Library. On Sundays he played an active role as a member of St Paul's Church, itself a centre of musical excellence. The freehold to his pew would have cost him £5.

Henrietta Lockhart/Laura Cox

202

202 Sir Joshua Reynolds, *A Portrait of Dr Ash*, 1788
   oil on canvas, 241.3 × 147.3 cm (see p. 191).
   The former United Birmingham Hospital Trust Fund, on loan to BMAG, 1994 Q 13

John Ash was an eminent physician and co-founder of the Birmingham General Hospital 'for the relief of the sick and lame.' He helped design the building and was appointed senior physician in 1779. Shown wearing the gown of a Doctor of Medicine, he holds the ground plan of the hospital, which is seen in the distance. Behind him is a statue representing Benevolence. An enthusiastic botanist, mathematician and social reformer, Ash embodied the spirit of scientific humanism that characterised the Midlands Enlightenment. BF

203 Printed circular, 1780s
   BAH, MS 3782/12/100/1

Following the foundation of the General Hospital, there was a campaign to establish a dispensary which could provide outpatient care for 'the provident portion of the labouring classes who may not be able to pay a surgeon to obtain medical aid'. This printed circular sets out the case for such an institution and proposes how it might be run. In 1793 Matthew Boulton became the treasurer of the newly formed General Dispensary. He was a staunch supporter of the cause, stating 'If the funds of the institution are not sufficient for its support, I will make up the deficiency.' SC

204 Tympanum from the General Dispensary, 1806–8
   stone, height 137.2 cm.
   BMAG, 1968 P 28

This sculpture of the Greek goddess Hygieia stood above the entrance to the General Dispensary in Union Street. This new building was designed by Birmingham architect William Hollins and opened in 1808. It replaced the original Dispensary in Temple Row which had outgrown its capacity within a few years of opening in the early 1790s. SC

205 King's Head Inn sign, date unknown
   paint on wood, height 141 cm.
   BAO (see p. 51)

Following the success of a campaign by Boulton, the Birmingham Assay Office opened for business on 31 August 1773. This inn sign is a reminder of its first premises, which were two rooms, rented from the landlord, above the (now vanished) King's Head Inn on New Street. Initially the Assay Office was open only one day each week; the business grew rapidly as the Birmingham

207

precious metals trades expanded, and the Office moved several times before the present one was built in 1877. It is now the busiest Assay Office in the world. SB

206 The Birmingham Assay Office Plate Register, 1773–92
BAO (see p. 50)

The Plate Register was where all the work submitted to the Birmingham Assay Office for assay and hallmarking was recorded. Boulton & Fothergill were the first customers when the Assay Office opened on 31 August 1773. They submitted over one hundred articles in silver, weighing a total of over eight hundred troy ounces. SB

207 Coffee pot, burner and stand, Boulton & Fothergill, 1769–70
silver, cane and ebonised wood, height 37.8 cm.
BMAG, 1996 M 1.

This coffee pot and stand with seated sphinxes flanking the burner are of exceptional quality and rarity. The design is attributed to James Wyatt; it bears a close resemblance to a Wyatt drawing of a coffee pot and stand in the Vicomte de Noailles album. Identical sphinx supports are also to be seen on an ormolu perfume burner, the drawing for which is in the Boulton & Fothergill pattern books. The coffee pot is one of the few surviving pieces of Soho silver hallmarked in Chester. FS

208 Tea vase, Boulton & Fothergill, 1773–4
silver and ivory, height 49.5 cm.
BMAG, 1965 M 68

Shaped like an urn, this tea vase is believed to be the first piece of silver to be assayed at the Birmingham Assay Office when it opened on 31 August 1773. It can be seen at the top of the list of Boulton & Fothergill items on the first page of the first Plate Register (Cat. 206, and p. 50). FS

208

209–210 Two passes for the New Street Theatre, 1774

silver and copper, diameter 32 mm.

BMAG, 1885 N 1541.151 and 2003.356

The New Street Theatre opened its doors for the first time on 20 June 1774. Like most provincial theatres of the time, it only staged plays during the summer season when the London theatres were closed and their companies went to work in other towns. Passes like this acted as season tickets for the year's performances. The men named on these two specimens, Joseph Green and Thomas Faulconbridge, were among the twenty-eight proprietors, the shareholders who built and owned the theatre as a speculative venture. Matthew Boulton was another proprietor, as were his business partner, John Fothergill, and Samuel Aris, publisher of the local newspaper *Aris's Birmingham Gazette*. DS

211 Playbill for the New Street Theatre, 2 August 1799

BMAG, 1973 F 71

Throughout his life Boulton enjoyed the theatre. In July 1799 Count Woronzov, the Russian ambassador, stayed at Soho House while he and Boulton discussed a new mint which Boulton was going to supply to St Petersburg. On Wednesday 31 July Boulton took the Count to see the famous comic actor Tom King perform in *The Clandestine Marriage* and *Blue Beard; or, Female Curiosity*. This playbill is for a performance just two days later, when the latter play was again performed. DS

Third Night of Mr. KING's Engagement.

Never acted here.

THEATRE, BIRMINGHAM.

This present FRIDAY, August 2, 1799, will be presented,
The celebrated COMEDY of The

MAN of the WORLD.

(Written by the late CHARLES MACKLIN, Esq.)

The Part of Sir Pertinax Macsycophant by Mr. KING,

Egerton, — Mr. M'CREADY,
Sidney, — Mr. HARLEY,
Melville, — Mr. POWELL,
Lord Lumbercourt, — Mr. JOHNSON,
Counsellor Plausible, — Mr. EGERTON,
Serjeant Eitherside, — Mr. STANWIX,
John, — Mr. SURMONT,
Tomkins, Mr. QUANTRILL, Sam, Mr. WILKINS.

Lady Macsycophant, — Mrs. GILBERT,
Constantia by a LADY (being her third Appearance here)
Betty Hint, — Mrs. M'CREADY,
Nanny, — Mrs. WHITMORE,
And Lady Rodolpha Lumbercourt, Mrs. JOHNSON.

To which will be added (Sixth Time) A GRAND MUSICAL ROMANCE, called,

BLUE BEARD;
Or, FEMALE CURIOSITY.

The MUSIC by Mr. KELLY.

The SCENERY and MACHINERY, particularly the DISTANT VIEWS of

BLUE BEARD's PROCESSION,
ILLUMINATED GARDEN,
BLUE CHAMBER,
TRANSPARENCIES,

Abomilique's Palace, and the Sepulchre,

By Mr. WHITMORE and NUMEROUS ASSISTANTS.

The Elephant, Camels, Palanquins, Banners, &c.

By Eminent ARTISTS, under the Direction of Mr. WHITMORE.

Abomilique, — Mr. HARLEY, Third Sphai, — Mr. DAY,
Selim, — Mr. TOWNSEND, Fourth Sphai, — Mr. WILKINS,
Shacabac, — Mr. FARLEY, Hassan, — Mr. QUANTRILL,
Ibrahim, — Mr. JOHNSON, Beda, — Miss SIMS,
First Sphai, — Mr. STANWIX, Irene, — Mrs. POWELL,
Second Sphai, — Mr. EGERTON, And Fatima, — Mrs. ILIFF.

The DANCE by Miss BRUGUIER.

The CHORUSSES by Messrs. Townsend, Stanwix, Day—Mrs. Gilbert, Mrs. Whitmore, Mrs. Quantrill, Miss Sims, Miss Bruguier, and several *Resident Vocal Performers*.

The Farce of the *Critic* is in Preparation.—Due Notice will be given of the next Representation of *Lovers Vows*.

212

212 Lithograph of the New Street Theatre, 1805
ink on paper.
BMAG, 1965 V 221.41

This picture shows the theatre as it looked just before Boulton died. He had been one of the original proprietors who agreed to build the theatre in August 1773, owning one of the thirty shares, which cost him £50. He took an active part in the attempts to make the theatre an officially licensed Theatre Royal, failing in 1777–9, but succeeding in 1807. From time to time he entertained visiting performers at Soho House, including Mrs Sarah Siddons.   DS

213 J. D. Pedley, *A Bird's Eye View of Vauxhall Gardens*, c. 1850
watercolour, 60.5 × 97.7 cm.
BMAG, 1979 V 527

Birmingham's pleasure gardens, located within the grounds of the old Duddeston Hall, were renamed Vauxhall Gardens in 1758 after their London counterpart. Commercial pleasure gardens were popular in the eighteenth and early

213

223

nineteenth century. During the summer months they hosted evenings of music and fireworks. Catherine Hutton commented that the events in Birmingham were 'upon a smaller scale, and in a lower style', a reference to the many tradesmen and their families who attended.

In 1800 James Bisset describes the gardens in his poetic Survey of the town.

VAUXHALL:
A rural spot, where tradesmen oft repair
For relaxation, and to breath fresh air:
The beauties of the place attractive prove,
To those who quiet and retirement love;
There, freed from toils and labours of the day,
Mechanics with their wives, or sweethearts, stray;
Or rosy children, sportive, trip along,
To see rare Fireworks – or hear a Song:
For oft, in Summer, Music's secret pow'rs,
Woos thousands to Vauxhall, to pass their hours. JC

214 John Rawstorne (architect) and Francis Tukes (engraver), *Perspective View of The Crescent now Erecting near the Town of Birmingham,* 1792
engraving, 42.0 × 63.0 cm.
BMAG, 1979 V 550

215 Charles Norton, *Proposals with the Plan and Specifications for Building The Crescent in Birmingham,* 1795
paper-bound pamphlet.
BMAG library

216 Edward Richard Taylor, *The Crescent, Cambridge Street,* late nineteenth century
watercolour, 28.7 × 19.4 cm.
BMAG, 1915 V 99

Throughout the eighteenth century speculative builders in Birmingham developed schemes of varying levels of ambition and success. Probably the most ambitious project in Birmingham was Charles Norton's Crescent. Initially proposed in 1788, the ill-fated scheme was re-launched in 1795.

The Crescent, designed by architect John Rawstorne, promised to be an 'ornament' to the town with an elevated position and 'extensive prospects'. The design was a Palladian-style crescent of twenty-three houses with a chapel at its centre. Coach houses and stables would be accessed at the back of the crescent, while the front aspect would only allow access to passenger vehicles.

The scheme failed due to the impact of a depression on the town's trade, and the growth of industry in the area. One critic in 1825 stated 'the houses have neither the advantages of town nor country, while they partake the inconveniences of both', citing 'the near neighbourhood of the canal-wharfs, with all the concomitant noises and a superabundance of vulgar language'. Only the flanking wings of the Crescent were ever completed. JC

217 Minutes of a meeting of the Street Commissioners, 17 December 1776
   BAH, Minute Book 1 of the Street Commissioners, 1776–85

(The Street Commissioners were established by an Act of Parliament (the 'Lamp Act' of 1769), which approved the setting up of a commission with the power to raise rates for the purpose of keeping the streets of the town clean, safe and free from obstructions. Among their other duties the Commissioners were also responsible for providing street lighting. In their meeting on 17 December 1776, the Commissioners agreed to divide Birmingham into twelve districts in an attempt to manage the town better. The districts included Digbeth, Park Street, High Town, Coleshill Street, Edgbaston Street, New Street, the Square, Bull Lane, Spiceal Street, Temple Row, Snow Hill and Mount Pleasant. Each district was assigned to a team of three Commissioners to manage.   JC

218 Notes made by Matthew Boulton on the criminal problems of Birmingham, their results and solutions, undated but ?1780s
   BAH, MS 3782/12/100/6
219 Printed letter inviting householders to a meeting to discuss the establishment of a 'Nightly Patrole' in the St Paul's–Snow Hill area of the town, 21 October 1789
   BAH, MS 3782/12/100/3
220 Printed report of the meeting, held on 22 October 1789
   BAH, MS 3782/12/100/4

Matthew Boulton was deeply concerned about the problem of crime in Birmingham and wrote several notes like this for himself on the subject. In this one he outlines problems with prostitution – 'Our Streets are infested from Noon Day to Midnight with prostitutes' – and other crimes. Among other things, he suggests that all watchmen should 'exert themselves and to take into custody all loose and disorderly women whose conduct on the streets have the appearance of prostituting' and 'pay particular attention to houses that sell alcohol or liquors'.

On 21 October 1789 local householders were invited to attend a meeting, to be held at the Bricklayers Arms in Great Charles Street, to establish a nightly patrol in the neighbourhood of St Paul's, Snow Hill and Lionel Street. The report of the meeting, dated 22 October, records the resolutions to form the patrol. A committee of twenty was appointed to run the scheme, the first name on the committee list being Boulton's. The committee was to decide how many night constables would be appointed. The constables would then be responsible for recording any persons and places of disorder, and for holding any 'disorderly' individuals at the 'general rendezvous' until the following day when they would be taken to the magistrate.   JC

221 Call for payment of £32, representing 8 per cent of a £400 investment in the Trent and Mersey Canal, 13 November 1769
   BAH, MS 3782/6/6/257

Boulton was a keen promoter of canals, seeing them as essential for the cheaper and easier distribution of his products. He invested in a number of canals, including the Birmingham Canal Navigation and, as here, the Trent and Mersey Canal. In such large undertakings it was common for investors to agree to pay a set sum for their shares, but only to produce the money in instalments, as funds were needed to cover construction and other costs.   DS

222 Receipt from John Galton to Matthew Boulton
   recording Boulton's payment of £100 on ten shares in the Birmingham Canal Navigation, 30 September 1771
   BAH, MS 3782/6/192/5

Boulton had bought shares in the canal but had not yet paid all the instalments of the purchase price when he used the shares as collateral in a deal with John Galton and his sister Mary. In consequence he was still responsible for meeting any calls for payment on the investment. This is a typical instance of how complicated his finances regularly became.   DS

223 Johannes Eckstein, *John Freeth and His Circle*, 1792
   oil on canvas, height 112 cm.
   BMAG, 1909 P 6

John Freeth, seated second from the left, was a poet and owner of the Leicester Arms, a pub in Bell Street, Birmingham. This was the meeting place of the members of the Birmingham Jacobin Club, pictured here. The club attracted a wide membership from merchants, industrialists and individuals who shared liberal Whig views.   SC

THE ANTI-SLAVERY DEBATE

The battle to end slavery went on from the 1770s until well after Boulton was dead. It is clear that he was generally against slavery in principle, but he was not a committed activist like his friend Josiah Wedgwood and other members of the Lunar Society. He bought books that set out the anti-slavery case (see no. 225) and his name is at the top of the list of those who are thanked by the former slave Olaudah Equiano (in a letter published in *Aris's Birmingham Gazette*, 28 June 1790) for their kindness to him on his recent visit to Birmingham and for subscribing to his autobiographical book, *The Interesting Narrative*. On the other hand, on 9 November in the same year John Dawson of Liverpool was writing to Boulton & Watt about the possible purchase of a steam engine for a sugar plantation in Trinidad, referring in his letter to the 'purchase of slaves which I am to have the supplying of', and the company carried on selling engines to West Indian plantations for years to come.

David Symons

224 Letter from John Divier, London, to Farmer & Galton, Birmingham, asking for prices for guns to export to Africa, 22 August 1768
   BAH, MS 3782/6/194/49

Farmer & Galton, a Birmingham-based business, supplied many guns to merchants involved in the slave trade. This request from a London merchant for four hundred guns of 'the corsest sort ... fit for Africa' shows how such a business profited from the trade, as the guns were often exchanged for enslaved Africans, who were then transported to the Americas.   CP

225 Bill from J. Johnson for the supply of books in February–March 1788
   BAH, MS 3782/6/194/49

Two of the books listed here, Newton's *Thoughts on Slavery* and *Considerations on Slavery*, show that Boulton was clearly aware of the growing momentum of the anti-slave trade campaign in 1788. John Newton, a former slave-ship captain and friend of William Wilberforce, published his account in that year, providing an influential, first-hand account of 'this disgraceful branch of commerce'.   CP

226 Samuel Galton's printed defence of his involvement in the gun trade, 1795
BAH, MS 3101/13/16/2

This lengthy response from Samuel Galton junior to his fellow Quakers is in response to criticisms that his gun manufactory business promoted war and underpinned the slave trade. Denying that guns promote war, Galton also argues that the consumption of West Indian products, such as rum, tobacco and sugar, are 'the very Ground and Cause of Slavery' rather than his trade with Africa:

> The Censure, and the Laws of the Society, against Slavery, and Oppression, are as strict and as decisive, as against War – Now, those who use the produce of the labor of Slaves, as Tobacco, Rum, Sugar, Rice, Indigo, and Cotton, are more intimately, and directly the Promoters of the Slave Trade, than the Vender of Arms is the Promoter of War; – because the Consumption of these Articles, is the very Ground and Cause of Slavery; – but the Manufacture of Arms is not the Cause, but only a consequence of War. Such of you as do not concede these Luxuries of Life to your Principles – Can you, consistently, require a Sacrifice from me, of a Concern in which my Property is so involved, and by which my Family would be so extensively injured? CP

THE AMERICAN REVOLUTION

Boulton's attitude to the American Revolution was somewhat ambivalent. The more radical of the Lunar Society members – Wedgwood, Whitehurst, Darwin and Priestley – openly supported the colonists in their revolt, but Boulton's views wavered, depending on the effect he perceived events were having, or were likely to have, on Birmingham's trade. Initially sympathetic to the colonists, when he realised that the products of an independent America might pose a threat to British businesses, he became a staunch supporter of the government, helping to organise a petition in Birmingham in early 1775 that called on ministers to take a hard line with the rebellious colonists. Once the war was lost and American independence had been conceded, he adjusted without difficulty to the new situation, angling for a contract to strike the new nation's coinage.

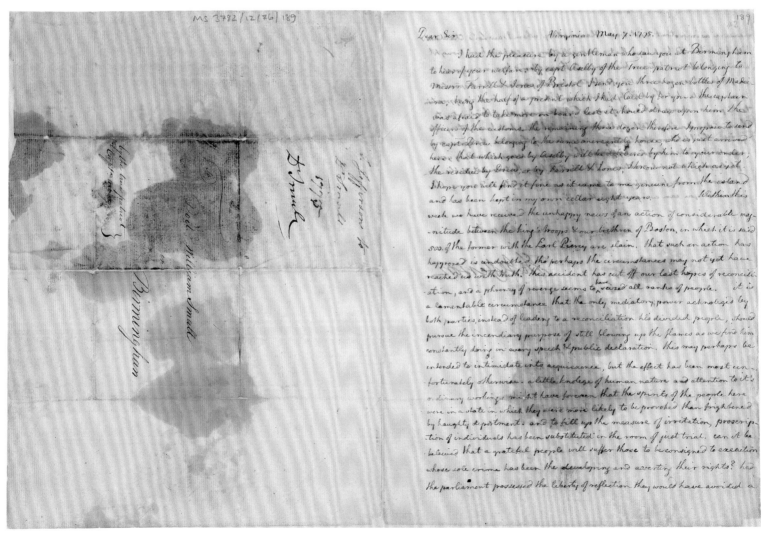

227 front

When that deal fell through, he happily supplied over twenty million copper blanks to the Philadelphia Mint, where they were struck into one-cent and half-cent coins, and produced medals for the US government (see no. 229). Boulton's attitude to this and other serious questions of the day (anti-slavery, the French Revolution) seems to have been one of steady pragmatism: principles are principles, but business is business.

David Symons

227 Letter from Thomas Jefferson, Virginia, to Dr William Small, Birmingham, 7 May 1775
BAH, MS 3782/12/76/189

Thomas Jefferson had been a student of William Small's when Small was Professor of Natural Philosophy at William and Mary College in Virginia from 1758 to 1764. Jefferson begins this letter by explaining that he is sending Small three dozen bottles of Madeira via Captain Aselby of the ship *True Patriot*, of Bristol. He then goes on to talk of reports of 'an action of considerable magnitude between the king's troops & our brethren of Boston', which presumably refers to the fighting at Lexington and Concord on 19 April 1775, the first battles of the American Revolution. The letter is a reminder of the close and friendly relationships that existed between individuals in Britain and the North American colonies despite the distances involved. When Jefferson wrote this letter he was clearly unaware that his old tutor and friend had died the previous February. DS

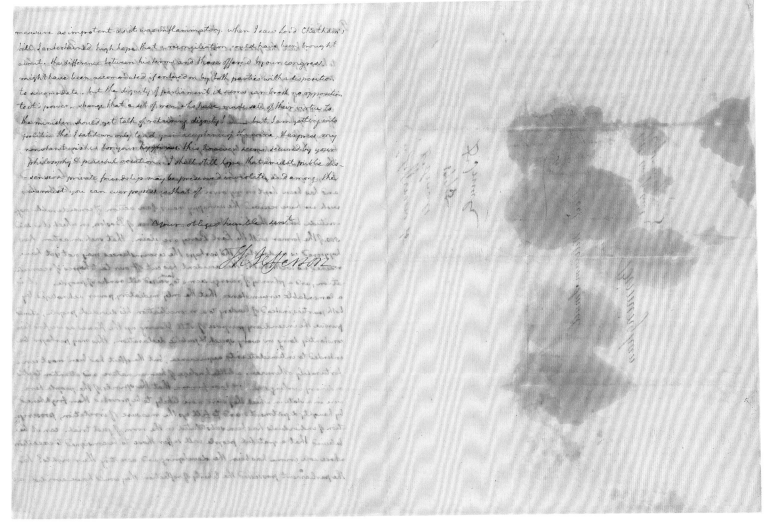

227 back

THE AMERICAN REVOLUTION

229

228 Josiah Wedgwood, *Bust of George Washington*, c. 1795

black basalt stoneware, height c. 41 cm.
BMAG, 1885 M 2717

Like the British in general, the members of the Lunar Society were divided in their attitudes to the American Revolution. Josiah Wedgwood was favourable to the cause of American independence, but Matthew Boulton's attitude wavered.　DS

229 George Washington medal, dated 1796 but struck at Soho 1798/9

copper, diameter 48 mm.
BMAG, 1885 N 1536.20

This medal was ordered by Rufus King, the American ambassador to Britain, towards the end of George Washington's second term as President (1793–7). Three different medals were produced, each with a design extolling the virtues of a settled agricultural life. This specimen depicts stock-raising, while the others showed sowing wheat and spinning yarn. The medals were used as presentation pieces to American Indian leaders.　DS

THE FRENCH REVOLUTION

Like many Britons, Boulton seems at first to have generally welcomed the French Revolution as bringing liberty to the French people in place of the tyranny of the *Ancien Régime*. But news of the massacre of prisoners in the Paris jails in September 1792, and the executions of Louis XVI and Marie Antoinette in January and October 1793 respectively (see nos. 273–274), caused a general revulsion at the revolutionaries, which Boulton fully shared. He wrote to his daughter of his horror at the turn of events and his belief that French refugees, of whom there were many in Cornwall, should be given every assistance.

David Symons

230 Thomas Rowlandson, *Reform advised. Reform begun. Reform compleat.*, 1793

etching with hand-colouring, 44.2 × 27.7 cm.
British Museum, 1931.0226.17

This print is a satirical response to contemporary calls for parliamentary reform, notably by the Society of the Friends of the People, formed in April 1792. It imagines dire consequences for England from the forces of reform, caricatured as three lean and belligerent Frenchmen who reduce John Bull from well-fed comfort to starvation and rags. It reflects how an initial sympathy for the changes in France (felt in many quarters) turned to revulsion after the execution of Louis XVI on 21 January 1793. VO

231 Tickets to a dinner held in Birmingham on 14 July 1791

BAH, MSS IIR 10 73499

Birmingham's new 'Constitutional Society' chose the anniversary of the storming of the Bastille for its inaugural dinner for 'any friend to freedom'. The dinner, attended by eighty people, provoked a riot which saw the destruction of the New Meeting House, as well as Dr Joseph Priestley's home and a number of other properties, an episode that came to be known as the 'Priestley Riots', although Priestley himself was not at the dinner. Priestley was caricatured in the newspapers as 'Gunpowder Joe' and 'Dr Phlogiston, the Priestley Politician or Political Priest'. HL

230

231

232 Report on the trials of the Priestley rioters, 29 August 1791

BMAG, 1970 F 1097

During the riots on 14 July 1791, thirty-one buildings were attacked and nine people died. Official reaction was mixed; the King expressed support for the rioters' actions. Of seventeen people prosecuted, just four were convicted, two being hanged and two pardoned. This newspaper report is based on notes taken in court.   HL

233 Two muskets made by Robert Wheeler, c. 1790, and supplied to Soho Manufactory

wood, brass and steel with flint and leather, length 176 cm.

BMAG, 1990 S 04181.00277 and 00280

These are two of a number of flintlock muskets that were bought by Matthew Boulton for the defence of the Soho Manufactory during the Birmingham riots of 1791. They were supplied by Robert Wheeler, a significant gun maker, who led moves for greater quality control in the gun trade.   JA

234 Unknown artist, *Joseph Priestley's house after the 1791 riot*, 1799

watercolour and ink on paper.

BMAG, 1997 V 87

235 Inventory of Priestley's property destroyed by the rioters, 1792

ink on paper.

BAH, MS 399801/IIR 30 1746

Joseph Priestley's home at Fair Hill, in the Sparkbrook area of Birmingham, was wrecked by rioters on 14 July 1791, and the contents destroyed. This inventory, prepared for the court case in 1792, gives us an insight into the range of specialist apparatus that Priestley had gathered for his ambitious scientific experiments. Distrust of radicals intensified during the 1790s, and in 1794 Priestley fled to America, dying in Northumberland, Pennsylvania, in 1804. The forlorn ruins of his house and laboratory at Fair Hill remained as a stark reminder of the violence that had erupted in Birmingham.   HL

236

WAR WITH FRANCE

The Napoleonic Wars mainly affected Birmingham by the disruption they caused to trade, and hence to manufacture. However, the ongoing threat of French invasion from the mid-1790s saw the creation of units of volunteers, pledged to resist any attack by force of arms (see nos. 238–239). Boulton was too old to serve, but he played his part by paying for the firearms for a squadron of the Staffordshire Volunteer Cavalry in 1795. When fresh volunteer units had to be raised in 1803, after the abortive Peace of Amiens (see nos. 240–242), his son Matthew Robinson Boulton served as an officer in the Loyal Birmingham Volunteers (catalogue nos. 243–245).

David Symons

236 Trafalgar medal, Soho Mint, 1806

silver, diameter 47 mm.

BMAG, 1885 N 1536.28

On 21 October 1805 Lord Nelson won a crushing victory over the combined Franco-Spanish fleet at Trafalgar, but was himself killed during the battle. Boulton had met Nelson when the latter visited Soho in 1802, so he approached the Admiralty and offered to supply (at his own expense):

> every sailor concerned in the late action at Trafalgar with a medal representing the battle on the reverse and the portrait of the gallant commander on the obverse.

The Admiralty agreed and the resulting medal, issued the following year, shows the moment when the *Royal Sovereign* began the attack on the enemy fleet. Specimens of the medal were struck in gilt, silver and tin. In addition to more than fourteen thousand medals distributed to the sailors and marines who fought at Trafalgar, Boulton gave examples to the King and Queen, the Tsar of Russia, senior naval officers, Lady Emma Hamilton and various of his own friends.   DS

236a

236a Catherine Andras, *Horatio Nelson*

portrait plaque, 265 × 225 × 35 mm, wax glass, wood, 1805.

National Maritime Museum, D6112.

Andras took the impression for this wax relief during Nelson's last shore leave in September 1805, just weeks before his death at Trafalgar. Lady Hamilton was said to be particularly pleased with the likeness.

236b

236b Nicholas Pocock, *The Battle of Trafalgar, 21st October 1805, The Beginning of the Action*
c. 1808, oil on canvas, 71.2 × 101.6 cm.
National Maritime Museum, BHC0548
Nicholas Pocock's painting of the Battle of Trafalgar is one of a pair showing the beginning and the end of the action. The battle scene is shown from the north-east.

237 Thomas Rowlandson, *The Wonderfull Charms of a Red Coat & Cockade*, c. 1785–90
pen and watercolour over pencil on laid paper, 25.8 × 21.1 cm.
BMAG, 1953 P 354
Rowlandson made many humorous drawings of mismatched couples, in this case a young woman of fashion and her beau, a much older officer. The title is a satirical comment on the supposed attractions of a man in uniform, which was perhaps one of the reasons for joining the volunteer regiments in the 1790s. VO

238 Edward Rudge (illustrator) and ?Forbes (engraver), *Three soldiers of the Birmingham Loyal Association*, 1797
coloured engraving.
BMAG, 1939 V 851
This engraving portrays a grenadier, a line infantryman and a light infantryman of the Birmingham Loyal Association, a volunteer unit formed in 1797 in response to the perceived threat of French invasion. Boulton was a supporter of the volunteer movement. In 1795 he bought sixty carbines and holsters, together with a number of pistols, for a troop of the Staffordshire Volunteer Cavalry. DS

239 Medal commemorating the presentation of colours to the Birmingham volunteer regiments, 1798
copper, diameter 41 mm.
BMAG, 1885 N 1541.126
This medal was produced by J. S. Jorden to mark the presentation of colours to Birmingham's two volunteer units, the Loyal Birmingham Light Horse Volunteers (cavalry) and the Birmingham Loyal Association (infantry), formed in 1797, at a well-attended public ceremony held on 4 June 1798. DS

237

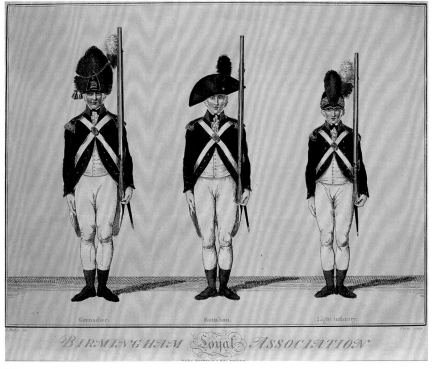

238

240

240 Medal presented to members of the Birmingham Loyal Association to mark the Peace of Amiens, Soho Mint, 1802
   silver, diameter 47 mm.
   BMAG, 1984 N 27

In March 1802 Britain and France signed a peace treaty at Amiens to end the war they had been fighting since 1793. With the peace, the local volunteer regiments were disbanded. The town of Birmingham presented silver medals, made by Boulton, to everyone who had served in the Birmingham volunteers. This specimen was given to William Mason of the sixth company of the Birmingham Loyal Association. Sadly the peace soon broke down. Fighting resumed in 1803 and continued until the final defeat of Napoleon at Waterloo in 1815. DS

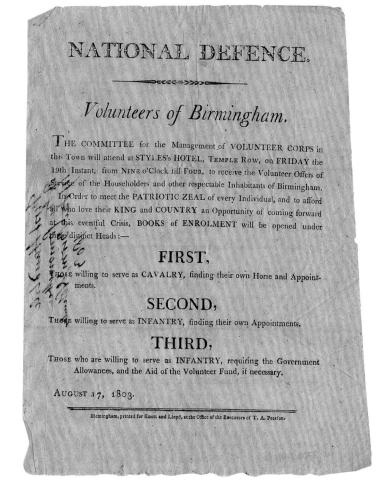

243

241 Drawing of Soho Manufactory illuminated for the Peace of Amiens, 1802
   ink on paper, 46.5 × 28.5 cm (unmounted).
   BAH, MS 3782/12/102/11

This shows the Soho Manufactory, illuminated with hundreds of coloured oil lamps to celebrate the Peace of Amiens in 1802. It is sometimes said that gas lighting, developed by William Murdock for Boulton & Watt for use in the Soho Manufactory and at Soho Foundry, was used to produce these illuminations, but this is not in fact the case. FT

242 Order for coloured glass lamps for the Soho illuminations to be supplied by Messrs. Shakespear & Johnson of Birmingham, 26 November 1801
   BAH, MS 3782/12/102/10a

The purple, green, crimson, yellow and white lamps used in the Soho illuminations to mark the Peace of Amiens were supplied by Shakespear & Johnson, who were makers of cut and plain glass and glass toys, of New Town Row, Birmingham. Over 2,600 lamps cost about £16, and the total expenditure for the celebration was nearly £60. FT

243 Handbill calling for the formation of a new volunteer force in Birmingham, 17 August 1803
   BAH, MS 3782/18/6/1

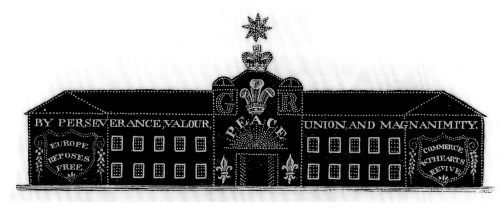

241

Following the breakdown of the Peace of Amiens, Britain went to war with France again in 1803. This call for a new volunteer force led to the formation of the Loyal Birmingham Volunteers, an infantry regiment composed of three battalions, each of five hundred men. Boulton's son, Matthew Robinson Boulton, became a major in the First Battalion.   DS

244 Letter from Amelia Alston to Matthew Boulton, 31 October 1803
BAH, MS 3782/13/53/117

Amelia Alston was a family friend of the Boultons. Here she teases Matthew Boulton that his son's new role in the Loyal Birmingham Volunteers has scared the French off invading:

> Since my arrival in London … I have heard nothing of France nor of ffrench [sic] men … I imagine Bonaparte has heard of your sons progress in the military art and begins to waver in resolution.   DS

245 Receipt for money collected for the Birmingham Loyal Volunteers, 1804
BAH, MS 3782/18/6/9

This receipt was issued by Dickinson, Goodall and Co., who were the owners of the Birmingham Bank, one of a number of private banks active in the town. Acting on behalf of Major Matthew Robinson Boulton, Matthew Boulton's son, Thomas Pemberton (the regimental paymaster of the Birmingham Loyal Volunteers) has paid in £262 10s. 0d. collected by public subscription on behalf of the Volunteers.   DS

246 John Phillp, Design for cane head terminal, c. 1803
pen and ink, 22.6 × 18.5 cm.
BMAG, 2003.0031.172

247 John Phillp, *Handsworth Troop of Horse drilling*, c. 1798
pen and ink, 188 × 230 mm.
BMAG, 203.0031.161

Phillp's first drawing shows a design for a cane head terminal for the Loyal Birmingham Volunteers, the defence force which was formed in 1803. M. R. Boulton was an officer in the force, but James Watt junior declined this role, offering to serve in the ranks in whatever position was thought useful.

The Handsworth Volunteer Cavalry, shown drilling in the second drawing, were formed in 1798 and John Phillp served as a lieutenant.   VL

247

STEAM

As an eighteenth-century polymath, Matthew Boulton would have been interested in steam purely as a physical phenomenon which he discussed with his friends in the Lunar Society. As an industrialist, however, he recognised it as a means of powering his factory and producing steam engines to be made and sold to other industrialists. He would also, at first, have realised its serious shortcomings because early engines were of poor power and very low efficiency. Then in the late 1760s he heard of James Watt's new design of engine with the potential to revolutionise the use of steam engines because it had double the power and three times the efficiency of earlier engines. Boulton and Watt were soon collaborating on the steam-engine patent and discussing how Boulton could be involved in the steam-engine business (see pp. 63–70).

At first their steam engines were used to pump water out of mines, to provide a water supply and sometimes to feed over a waterwheel to drive machinery. The engines ran too irregularly to drive directly most factory machinery, but as the pumping-engine design became more reliable Boulton pressed Watt to turn his attention to powering factories. Through the 1780s the rotative engine was developed into a well-controlled and efficient factory power unit suitable for use in a variety of industrial applications from textile mills to ironworks. At Soho, Boulton planned and built the first steam-powered mint in the world (see pp. 104–5).

In the 1790s, Boulton and Watt took their sons into partnership and planned for the time when they could no longer obtain income from the patent licences. Boulton's Soho Manufactory had been supplying increasing numbers of components for the engines being built, but in 1795 they opened Soho Foundry (see pp. 106–7) to supply complete sets of components for steam engines, thus completing their transformation from engine design consultants to engine manufacturers. Boulton's name remained associated with the steam-engine business until 1848 and Watt's until after 1900, while other firms slowly took over the lead in exploiting steam power.

Jim Andrew

248

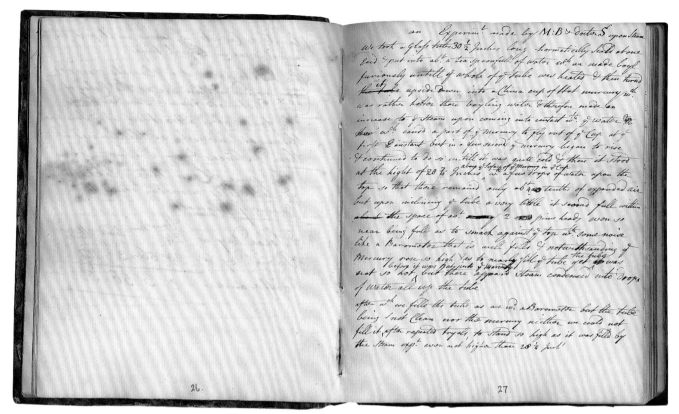

250

248 Sir William Beechey, *Portrait of Matthew Boulton*, 1810 copy of a 1798 portrait
   oil on canvas, 124.7 × 99 cm.
   BMAG, 2003.0007.44 (see p. 206)

The portrait of Boulton painted by Sir William Beechey in 1798 is an indication of Boulton's growing status, as Beechey was a senior member of the Royal Academy. Once again Boulton is shown holding a medal, indicating how important this branch of his business was to him.

This 1810 copy of Beechey's portrait of Boulton was ordered for James Watt shortly after Boulton's death and would originally have hung at Watt's home, Heathfield. Later, it hung at Aston Hall opposite Beechey's portrait of Watt when his son James Watt junior lived there.   VL

249 Sir Thomas Lawrence, *Portrait of James Watt*, 1812 (see p. 207)
   oil on canvas, 142 × 112 cm.
   BMAG, 2007.0889.

This portrait of James Watt was commissioned by Watt's son, James Watt junior. X-ray analysis of it has revealed an image of a steam engine on the book to Watt's right which was overpainted later. It is not known why this was painted out, but it may have been at Watt's own intervention.   LC

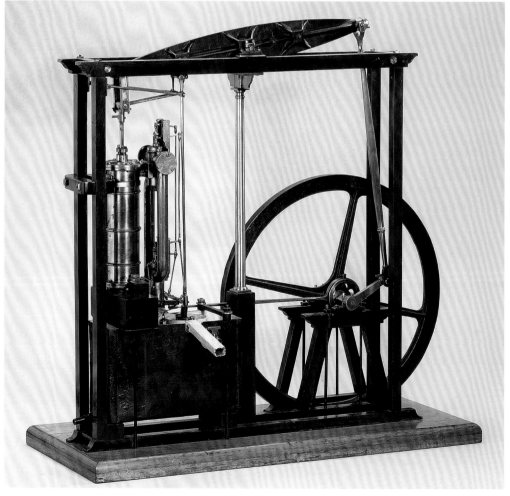

251a

250 'An Experiment made by MB & Doctor S upon Steam', Notebook 1, General, 1751–9
BAH, MS 3782/12/108/1

Boulton was interested in the power and potential of steam even before he met James Watt. This entry from his earliest notebook describes an experiment involving mercury and a steam-heated glass tube. The identity of 'Doctor S' is a puzzle, for although Boulton and his friend Dr William Small shared a common interest in steam, so far as is known they had not met at the time of this experiment. DS

251 James Watt, Sheet of drawings relating to steam-engine design, 1774–5
BAH, MS 3219/4/221/1

This sheet shows two ideas drawn for the Committee of the House of Commons, when James Watt was renewing his application for a steam-engine patent. The steam-wheel at the left was a first attempt to create rotary motion. The steam-wheel used a circular chamber fitted with valves, on a horizontal shaft turned by steam pressing on a weight of mercury. On the right are drawings of the double engine, which injects steam just above atmospheric pressure to drive the piston, helped by a partial vacuum and a system of valves. The double engine included a separate condensing chamber and gave greater efficiency in fuel consumption. Watt developed this engine with Boulton's assistance. It replaced Newcomen engines in areas where coal was expensive, and went on to be used where natural power sources such as wind and water were absent. ST

251a Model of a rotative beam engine, c. 1805
The Science Museum, 1861–0047; image: Science Museum/Science & Society Picture Library.

251b Apprenticeship indenture of William Tongue, 1797
The Science Museum, 1933–509

This 1:8 scale model, accompanied by the apprenticeship indenture of the apprentice who made it, represents the rotative steam engine as Watt left it on the expiry of his patents in 1800. William Tongue was an apprentice with Boulton & Watt from 1797 to 1804. His indenture, signed by Matthew Boulton and James Watt, shows that in lieu of 'Meat, drink, lodging and other necessaries' he was paid a starting wage of eight shillings a week for the first year, rising to fourteen shillings a week for the seventh and final year, and was to be instructed in 'the Art of Manufacturing and erecting Engines'. SM

251b

252 Fragment of a cylinder for a steam engine, 1797
Cast iron, 60 × 17.5 cm.
Tyne and Wear Museum Service, TWCMS: B2902

This is part of the large steam cylinder for a Boulton & Watt engine installed in 1797 at Hebburn Colliery Pit B, Tyneside. The cylinder was supplied by the Coalbrookdale Company at a time when John Wilkinson was no longer trading and Boulton & Watt's foundry could not yet supply such large cylinders. JA

253 Modern model of the lap engine
metal, plaster and timber, 96 × 120 × 82 cm.
BMAG, 1950 S 00022, at Thinktank

This is a model of the engine built in 1788 to drive the polishing machines at Boulton's Soho Manufactory. It was one of the first fully-developed Boulton & Watt rotative engines, with a double-acting cylinder, sun and planet motion, parallel motion and a governor to control its speed. JA

253

THE SOHO MINT

There were effectively three Soho Mints (see pp. 81–98 and pp. 104–5). The first, constructed in 1787–9, resulted from Boulton's experiences while producing a coinage for the East India Company in 1786, from his conviction that steam-power should be applied to the manufacture of coins and from his certainty that he was about to win a contract to strike a new copper coinage for Britain. Built at a cost of about £8,000, the First Soho Mint housed eight coining presses set in a circle and powered via an overhead wheel. Technical problems were gradually solved and the millions of coins, medals and tokens produced established the mint's reputation nationally and internationally.

In 1797 Boulton won the long-sought copper coinage contract from the British government. This convinced him to construct the second Soho Mint, because striking the large 'cartwheel' coins tested his existing equipment to the limit. Once the second mint was up and running, the first mint was demolished.

The second mint incorporated a series of technical improvements which made the coining process quieter and more efficient, and put less strain on the machinery (which was operated by twelve-year-old boys). This second mint was responsible for the British copper coinages of 1799 and 1806–7, and the Bank of England silver dollars of 1804, as well as millions of other coins. However, production slowed and for nearly a decade from 1813 virtually nothing was struck at the Soho Mint. Finally in 1824 Matthew Robinson Boulton sold it to the East India Company, and it was dismantled and shipped to Bombay (Mumbai).

The story of the third Soho Mint is strictly outside our remit here, but can be quickly told. Having disposed of the second mint, in 1826 Matthew Robinson Boulton was tempted again by the possibility of lucrative contracts from South America to build a new, smaller mint, with just four coin presses. Unfortunately the expected deals fell through and, while the third mint attained a modest output in the 1830s and 1840s, it never reached anything like its full potential. In 1850, eight years after Matthew Robinson Boulton's death, all the machinery was auctioned off. Much of it was bought by an enterprising Birmingham man, Ralph Heaton, who used it to set up his own mint, Heatons, later the Birmingham Mint, which continued to mint coins in Birmingham until it ceased trading in 2003.

David Symons

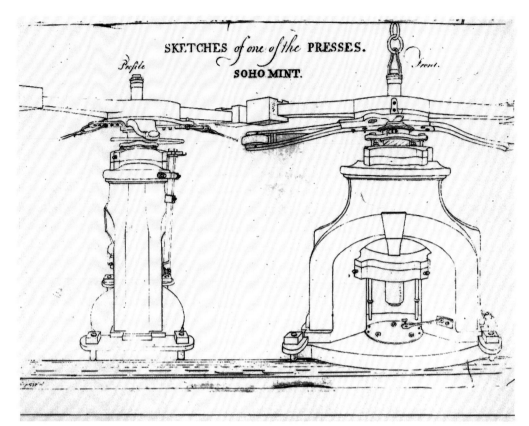

254

255

254 John Phillp, Three drawings of machinery at the Soho Mint: Side and front views of a coin press in the original Soho Mint, 1797
pen and ink on paper, 34 × 49.5 cm.
BMAG, 2003.0031.184

255 Side view of a coin press from the Soho Mint as rebuilt in 1798–9, 1799
pen and colour wash on paper, 24.6 × 20 cm.
BMAG, 2003.0031.149

256 A ceiling baffle, a vacuum trumpet, and decorative plates from a coin press, 1799
coloured drawing on paper, 48.3 × 62.0 cm.
BMAG, 2003.0031.68

The machines shown in Phillp's drawings enabled Boulton to make every coin of a particular denomination identical in appearance, size and weight, which had not been possible before using only man-powered machinery. Boulton played a crucial role in the modernisation of coinage, and Soho Mint products were distributed around the world. ST

THE SOHO MINT

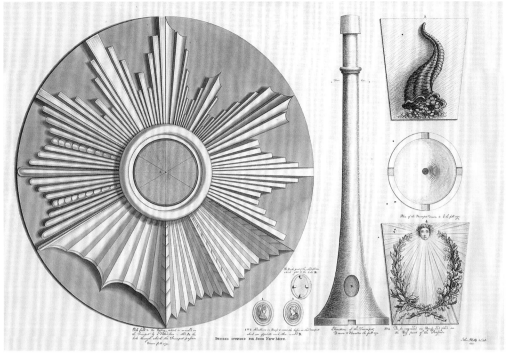

256

257 Medallic scale medal, struck at the Soho Mint in 1803 but dated 1798

bronzed copper, diameter 42 mm.
BMAG, 1885 N 1541.238

The die for this medal was engraved for Matthew Boulton by Rambert Dumarest. It indicates the rates of production that Boulton claimed could be achieved for coins at the Soho Mint. The reverse has a series of concentric rings, each intersected by a number representing the number of coins of that diameter that can be struck in one minute, from 920 per minute at the smallest near the centre to 460 per minute at the largest on the outermost ring. Boulton produced this medal to counter claims by his former engraver, Jean-Pierre Droz, that he and not Boulton was responsible for the improvements in coining machinery at Soho. Translated from the (much-abbreviated) French, the inscription on the reverse reads:

> Matthew Boulton erected at Soho, England in 1788, a steam-powered machine to strike coins. In 1798 he set up a much better one with 8 new presses. These circles and numbers indicate the diameter and number of pieces struck per minute by 8 children without fatigue, of the smallest or greatest volume, or of 8 different combinations. One can increase efficiency to the necessary degree. ST

258 John Phillp, Two drawings with designs for halfpennies, c. 1802

ink on paper, 25.8 × 42 cm and 21.2 × 27.3 cm.
BMAG, 2003.0031.128 and 129

These drawings show John Phillp experimenting with radically new designs for halfpenny coins. Instead of the traditional figure of Britannia, he tries out a series of symbols like the royal coat of arms, the Garter Star, crowns, roses, shamrocks and thistles. The number thirty-eight (in one case forty) indicates how many coins were to be minted from one pound of copper. No coins were ever produced with these designs. DS

£ s. d.: MONEY IN MATTHEW BOULTON'S BRITAIN

In the eighteenth century Britain was still using its traditional currency system, which survived until decimalisation in 1971. In this system:

12 pennies or pence (d.) = 1 shilling
20 shillings (s.) = 1 pound (£)

meaning that there were 240 pennies in a pound.

The penny was subdivided into 2 halfpennies (½d.) or 4 farthings (¼d.), which were the lowest-value coins.

The standard gold coin was the guinea, worth 21s.

Sums of money could be written in a number of ways. Thus:

4s. 6d. was the same as 4/6.
£1 5s. 3½d. was the same as 25s. 3½d. or 25/3½.
£3 3s. 0d. was the same as 3 guineas.

In Birmingham, a town with a high proportion of well-paid, skilled metalworkers, male weekly wages in the 1770s varied from a minimum of 7s. to as much as £3 in a few exceptional cases. A fairly average wage would have been in the region of 15s to 20s. Women could expect to earn less than this, their wages varying from 7s. to a maximum of £2 for the very best paid. A live-in domestic servant might earn as little as £2–£3 a year, but would have their food, clothing and lodging provided for them.

David Symons

257

Trade tokens struck at Soho Mint
259 Parys Mine Co., Anglesey, halfpenny token, 1791
    copper, diameter 28.5 mm.
    BMAG, 1885 N 1541.155
260 John Wilkinson, ironmaster, halfpenny token, 1792
    copper, diameter 28.5 mm.
    BMAG, 1885 N 1536.86
261 Cronebane halfpenny token, 1789 (see p. 92)
    gilt copper, diameter 29 mm.
    BMAG, 1885 N 1536.94
262 Glasgow halfpenny token, 1791 (see p. 92)
    bronzed copper, diameter 28.5 mm.
    BMAG, 1885 N 1536.87
263 Southampton halfpenny token, 1791
    bronzed copper, diameter 28 mm.
    BMAG, 1967 N 906
264 Inverness halfpenny token, 1793
    bronzed copper, diameter 28.5 mm.
    BMAG, 1885 N 1536.96
265 Bishop's Stortford halfpenny token, 1795 (see p. 92)
    gilt copper, diameter 29 mm.
    BMAG, 1885 N 1536.90

In response to the national shortage of small change, the Soho Mint struck large numbers of tokens for the industrial giants Thomas Williams 'the Copper King', who ran the Parys copper mine and associated works on Anglesey, and 'Iron Mad' John Wilkinson, who made parts of the steam engine and mint castings for Boulton. These tokens were widely distributed and used as small change throughout the country. Various other industrialists and merchants also commissioned Boulton to make tokens for use in their local areas. The gilt specimens here were struck as presentation pieces and for sale to collectors; the tokens for everyday use were all made out of copper.   ST

266 Gilt 5s. truck ticket for Pen-y-Darran Ironworks, Soho Mint, 1800
    gilt copper, diameter 34 mm.
    BMAG, 1968 N 170
267 Gilt 2s. 6d. truck ticket for Pen-y-Darran Ironworks, Soho Mint, 1800
    gilt copper, diameter 32 mm.
    BMAG, 1968 N 171

The Pen-y-Darran Ironworks were established in 1784 by Jeremiah, Thomas and Samuel Homfrey (H & Co.), the sons of Francis Homfrey, in Merthyr-Tydfil. The copper tickets were used to pay workers and were to be spent in the company shop. The ironworks was used mainly for manufacturing weapons and ammunition.   ST

The marriage of the Prince of Wales
268 Medal marking the marriage of the Prince of Wales to Caroline of Brunswick
    Soho Mint, 1795
    Formerly in the possession of the Watt family, with its original metal shell and paper wrapper, copper, diameter of medal 48 mm.
    BMAG, 2003.0035.2.1–3 (see p. 97)
269 Original obverse die for the same medal
    1795
    steel, height 58 mm, diameter 75 mm.
    BMAG, 2004.0182.1 (see p. 97). (Die presented by the National Art Collections Fund.)
270 Letter from Matthew Boulton to Matthew Robinson Boulton, 12 February 1795
    saying he has met the Prince of Wales, who has agreed to sit for a portrait for a medal for his wedding.
    BAH, MS 3782/13/36/118

In January 1795 Conrad Heinrich Küchler, Soho's leading die engraver, suggested to Boulton that they should produce a medal to mark the marriage of the Prince of Wales to Princess Caroline Amelia Elizabeth of Brunswick, which was to take place on 8 April 1795. Unfortunately Boulton was not happy with Küchler's first attempt at the Prince's portrait:

> I am sorry to say that the Prince of Wales strikes me as the head of a Citizen Alderm^n for the …. Hair has too much the appearance of a Wig and it is not a good Sembl of the Prince…

266

Küchler did not complete the dies until 10 May, a month after the wedding had taken place, and Boulton decided against issuing any medals so long after the event. The royal couple separated after a year, and it seems that the surviving medals were only struck for collectors, in about 1810, on Matthew Robinson Boulton's instructions. To make matters worse, the wrong date, 1797, appears on the medal by mistake.   ST

Commemorative medals from the Soho Mint
271 Medal celebrating the recovery of George III from illness
    1789
    silver with gilt edge, diameter 34 mm.
    BMAG, 2003.0035.1.1 (see p. 96)
272 Medal on the victory of Lord Cornwallis over Tippu Sultan, the ruler of Mysore in India
    1793
    bronzed copper, diameter 48 mm.
    BMAG, 1974 N 27 (see p. 87)
273 Medal on the execution of Louis XVI of France
    1793
    copper, diameter 51 mm.
    BMAG, 1885 N 1541.5
274 Medal on the execution of Queen Marie Antoinette of France
    1793
    copper, diameter 48 mm.
    BMAG, 2000 N 7.2 (see p. 96)
275 Medal commemorating the naval victory by Admiral Lord Howe over the French at the Glorious First of June
    1798
    copper, diameter 48 mm.
    BMAG, 1885 N 1541.54 (see p. 87)
276 Medal marking the restoration of King Ferdinand of the Two Sicilies to his throne by Nelson
    1799
    gilt copper, diameter 48 mm.
    BMAG, 1885 N 1541.58 (see p. 96)
277 Medal on the Union of Britain and Ireland
    1801
    gilt copper, diameter 48 mm.
    BMAG, 1885 N 1541.50

Boulton made a variety of medals at the Soho Mint to celebrate contemporary events, including the recovery of George III from his first bout of madness, the executions of Louis XVI and Marie Antoinette in France, and battles at sea and on land. He also marked the Union of Britain and Ireland in 1801. Produced as commercial ventures, these medals were sold in a variety of metals, silver and gilt copper for the better-off, and cheaper versions in tin.   ST

284

Hafod Friendly Society

278 Design drawing by J. Phillp for Hafod Friendly Society medal
1798, ink on paper (with spelling errors corrected), 12.5 × 16.8 cm.
BMAG, 2003.0031.111 (see p. 98)

279 Hafod Friendly Society medal
1798, bronzed copper, diameter 41 mm.
BMAG, 1885 N 1541.105 (see p. 98)

The sketch shows the corrections made to the Welsh inscription, which translates as the three traditional Welsh virtues: sobrwydd (sobriety), diwydrwydd (diligence) and brawydgarwch (fraternity). The medal was struck for the Hafod Friendly Society run by Thomas Johnes, who had created a model agricultural establishment and parkland at Hafod in Cardiganshire. ST

Imperial coins struck at the Soho Mint

280–2   Sumatra (Bencoolen)
1786, copper 1, 2 and 3 keping, diameters 20–27.5 mm.
BMAG, 1885 N 1541.183–184 and 2007.1261 (see p. 90)

283 Sumatra
1804, copper 4 keping (formerly in the possession of the Watt family) with its original metal shell and paper wrapper, diameter 31 mm.
BMAG, 2003.0035.3.1, 4 and 7 (p. 93)

284 Sri Lanka
1802, copper 1/48th rupee (formerly in the possession of the Watt family) with its original metal shell and paper wrapper, diameter 30 mm.
BMAG, 2003.0035.7.1, 4 and 7

285 Bombay
1804, bronzed copper-proof double pice, diameter 30 mm.
BMAG, 1885 N 1541.193 (see p. 93)

286 Sierra Leone
dated 1791 but struck in 1792
bronzed copper proof half dollar, diameter 30.5 mm.
BMAG, 1885 N 1541.173 (see p. 93)

287 Gold Coast
1796, bronzed copper proof 1 ackey, diameter 31 mm.
BMAG, 1885 N 1536.60

288 Bahamas
1806, copper penny, diameter 28 mm.
BMAG, 1885 N 1536.69

289 Bermuda
1793, bronzed copper proof penny, diameter 30 mm.
BMAG, 1885 N 1526.85

290 Isle of Man
1798, gilt copper proof penny, diameter 33 mm.
BMAG, 1885 N 1541.211 (see p. 93)

From 1786 the Soho Mint struck many millions of coins for customers across the world. The East India Company ordered coins for its possessions in India and the East Indies. Other coins were produced for British colonies in Asia, Africa and the Caribbean, as well as for places closer to home like Ireland and the Isle of Man. Boulton even hoped to strike coins for the newly-independent USA, but this did not happen. Instead, however, between 1797 and 1807 he made more than twenty million copper blanks for half-cent and one-cent coins and shipped them across the Atlantic to the US Mint in Philadelphia. ST/DS

277

288

289

British regal copper coins – New and old, fakes and forgeries

294–296 Two halfpennies and a farthing
  1771–4, Copper, diameters 23–28.5 mm.
  BMAG, 1969 N 602, 611 and 617 (see p. 91)

297 A very worn halfpenny of George II
  (r. 1727–1760), copper, diameter 27 mm.
  BMAG, 2008.1413 (see p. 91)

298–301 Four forged halfpennies
  1770s or later, copper, diameters 27–27.5 mm.
  BMAG, 1932 N 107.13; 1971 N 461; 2007.2145 and 2147

302–306 Five 'evasive' halfpennies
  1770s or later, copper, diameters 26.5–27.5 mm.
  BMAG, 1885 N 1526.258 and 422; 1930 N 447.12; 1964 N 2139; 2007.2172 (see also p. 91)

Denmark and Russia

291 Pattern for a proposed, but never issued, coinage for Denmark
  1799, copper, diameter 39 mm.
  BMAG, 1885 N 1536.72

292–293 Two sample pattern Russian roubles
  1804, copper, diameter 41 mm.
  BMAG, 2007.2035 and 1885 N 1536. 43

The Danish piece is one of a number that were struck, in five different sizes, as samples to show the Danish government what could be produced with the minting machinery they were negotiating to buy from Boulton for the Copenhagen Mint. Unfortunately the events of the Napoleonic wars, including Nelson's bombardment of Copenhagen in 1799, made it difficult for Boulton to complete the contract. The Danish Mint was finally functioning by 1809.

The Russian pieces were not intended for circulation, but are mint samples, produced to help train staff at the mint in St Petersburg in how to operate the minting machinery that Boulton had supplied. ST

The Royal Mint produced a substantial issue of copper coins in the years 1770–5, but then struck no more for the next twenty years, leading to a growing shortage of low-value coins. Cat. nos 294–6 show what the coins looked like when they were freshly made, but they did not stay like this for very long once they entered circulation. In fact a lot of the circulating copper consisted of old, worn coins. Some (297), were worn virtually flat after thirty years or more in use. Forgeries (298–301) were even more common. These were of very variable quality. Some were quite convincing, while others were frankly awful, but people used them because there was nothing better available. Most of these forgeries were made in Birmingham. Some manufacturers evaded the laws on forgery by making pieces that looked like real coins, but did not exactly reproduce the correct designs or legends. These are known as 'evasive' halfpennies (302–306). One example here replaces King George III's name with GOD SAVE THE KING, while another replaces the normal reverse legend of BRITANNIA with BRITONS HAPPY ISLE. DS

Droz and Pingo

307–308 Two examples of a pattern halfpenny
from dies engraved by Jean-Pierre Droz at
the Soho Mint, 1788
copper and gilt copper, diameter 36 mm.
BMAG, 1969 N 637 and 1969 N 651 (see p. 91)

309–310 Pattern halfpenny and farthing
from dies engraved by Lewis Pingo at the
Royal Mint, 1788
copper, diameters 35 mm and 28 mm.
BMAG, 1969 N 622–623 (see p. 91)

In 1788 Boulton was trying to obtain a contract to make the regal copper coinage for Great Britain. He commissioned the Swiss engraver Jean-Pierre Droz to make some specimen coins so that he could show his designs to the Government. Lewis Pingo, the chief engraver at the Royal Mint, was also making pattern coins. The designs of both show the head of George III and Britannia, but the Droz coin is more delicately engraved. However, Boulton soon became disenchanted with Droz: 'He is the vainest and most conceited Charlaton I ever knew'. ST

311 A record of Matthew Boulton's appearance before the Committee of the Privy Council on Coin, London, 1789
BAH, MS 3782/12/97/7

The Committee had been set up in 1787, charged with carrying out a thorough examination of the nation's coinage. Here we see Boulton being questioned, as a businessman well-versed in metal-working and coining, on the practicality of using a cheaper alloy in place of pure copper for low-value coins, and on the likely costs involved in striking a new coinage. DS

312 Royal proclamation declaring the 'cartwheel' coinage legal tender, 26 July 1797
printed notice on paper, 41.5 × 33 cm.
BAO

By this proclamation the 'cartwheel' pennies and twopence pieces (so-called because of their size and their bold rims) were made legal tender for Great Britain up to the value of 1s., to provide 'an Immediate Supply of such Copper Coinage as might be best adapted to the Payment of the Laborious Poor'. ST

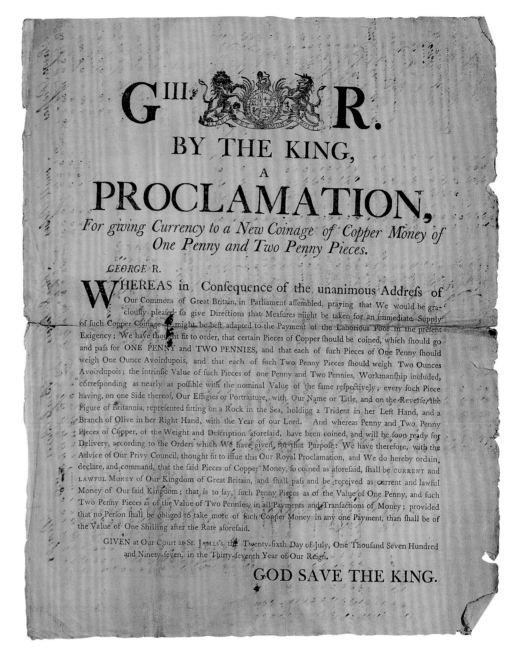

312

313 Royal licence for Boulton to strike the 'cartwheel' coinage, issued under the privy seal, 9 June 1797
four vellum sheets with a wax seal in a tin container, all in the original decorated wooden box, diameter of seal 15.0 cm.
BAH, MS 3782/17/4

This is the official licence instructing Boulton to produce coins at a rate of twenty tons a week, to a total of 480 tons of pennies and twenty tons of twopence pieces, with the first delivery on 3 July 1797. They were to be wrapped in paper packets of two shillings, which would then be packed in wooden casks. Matthew Boulton was to ensure that the copper was rolled, hot-scaled and measured, then cold-rolled in polished rollers exactly to the size and weight required, then cut out into blanks and coined 'with such dies as shall be approved'. ST

Cartwheel coinage struck at the Soho Mint, 1797
314 Proof twopence
gilt copper, diameter 42 mm.
BMAG, 1969 N 726
315 Twopence
copper, diameter 40 mm.
BMAG, 1969 N 736 (see p. 95)
316 Proof twopence
copper, diameter 41 mm.
BMAG, 1969 N 728
317–318 Two pennies
copper, diameter 36 mm.
BMAG, 1969 N 775–776 (see p. 95)
319 Pattern penny
with a helmeted Britannia, pewter, diameter 36 mm.
BMAG, 1969 N 744

Between June 1797 and 1799 some 1,250 tons of copper were minted into 'cartwheel' coins at Soho, a total of 43,969,204 pennies and 722,180 twopence pieces. This temporarily alleviated the shortage of small change. However, the coins were very heavy (the penny weighed 1 oz./28g., the twopence double this) and, as the price of copper rose, many were melted down. For the first time on a British coin Britannia was shown with a trident, symbolising Britain's sea-power. The pewter penny with a helmeted figure of Britannia is a 'pattern', a sample piece trying out a possible design that was not used. ST

Pattern guineas
320–321 Two pattern guineas
produced at the Soho Mint, 1791
gilt copper, diameter 24 mm.
BMAG, 1971 N 484 and 1930 N 190
322–323 Two pattern guineas
produced at the Soho Mint, 1798
gilt copper, diameter 24 mm.
BMAG, 1885 N 1541.158 and 1930 N 204

These gilt pattern guineas were made by Boulton to show how his minting machinery could produce coins for the government more effectively than the Royal Mint. He wanted to sell his process and did not intend to mint gold coins himself. The reverse features a crown above the royal coat of arms, both with and without a 'cartwheel' rim. These designs were never used for real coins. ST

British coins struck at the Soho Mint in 1799 and 1806–7
324–325 Two halfpennies
1799, copper, diameter 30 mm.
BMAG, 1969 N 864–865
326–327 Two farthings
1799, copper, diameter 23 mm.
BMAG, 1969 N 884–885
328 Pattern halfpenny with crowned bust
1799, gilt copper, diameter 30 mm.
BMAG, 1969 N 842
329–330 Two pennies
1806, copper, diameter 33.5 mm.
BMAG, 1969 N 927 and 1969 N 931 (see p. 95)

331–332 Two halfpennies
1806, copper, diameter 28 mm.
BMAG, 1969 N 962–963
333–334 Two farthings, 1806
copper, diameter 21 mm.
BMAG, 1969 N 978 and 1969 N 975

Boulton began to strike a second copper coinage, of halfpennies and farthings, in November 1799. The agreement stipulated that 550 tons of copper should be turned into coins, with 36 halfpennies or 72 farthings being struck from each pound of copper. Over 42 million halfpennies and 4 million farthings were made in 8 months. The coin showing the king wearing a crown is a 'pattern' with a design that was considered but not used.

By 1805 small change was again becoming scarce and a further coinage was ordered. This time pennies, halfpennies and farthings were struck, over 165 million coins being made at Soho in 1806 and 1807. ST

Countermarked and counterfeit dollars
335 Countermarked Spanish dollar
1797, silver, diameter 39 mm.
BMAG, 1971 N 602
336 Counterfeit of a countermarked Spanish dollar
1797, silvered base metal, diameter 39 mm.
BMAG, 1935 N 547.586

In March 1797, in an attempt to put a large amount of silver coin into circulation quickly in the face of a financial crisis, the government had over a million Spanish silver dollars countermarked with a small punch showing the king's head. These were then put into circulation with a value of 4s. 9d. Unfortunately it proved too easy to forge both the coin and the countermark and the coins had to be recalled in September 1797. DS

335

336

Dollars for the Bank of England

337 Bank of England dollar
1804, silver, diameter 40.5 mm.
BMAG, 1932 N 285.434 (see p. 95)

338–339 Two proof Bank of England dollars
1804, bronzed copper, diameter 40.5 mm.
BMAG, 1885 N 1536.130 and 1885 N 1541.17

340 Pattern Bank of England dollar
1804, bronzed copper, diameter 40.5 mm.
BMAG, 1885 N 1536.140

In 1804 the government tried once more to solve the recurrent problem of the lack of silver coins in circulation. Just as in 1797, they initially tried countermarking Spanish silver dollars with a small head of King George III, but once again this failed. Boulton, however, had been experimenting and showed that his coin presses could take the Spanish coins and strike completely new designs on them. So impressive were the results that he was commissioned to produce over one million overstruck dollars in 1804 alone. The overstriking did not remove all traces of the original coin; on this specimen they are quite clear below the king's bust on the obverse and around the date on the reverse. Characteristically, Boulton argued that the fact such traces were visible would be an added safeguard against forgery.

Proofs are coins struck on specially prepared flans from carefully polished dies. At Soho they were often struck in different metals to the finished coin or medal. The pattern carries a different reverse design to the one finally used. DS

341 Soho Mint record book – 'British Coinage 1798–1800
Consignments and Weights. Book No. 1 (1797–98)', c. 1800
BAH, MS 3782/3/71

This book records the dispatch from the Soho Mint of wooden casks full of 'cartwheel' pennies between 21 July 1797 and 30 October 1798. In that time no fewer than 3,018 casks are recorded as being sent out, each containing 6,000 coins (£25-worth) and weighing over 370 lbs. The first page shows that the very first coins sent out went to Boulton's old friend Sir Joseph Banks, who was sent one guinea's-worth (252 coins). The next shipment was to the Treasury and consisted of £4,000 worth of coins (960,000 of them) in 160 casks. It is noteworthy that whenever possible, shipments were sent by canal, the cheapest and easiest way to move such heavy merchandise. DS

342–344 5 sols and 2 sols tokens, and a medal depicting Jean-Jacques Rousseau, minted at the Soho Mint for Monneron Frères of Paris, 1791–2
copper, diameters 32 to 38.5 mm.
BMAG, 1885 N 1541.4; 1885 N 1536.97; 1885 N 1541.10 (see p. 92)

In 1791 France, like Britain, was suffering from a shortage of small change. Monneron Frères, a partnership of brothers with whom Boulton had already had dealings, decided to issue their own tokens, like the British trade tokens, and asked Boulton to supply them, along with a number of medals depicting great men and events of the French Revolution. The project ran into all sorts of problems. First the Monnerons experienced serious financial troubles, then the French government banned the import and issuing of such tokens. DS

345 Copy specification for material for a new mint being sent to Denmark, 5 June 1805
BAH, MS 3782/3/137

Boulton first discussed the possibility of supplying the Danish government with a complete steam-powered mint in 1796. However, it took another ten years before the machinery was finally installed in Copenhagen, and the first coins did not come off the presses until 1809. DS

347

THE SOHO FOUNDRY

The Soho Foundry was established by the second generation of Boulton and Watt, and in particular by James Watt junior, as the first purpose-built and integrated steam-engine manufactory in the world. The Foundry, in Smethwick, was first laid out in 1795–6, about a mile to the west of the Soho Manufactory. For the first time all the parts making up a steam engine, including the vital cylinder, were cast or forged on one site, processed to fit accurately and then assembled to make a complete product. A location on the Birmingham Canal was essential in order to transport the finished engines more easily to customers. The original plan of the Foundry was supplied by William Wilkinson following the dissolution of his partnership with his brother, John, in 1795. They had previously supplied most of the cylinders for Boulton & Watt engines. The collapse of the Wilkinson partnership and the expiry of Watt's patent in 1800 prompted the building of the Foundry, allowing Boulton & Watt to get ahead of any competitors.

The Soho Foundry, which was the first factory in the world to be lit by gas supplied from fixed installation, grew to about three times its original size during the first half of the nineteenth century, and a great spoil heap of ash from the furnaces advanced relentlessly northwards. Buildings were constructed alongside the ash bank, causing unexpected changes in level across the site.

In 1848 the firm of Boulton & Watt folded on the death of James Watt junior and it was taken over by James Watt and Co., which, despite the name, had no Boulton or Watt family connection. In 1895 James Watt and Co. was dissolved and the site purchased by W. & T. Avery Ltd, weighing-machine manufacturers, who still own it today under the name of Avery Weigh-Tronix. Unlike the Soho Manufactory and Soho Mint, sufficient Soho Foundry buildings survive from the Boulton & Watt era to have been listed Grade II*, though they are currently in poor condition and the subject of restoration proposals.

George Demidowicz

346 Plan of the Soho Foundry, mid-1795
ink on paper, 44.5 cm × 63.2 cm.
BAH, MS 3147/5/1453

The Soho Foundry was built at Smethwick, conveniently alongside the canal, on land bought by James Watt to provide a suitable site for casting engine parts and building Boulton & Watt engines. William Wilkinson (previously manufacturer of engine cylinders for Boulton & Watt) sent a proposed layout for the new Foundry in August 1795; it exploited a fall in the land so that cylinders newly cast in the main building (L) could be transported on the level to the boring mill (I). The layout was quickly modified and extended, and this plan shows the buildings that had been constructed by early 1796. (North is at the bottom of the plan.) GD

347 John Phillp, *View of the Soho Foundry*, c. 1799
pencil on paper, 98 × 130 mm.
BMAG, 2003.31.18

This is one of only two sketches of the Soho Foundry at this date, the other being by William Creighton. This image highlights how close the site was to the Birmingham and Wolverhampton Canal, an important consideration when selecting the site as it made the transportation of bulky engine parts easier. VL

348 Matthew Boulton's notes for his speech at the 'consecration' of the Soho Foundry, January 1796
BAH, MS 3782/13/37/19

After the main Foundry roof was covered, a rearing feast was organised on 30 January 1796 for all the engine-smiths and workmen involved in the construction. After a meal that included two fat sheep, rounds of beef, legs of veal and gammons of bacon followed by 'innumerable plumb puddings', Matthew Boulton delivered the main address, though in truth he had been little involved in developing the first purpose-built manufactory of steam engines in the world. He sprinkled the walls with wine and named it 'Soho Foundry', and expressed the hope that the name would endure forever. GD

349 Piece of cast iron from William Murdock's collection, late eighteenth century
Lapworth Museum, University of Birmingham, BIRUG 17.162

This piece of cast iron is labelled as being 'from Mr Boulton's press'. It was preserved in the geological collection of William Murdock, which now belongs to the Lapworth Museum at the University of Birmingham. Murdock was a Scottish engineer who worked for Boulton & Watt in Cornwall for a number of years from 1779 as a steam-engine erector. While there he, like Boulton, acquired many mineral samples. DS

SOHO PROJECTS

The steady flow from Soho of toys, jewellery, silverware, ormolu, clocks, coins, medals, steam engines and minting equipment, has been demonstrated in the foregoing pages. Yet with all this, Boulton was still ready to try a new venture when opportunity presented, as represented by the following entries. Some employment practices at the Manufactory were also ahead of their time and were watched with interest by others.

350 View of Soho Manufactory, from James Bisset's *Magnificent Directory*, engraved by Francis Eginton junior, 1800 (see p. 29)
line and stipple engraving, 12.7 × 19.9 cm.
BMAG, 1997 V 1.32

This plate from Bisset's *Magnificent Directory* emphasises Boulton the businessman, important in a guide that listed many of the manufacturers in Birmingham. The list of eight firms under the image makes clear the size and remarkable diversity of Boulton's enterprise, and makes him stand out from the other entries. VL

MECHANICAL PAINTINGS

One of Boulton & Fothergill's most intriguing enterprises was the production of mechanically duplicated paintings, known as polygraphs. The process, invented by Francis Eginton, enabled original historic and contemporary paintings to be reproduced on canvas in editions through a combination of mechanical and manual techniques. To the untrained eye they would be identical to the original, but at a fraction of the cost. The pictures could be framed as canvases or inset into furniture. They were also particularly suited to interior design schemes with images used as over-doors or ceiling panels. Mrs Elizabeth Montagu purchased mechanical paintings from Soho for the decoration of Montagu House in Portman Square, her new London house designed by James 'Athenian' Stuart and completed in 1781. For the original pictures, Boulton enlisted the support of some of the leading artists of the day, including Sir Joshua Reynolds, Benjamin West, Philippe de Loutherbourg and Angelica Kauffman.

Production began commercially in 1776 and Boulton was tight-lipped about the techniques. In a letter to Sir Watkyn Williams Wynn he says vaguely, '... by some peculiar Contrivances I am enabled to make better Copies of good originals than can be done otherwise without much greater expense.' There is still much speculation about the process involved and there is as yet no conclusive evidence. It seems probable that a reduced copy of the painting was etched in aquatint on a copper plate. The outline, main tonal and colour areas of the picture would be printed on paper which was then laid onto a primed canvas and transferred by a rolling press. The paper was then peeled off and the basic image in its 'dead coloured state' was given to a painter, or team of painters, for finishing. The manual addition of highlights and details helped to disguise the underlying print. Varnish was then applied to seal the surface and mimic the appearance of a true oil painting. Essentially, they were elaborate hand-coloured prints on canvas. A similar technique was being used by the miniaturist Joseph Booth in the early 1780s in London, where a Polygraphic Society was founded to promote the medium. He recruited many of the same artists, who seemed eager to exploit the potential of this new and impressive medium of mass production.

At Soho interest began to wane due to lack of orders and rising costs. Eginton left after a dispute with Boulton, but some of the equipment was transferred to him and for a while he maintained production independently, but polygraphic reproduction was short-lived and its trade secrets were soon forgotten.

Brendan Flynn/Martin Ellis

351

351 James Millar, *Portrait of Francis Eginton*, c. 1796
oil on canvas, 89.8 × 73.6 cm.
BMAG, 1912 P 24

Eginton was originally in charge of the Boulton & Fothergill japanned ware department and also worked as a designer in silver and ormolu. The invention of mechanical painting at Soho is usually credited to him. He fell out with Boulton and left Soho in 1780, but continued the production of mechanical paintings independently, as a sideline to his stained glass business.   BF

352–353 Francis Eginton, *Winter* and *Summer*, mechanical paintings after original paintings by Philip de Loutherbourg, c. 1778
polygraphic reproductions in ink and paint on canvas, both 87.6 × 123.2 cm.
BMAG, 1885 P 2591 and 1885 P 2589

Eginton probably produced these pictures for Boulton & Fothergill. De Loutherbourg had exhibited the original paintings of *Winter* and *Summer* at the Royal Academy in 1776. They depict contrasting and light-hearted scenes of London life and were designed to hang as a pair. The trees, positioned to right and left respectively, create a balanced overall composition. They made an ideal subject for Eginton's polygraphic process. In *Winter*, the seated man being fitted with skates is a portrait of de Loutherbourg himself. The group behind him includes his wife, Lucy Paget, Captain Cook's draughtsman, John Webber, and the famous French choreographer, Jean Noverre. BF

354 Letter and copy invoice concerning mechanical paintings, 1791
BAH, MS 3782/6/194 27a and 27b

The manufacture of mechanical paintings at Soho had ceased by 1782, but Francis Eginton continued production independently. This copy invoice and its covering letter are from Boulton's accountant, John Hodges. They show that Eginton was supplying mechanical paintings to Soho for resale, at least until 1791. They also mention works sent to him for repair. The list indicates the popularity of Angelica Kauffman's works which accounted for over half the stock list. BF

352

353

355

nesses and individuals. Watt's machines provided the first effective document-copying process, saving an enormous amount of labour, and the principles remained in use throughout the nineteenth century.

Letters were written with a slow-drying ink, and when finished a sheet of moist tissue was placed on top of the page. Letter and tissue were then pressed together using a roller, making the ink soak through the tissue. The tissue paper formed the copy, in reverse, which could be read from the other side. Many such copy documents are to be found in the Archives of Soho, many of them still clearly legible. LC

358 Poster with the rules of the Soho Insurance Society, c. 1792 (see p. 26)
ink on paper, 71 × 50.5 cm.
BAH, MS 3147/8/4/

The Soho Insurance Society was established by Matthew Boulton and run by a committee from the workforce, and is a very early example of such a scheme. Workers' contributions, based on their level of pay, were used to create a fund to support those who were sick, and when necessary to help with funeral expenses. The lowest-paid (those earning less than 2s. 6d. per week) did not have to contribute to the fund but could receive assistance from it. In addition to the standard rate of sick pay (which varied according to the sick person's wage-rates), extra hardship payments could be made where it was deemed appropriate, but allowances were not paid if the sickness was a result of drunkenness, debauchery or fighting.

Although this poster is dated c. 1792, the Society was well-established by then, for in 1782 the Manchester physician Dr Thomas Percival wrote asking for information on how it operated. At the time of Percival's letter, Boulton's manager John Hodges wrote that there was currently £120 in the funds, which included donations made by visitors to Soho. The rules show that fines for fraudulent applications to the fund and failure to attend committee meetings were also added to the funds. FT

355 Letter-copying machine, by James Watt & Co., 1780s
wood and metal, 44.4 × 29.2 × 13 cm.
BMAG, 1994 S 4400

356 Four-page advertisement for 'a new method of copying letters'
1780, BAH, MS 3147/20/1

357 Two pamphlets about the use of the copying machine
1780s, BAH, MS 3147/20/2 and 3

Soho generated a huge amount of correspondence, technical drawings and specifications, all of which needed copies to be kept. This meant handwriting or drawing everything twice, until Watt designed copying machines. He patented the invention in 1780. Watt carried out much research and development to produce the special ink required to make the process work.

Boulton & Watt used the machines and they were manufactured at Soho for sale to other busi-

359 Order from the Soho Insurance Society for payments to sick workers, 23 February 1805
copy press paper.
BAH, MS 3147/8/46

This is an order for payments amounting to £1 10s. 0d., to six workers at Soho Foundry who were sick that week. For each day he was off sick a worker received three-quarters of his normal daily pay. The daily wage rates recorded on this order range from 2s. 8d. to just 8d. This is a file copy of the original order produced using a copying machine. FT

361

SOHO AND THE WORLD

The Soho Manufactory was no ordinary button and buckle workshop such as could be found tucked in the back alleys of Birmingham in the middle decades of the eighteenth century. Erected on Handsworth Heath and visible from some distance, it was destined to attract attention and comment from the outset. Lord and Lady Shelburne were among the first visitors in 1766. They toured the industrial premises, purchased watch chains and trinkets at 'an amazing cheap price' and afterwards drank tea with Mrs Boulton at Soho House overlooking the Manufactory. In the 1770s Soho became an established venue on the summer season touring circuit of the aristocracy and gentry. Matthew Boulton would view every sightseer as a potential customer and he took steps to enhance the attractions of his manufactory. A show gallery and tea room were added, and his garden was progressively transformed into a park replete with walkways, a waterfall, statues and a miniature temple.

Foreign visitors followed hard on the heels of England's leisured classes, particularly once news of the steam-engine developments there spread on the continent of Europe. To Matthew Boulton's immense satisfaction, Soho was listed in guide books as a semi-obligatory stopping-place on the itinerary of the technological Grand Tour. Contingents of visitors arrived from every corner of Europe and the western world. However, by the late 1780s, when this flow of visitors was approaching its zenith, Boulton and his partners were finding that the duties of hospitality were becoming rather irksome. Sightseers, moreover, were not always what they appeared to be. Mixed in with those arriving from countries such as France, Prussia and the Scandinavian states were a number of industrial spies. In common with other Birmingham businessmen, Boulton became wary of showing the more innovative features of his manufacturing processes to all and sundry. Nevertheless, foreign visitors could nearly always obtain access to Soho on a letter of recommendation, even if Boulton preferred to escort them around his park rather than the buildings housing the steam-driven machinery. Not until the turn of the century would the Soho Manufactory be closed to visitors. Nevertheless, the adjacent Soho Mint buildings remained accessible to all comers until Boulton's death in 1809.

Peter Jones

360 J. Walker (engraver), Soho, Staffordshire, from The Copper Plate Magazine (later republished in The Itinerant), 1798–9
copper-plate engraving, 19.2 × 23.3 cm.
BMAG, 1965 V 221.80 (see p. 103)
This view shows Soho as a destination for visitors: a fashionable young couple leave in their carriage, having just seen the showroom, perhaps having made a purchase, while others stroll in the landscaped grounds. VL

361 Mr Boulton's Manufactory at Soho near Birmingham, c. 1795–1809
etching and aquatint, 23.2 × 35.9 cm.
BMAG, 2003.31.88 (see p. 223)
This is a later reworking, possibly by John Phillp, of the aquatint plate of Soho produced by Francis Eginton in 1773 (see p. 25). The original plate included John Fothergill in the caption, which would have needed to be changed following the end of the partnership. At the same time other areas such as the sky were also reworked. This later work was less skilled than Eginton's and was probably undertaken by someone with less experience in the aquatint process. The presence of a number of aristocratic coaches attest to Boulton's success in attracting well-off visitors and customers to the Manufactory. VL

362 Handwritten booklet listing visitors to Soho, c. 1790
BAH, MS 3782/13/52/1a
From the 1770s a steady flow of travellers from many parts of the world made their way to Birmingham for the sole purpose of visiting its workshops. The Soho Manufactory became a recognised stopping-place for these industrial tourists. It is likely that Matthew Boulton kept a detailed record of the individuals who received his hospitality, but the booklet on display is the only document of this type to have survived in a public collection. It contains the names of some of the more illustrious visitors between 1781 and 1790. In 1800 Boulton reluctantly called a halt to admitting visitors due to the disruption they caused and, more particularly, the instances of industrial espionage which had resulted (see p. 78). PJ

363 Letter of recommendation, 24 May 1783
BAH, MS 3782/13/52/4
Most travellers wishing to pay a visit to the Soho Manufactory obtained letters of introduction which were either sent in advance to Matthew Boulton or presented on arrival in Birmingham. Here the ironmaster John Alströmer recommends the Swedish natural philosopher Count Axel Fredrik Cronstedt. His work on mineralogy would have been well-known to Boulton and his Lunar associates. PJ

364 Letter of recommendation, 24 July 1783
BAH, MS 3782/13/52/7
The diplomat Alleyne Fitzherbert had come to prominence in 1782 during the negotiations to bring the American War of Independence to an end. Several months after this letter of recommendation was written, he would set out for Russia as Britain's Minister to the Court of Catherine the Great. PJ

365 Letter of recommendation, 1783
BAH, MS 3782/13/52/6
Matthew Boulton was fond of quoting from the Baron de Montesquieu's *Spirit of the Laws*, and he would have been hugely gratified to receive a visit from the great philosopher's grandson and heir, together with two other French army officers of noble birth. PJ

THE WORLD BEYOND SOHO

While what Elizabeth Montagu called 'the great world' beat a path to the Soho Manufactory, to gaze in awe at the machines and the mass of workers in action, Soho itself, in the person of Matthew Boulton, constantly looked outwards. He saw the world beyond Britain's shores both as a great marketplace and as a field of knowledge to be explored. Boulton's various book lists reveal that he bought sets of maps of different countries as well as travel books. He seems only to have travelled abroad three times himself, once to Holland (1779) and twice to France (1765 and 1786), each time writing detailed accounts of the experience and recording his impressions of the places and people encountered. But he corresponded with business and scientific associates in other parts of the world and, like many other eighteenth-century people, viewed with fascination the exotica that returning travellers brought back with them. He was anxious that his son should acquire a European polish, therefore Matt studied in both France and Germany. His father sometimes wrote to him in French to encourage the boy in his study of the language, in which Matt became proficient.
Shena Mason

366–367 Two 'Otaheite' or 'Resolution and Adventure' medals, 1772
silver and platina (a soft brass alloy), diameter 43.5 mm.
BMAG, 1981 N 17 and 1997 N 11 (see p. 90)
Boulton was always on the lookout for faraway trading opportunities. The Otaheite (modern Tahiti) medals were Boulton's first numismatic product. He was commissioned by the Admiralty to make two thousand 'brass' medals for Captain Cook to take with him to distribute as gifts on his second voyage to the Pacific (1772–5). Joseph Banks, who had been on Cook's voyage in 1768–71 and was to have accompanied Cook as the second expedition's botanist, added a private order for two gold and a number of silver examples for his own use. The medals bear the expedition's intended date of departure, March 1772, but the fleet actually did not sail until July. In the end, Banks did not go. DS

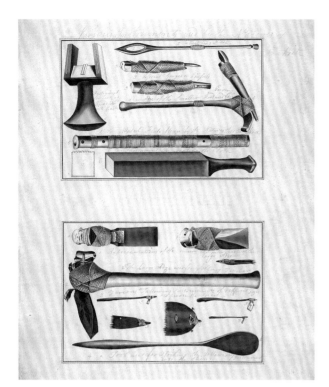

369

370

368 Letter from Matthew Boulton to John Scale, 31 March 1772
BAH, MS 3782/12/23/230
Boulton was in Stratford-upon-Avon when he wrote this letter to his manager, John Scale, at Soho, concerning the transporting of the sidereal clock (Cat. 170) from Soho to London for the Christie's sale in April. Towards the end of the letter, Boulton asks Scale to bring 'a few pairs of ye Otehite Ear rings … & also one of their queer Gods'. These may have been items that Boulton was making at Soho for Captain Cook to take as gifts or trading goods on his second voyage to the Pacific. DP

369 John Phillp, Drawings of objects from Otaheite, c. 1790–1810
ink on paper, 40 × 26.1 cm.
BMAG, 2003.0031.82
These detailed drawings of objects from Otaheite are by John Phillp. Boulton's friend Sir Joseph Banks had been to Tahiti with Captain Cook on his first circumnavigation and had brought items back with him; similar pieces were also exhibited in Birmingham entrepreneur James Bisset's museum, and Phillp's drawings may be of some of these. They demonstrate the contemporary interest in exotic cultures. DP

370 William Alexander, *The Emperor of China approaching his Tent in Tartary to receive the British Ambassador, Lord Macartney*, 1795
pen and ink and watercolour over pencil on paper, 32.8 × 50.8 cm.
BMAG, 1953 P 33
The British Embassy to China in 1792–3 was intended to establish diplomatic relations between the two countries and to foster trading links. Products made at Soho were among those carried by the embassy as examples of British manufacturing.

William Alexander was the official draughtsman to the mission, but he was not permitted to attend the meeting of Lord Macartney and the Chinese emperor K'ien Lung on 14 September 1793. This watercolour, made after his return to England, is based on eyewitness reports and existing sketches. VO

371 Letter from the Committee of the Privy Council on Trade, 19 July 1792
BAH, MS 3782/12/93/74

372 'A General List of Goods Manufactured at Birmingham and its Neighbourhood', 1792
BAH, MS 3782/12/93/75
The Directors of the East India Company decided to send examples of British goods to China with Lord Macartney's mission in the hope that this would lead to more trade with that country. This letter was sent to Matthew Boulton and Samuel Garbett, asking them to collect together 'Patterns of the principal Manufactures carried on in your Town, & Neighbourhood', and to send these to London. The list that was compiled as a result of this request covers two-and-a-half pages and provides a remarkable overview of what two leading businessmen considered to be the most important products of the town. Perhaps unsurprisingly, given Boulton's involvement, it begins with buttons, plated wares and coins. Boulton thought highly enough of his nephew, Zaccheus Walker junior, to recommend him as a suitable person to join the mission, but the suggestion does not seem to have been taken up. DS

373

LAST YEARS AND LEGACY

By 1800 Matthew Boulton was in poor health, having developed a kidney problem that caused him much pain, but despite recurrent bouts of illness he continued to take a lively interest in the business, his garden and his lifelong passion, science, or 'natural philosophy'. In January 1801 his daughter Anne had to plead with him to give up stargazing through his telescope while the weather was so cold. However, from 1802 he spent much of his time confined to the house, being obliged to receive Nelson in his bedchamber when the great man visited Soho. Although Boulton rallied from time to time – enough to organise the production of the Trafalgar Medal in 1805–6 – his health was now inexorably deteriorating. Matthew Boulton died at Soho House on 17 August 1809, shortly before his eighty-first birthday.

After his death James Watt wrote: 'had Mr B. done nothing more in the world than what he has done in improving the coinage, his fame would have deserved to be immortalized.' The introduction of the world's first mass-produced coinage was undoubtedly one of Boulton's greatest achievements. However, as Watt hinted, he did far more than that, and much of what he did has left us legacies today. He was part of an extraordinary generation at an exhilarating time, when the boundaries between 'art' and 'science' were as yet indistinct, and knowledge gained for its own sake and pleasure opened up infinite possibilities for exploration and progress.

The steam-engine developments, for which Boulton and Watt are best known, contributed significantly to the foundation of Britain's nineteenth-century wealth. Boulton was one of the pioneers of the factory age; the creation of the Soho Manufactory and the introduction of more efficient working practices transformed the world of manufacturing. Boulton's workers' insurance scheme was the model for other such schemes.

The canal network that threads its way through our countryside and cities reflects the vision of early canal investors like Matthew Boulton, who wanted to improve transport links and achieved the extraordinary feat of making Birmingham the most landlocked port in Britain. Without Boulton's triumph in securing an Assay Office for Birmingham, without his establishment of a culture of large-scale fine metalworking in the town, and without his vision and determination to compete with the established London trade, the story of silver in Birmingham would be very different. A new generation of silverware manufacturers, many of whom had been trained at the Soho Manufactory, emerged in the early nineteenth century. That they were able to produce confident, ambitious and beautiful silverwares was due to Boulton. They in turn laid down the industrial base which established Birmingham as a great silver city.

The Soho Manufactory was demolished in the 1860s but the decorative objects made there can be admired in museums and historic houses around the world. Boulton's Handsworth home, Soho House, is now a museum run by Birmingham Museums & Art Gallery. And in the remarkable collection known as the Archives of Soho, housed in Birmingham Archives & Heritage at the city's Central Library, and studied by scholars from all over the world, are to be found not only Matthew Boulton's life story but information on almost every aspect of eighteenth-century life.

Laura Cox / Chris Rice /
Martin Ellis / Shena Mason

373 Carl F. von Breda, *Portrait of Matthew Boulton*, 1792
oil on canvas, 145 × 120.5 cm.
BMAG, 1987 F 106

Boulton was asked by his children and his friend, the engineer John Rennie, to have his portrait painted. The Swedish artist Carl F. von Breda travelled to Birmingham in 1792, where he painted Boulton, James Watt and William Withering. Boulton is shown holding a medal, with the Soho Manufactory visible in the background. Boulton's daughter Anne told him in a letter that friends had said that his portrait 'does all but speak'. VL

374 Letter from Elizabeth Fowles (Manchester) to Matthew Boulton, 14 January 1806
BAH, MS 3782/6/135/57

A robbery in the counting house at the Soho Manufactory at the end of 1800 was foiled by Boulton and an armed band of workers, following a tip-off. The robbery was the subject of alarmed and excited correspondence between Boulton, his daughter and Charles Dumergue, at whose London home Anne was staying. The robbers were caught and tried at Stafford Assizes, where drawings of the Manufactory showing the route the thieves had taken were produced in evidence (see p. 2 and p. 102). Some of them, including one William Fowles, known as 'The Little Devil', were transported to New South Wales. Elizabeth Fowles was the 'Little Devil's' wife, left behind in England. Here she writes that her husband has died in Australia and she is in great need, and asks for Boulton's help. Boulton's personal clerk, William Cheshire, was instructed to ask Peter Ewart, a former Soho engineer who had become a master millwright in Manchester, to check the woman's story, writing 'Mr Boulton sincerely wishes the Widow Fowles had a part of the 150 Guineas her late husband stole from Soho, but as it is possible her case may merit some compassion … he would wish to extend some little relief to so unhappy a Being if her distresses are not fictitious, & her habits of life are tolerably orderly.' Ewart was asked to give Mrs Fowles a few guineas on Boulton's behalf if he was satisfied and to apply to Boulton for reimbursement. SM

375–377 Letters from Mrs Sarah Nicholson to Matthew Boulton, 30 June, 5 July and 17 July 1802
BAH, MS 3782/12/47/194, 205 and 223

'Sally' Nicholson ran the Piccadilly household for Boulton's friends, the Dumergues; at one time she may have been Miss Sophia Dumergue's nurse, as Sophia refers to her in her will as 'My more than mother'. Sally's affectionate letters to Matthew Boulton include these three explaining that, with Mrs Fitzherbert's agreement, she has arranged for him to borrow the Prince of Wales's wheelchair, which is to be sent to Soho. The third letter notes that 'the Prince is delighted you have it, & beg'd we would assure you from him if his House could produce any thing that was agreeable to you, it was at your command.' SM

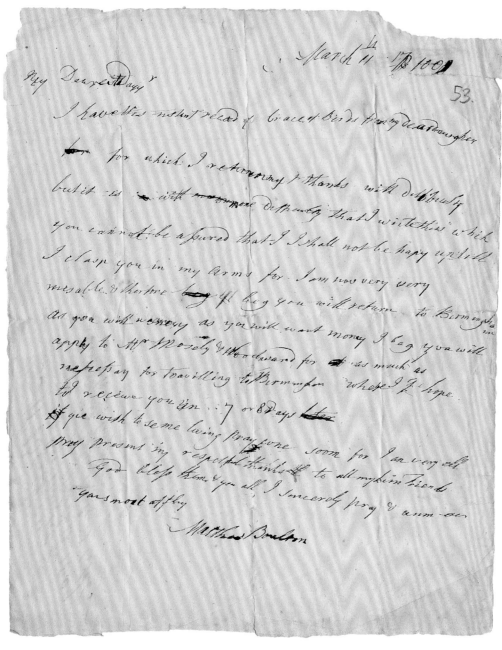

382

378 Draft letter from William Cheshire to a Mr Pigasse (Dublin), 21 April 1803
BAH, MS 3782/6/132/19
William Cheshire, Boulton's confidential clerk and bookkeeper, writes that Matthew Boulton's health is now so much worse that his physician has instructed him to give up business to his son. Boulton was seriously ill in 1802–3, probably with the kidney complaint that dogged him in his final years. SM

379 Matthew Boulton's diary for 1804
BAH, MS 3782/12/107/32
The notes in the front of the diary include one to write to Mr Rosini, Master of the Concert Rooms at Bath, about juice of nettles curing his nephritic complaints. Matthew Boulton was constantly in search of effective remedies for kidney problems. SM

380 Report by Dr John Carmichael on Matthew Boulton's health and treatment, 17 March 1807
BAH, MS 3782/12/52/28

381 Letter from Henry Cline, London, to Dr John Carmichael, 23 March 1807
BAH, MS 3782/12/52/29
Dr Carmichael of Birmingham, who at one time was engaged to Matthew Boulton's daughter, Anne, was one of several physicians who attended Boulton in his later years. In his report to the London physician Henry Cline he describes his patient's painful bladder condition, for which he is being given opium. Cline broadly approves Dr Carmichael's treatment regime and offers some additional suggestions. SM

382 Letter from Matthew Boulton to Miss Anne Boulton (Bath), 11 March 1809
BAH, MS 3782/14/76/53
Anne Boulton was staying in Bath with her brother when she received this letter from her father, endorsed on the back by Matthew Robinson Boulton 'Supposed to be the last letter written by him'. It includes the phrase 'If you wish to se [sic] me living pray come soon for I am very ill'. SM

Two letters to James Watt describing Matthew Boulton's funeral:

383 from James Watt junior, 25 August 1809
BAH, MS 3219/4/33/36

384 from John Furnell Tuffen, 29 August 1809
BAH, MS 3219/4/49/88

James Watt was in Glasgow when his old friend and partner died on 17 August 1809, so he was represented at the funeral by his son, James junior. In these two letters James junior and Tuffen, a banker and an old friend of Watt's, give him a full account of the proceedings. Watt junior writes: 'The remains of our excellent friend were yesterday committed to the Grave. The interval of weather was favourable, the procession well conducted, and the ceremony awful & impressive.' The lengthy procession was made up of ten coaches of mourning and numerous relations, friends, colleagues and Soho workers. Tuffen describes the scene:

> Early on Thursday morning the entrance to Soho & the road from thence to Handsworth Church was lined with spectators on foot, on horseback & in carriages, to the number it is said of at least 10,000 persons. Although the church was crowded in every part & multitudes remained without who could not gain admittance, the utmost stillness & solemnity prevailed, & the effect of the music was visible in almost every eye. In short nothing could be more appropriate or better conducted, & there never was perhaps a public funeral attended by so many real & respectful mourners. Thus my dear Sir has the Grave closed on one of our oldest & dearest friends, whose like, take him for all in all, we shall not see again. DS

385 John Phillp, Sheet of designs for Matthew Boulton's coffin plate, 1809
pencil on paper, 25.3 × 20.3 cm.
BMAG, 2003.0031.100

This sheet records ten variously elaborate designs for Boulton's coffin plate. It is not known which, if any, of these was finally used. VO

385

386

388

386 John Phillp, *View of Handsworth Church taken on the Spot*, 1796

pen and ink over pencil on paper, 32.3 × 46.6 cm.

BMAG, 2003.0031.38

Phillp's drawing depicts the parish church of St Mary's, Handsworth, where in 1809 Boulton was buried in a family vault where his daughter, son and daughter-in-law were also subsequently interred. The church contains monuments to William Murdock and James Watt, both of whom are also buried there, as well as to Boulton.   VO

387–388 Two examples of the memorial medal struck for distribution at Matthew Boulton's funeral, 1809

silver and copper, diameter 40.5 mm.

Silver: private collection; copper: BMAG, 1885 N 1541.91

This simple medal was produced between Matthew Boulton's death on 17 August and his funeral on 24 August. The Soho Mint archives record that 530 'tokens' were struck 'for distribution amongst Attendants and Workmen', but the existence of a silver version of the medal was a recent discovery. According to the owner's family tradition, the silver medals were presented to the men who carried the coffin at the funeral. In a letter to his father about the funeral, James Watt junior describes them as 'ten of the oldest men (who had been for 30 to 50 years in his [i.e. Boulton's] service), with ten others to relieve.'

Watt also describes the distribution of the medals to the Soho workforce, who attended the funeral en masse, 'in number about 430, and 60 to 70 women':

> It is impossible to conceive of any thing more proper and respectful than the conduct of the workmen has been throughout. After the ceremony, they retired to the different public houses where refreshment had been provided for them of cold meat, and when each received from Mr Boulton a jetton with his age & death on one side, and <u>in memory of his obsequies</u> on the other, after receiving these and drinking the memory of their departed benefactor standing & in silence, they all repaired to their respective homes, and not a Soho man was to be seen upon the road for the remainder of the day.   DS

389 Matthew Boulton's Will, dated 3 June 1806, proved on 15 September 1809

BAH, MS 3782/13/37/28

In his Will, Matthew Boulton leaves the bulk of his estate, including farms and other land in Warwickshire and Staffordshire, to his son, Matthew Robinson Boulton. He nominates Matt to be his successor in the business at Soho and appoints him sole executor of the Will. A codicil provides for his daughter, Anne, to receive all her mother's jewels, together with £1,500 to purchase furniture, an annuity of £200 for life and the interest on £5,000, in addition to a settlement made for her which included capital and rent-producing farms, all intended to make Anne financially independent of her brother. Matt is to inherit a diamond ring given to his father by Emperor Alexander I of Russia, while other bequests to various friends and good causes include sums ranging from £20–£500 to nephews and nieces, friends including members of the Fothergill family, servants and some of the longserving Soho workers. The General Hospital and the Birmingham Dispensary both receive £100.   SM

391

392

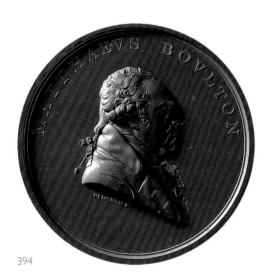

394

390 Obituary of Matthew Boulton from *The European Magazine and London Review* for September 1809
BAH, MS 3782/13/37/150

After Boulton's death this obituary was published with his portrait. The obituary celebrated Boulton's status as an entrepreneur as well as his life achievements:

> We feel a great degree of national pride in adding, to the long list of British worthies whose portraits have adorned our volumes, the name of one who, by his energetic mind, scientific acquirements, and liberal and benevolent sentiments, has done so much honour to… the commercial importance and prosperity of the country. LC

391 Memorial medal for Matthew Boulton, c. 1809–10 or later
copper, diameter 46 mm.
BMAG, 1966 N 693

This medal features a bust of Boulton based on a wax portrait produced in 1803 by Peter Rouw, a sculptor and modeller of gems and cameos. Below the bust is a small image of the Soho Mint. The long inscription on the reverse is copied from Boulton's funeral monument in St Mary's church, Handsworth. It was not made at Soho, indicating the respect in which Boulton was held by other medallists of the time. DS

392 Memorial medal for Matthew Boulton, struck at the Soho Mint, c. 1809–10 or later
copper, diameter 48 mm.
BMAG, 1885 N 1541.90

The obverse die used to strike this medal was engraved by C. H. Küchler. It was apparently never completed as it seems clear that there was supposed to be an inscription on the ribbon below Boulton's bust. It is possible that it was left unfinished when Küchler died in July 1810, less than a year after Boulton. DS

393–394 Two examples of a medal commemorating Matthew Boulton, designed by G. F. Pidgeon and struck at the Soho Mint, 1819, with protective metal shell (394a)
bronzed copper, diameter 64 mm.
BMAG, 1885 N 1536.14 and 2003.35.1–2

This large medal was commissioned by Matthew Robinson Boulton to commemorate the death of his father. Pidgeon was asked to engrave the dies in January 1812, but seems to have worked very slowly. The obverse die, with Matthew Boulton's

396

portrait, was ready by February 1815, but the reverse die was not finished until 1817. Matthew Robinson Boulton seems then to have decided to wait for the tenth anniversary of his father's death before striking the medals. The edge has an inscription dedicating the medal to Boulton's friends. The medal with its Soho Mint metal shell originally belonged to James Watt or his son James Watt junior. ST

395 'List of Persons to whom it is intended to present a Medal of the late Matthew Boulton Esq', compiled by Matthew Robinson Boulton, 30 September 1820
BAH, MS 3782/13/37/117

This list, compiled by Matthew Robinson Boulton, includes friends and colleagues of his father to whom a specimen of the 1819 memorial medal was to be presented. It indicates whether the individual received a bronze or gilt medal and when these were distributed. The range of people represented is amazing, from the illiterate smith John Busch, who had worked with Boulton in developing the Soho Mint, to the Earl of Dartmouth. ST

396 Sir Thomas Lawrence and Sir Martin Archer Shee, *Portrait of Matthew Robinson Boulton*, c. 1830
oil on canvas, 230 × 140 cm.
BMAG, 1987 F 331.

Boulton and Watt admitted their sons, Matthew Robinson Boulton and James Watt junior, as partners in the business in 1794. They were not great entrepreneurs but were able managers. They got the accounts in order and introduced a work/time analysis that streamlined production at the Soho works.

This dramatic portrait was unfinished when Sir Thomas Lawrence died in 1830 and was completed by Sir Martin Shee who succeeded him as president of the Royal Academy. BF

397

397 Photograph of the Soho Manufactory
c. 1860
Benjamin Stone Collection.
BAH, Local Studies Collection, Box 14, Print no. 33

The Soho Manufactory survived just long enough into the nineteenth century to be photographed. This photograph shows the principal building shortly before its demolition. Many of the details familiar from earlier eighteenth-century drawings can still be seen. A century old at the time the photograph was taken, the 'principal building' is still an impressive structure, if somewhat neglected and no longer set in its polite parkland landscape. SM

LATER GENERATIONS

At the time of Matthew Boulton's death in 1809 his daughter and son were aged forty-one and thirty-nine respectively. Both were unmarried. Matthew Robinson Boulton was running the Soho business with forty-year-old James Watt junior, who was also single. The two Boultons continued to live at Soho House. Anne Boulton had been engaged briefly to her doctor, John Carmichael, but that had fallen through.

In 1810 James Watt junior made it known to his parents and to Matt Boulton that he and Anne Boulton wished to marry. The Watts seem to have been in favour of the marriage but Matt was not, though in letters he is not explicit about his reasons. Without her brother's approval Anne does not seem to have felt able to go ahead with the marriage, though in a letter to her stepson Annie Watt suggests that Anne's regard for him was deep and longstanding. Anne remained single, and so did James Watt junior.

In 1815 Matt Boulton bought a country estate at Great Tew in Oxfordshire, initially for recreational use. Later it would become the family home. In 1817, at the age of forty-seven, he married twenty-two-year-old Mary Anne Wilkinson, niece of 'Iron-mad' John Wilkinson. Anne Boulton and the new Mrs Boulton were not on good terms, and shortly after the birth of her brother's first child Anne moved to Thornhill House, not far away. Her father had given her the house at the time of her engagement to Dr Carmichael, but James Watt junior had been renting it from her for some years; he moved to Aston Hall to allow Miss Boulton to take over Thornhill. She died there in November 1829 at the age of sixty-one, three days after suffering a 'paralytic seizure'.

Matthew Robinson Boulton and his wife had three daughters and four sons; Mrs Boulton died following the birth of the last child in 1829. Of Matthew Boulton's seven grandchildren, four had no children. Two of the girls, Katharine and Mary Anne, married and had families of which descendants are alive today. Their eldest brother, Matthew Piers Watt Boulton, had six children by two marriages, none of whom have descendants now living. Matthew Robinson Boulton died in 1842. By then most of the Soho businesses had been closed down or sold.

James Watt senior died in 1819. He was survived by his second wife, Annie, and his son James junior, the son of his first marriage. The children of his second marriage, Jessie and Gregory Watt, had both died of tuberculosis in 1794 and 1804 respectively. James Watt junior died in 1848.

Shena Mason

398 Catalogue of the Soho Mint auction sale, 1850
BAH, MS 3147/10/74

Over the five days from 29 April to 3 May 1850 the entire contents of the Soho Mint were sold in 707 lots. As the title page of this sale catalogue describes, the sale covered all of the machinery and plant of the Mint, including the coining presses, rolling mills and steam engine. Also sold was the 'extremely valuable collection of dies'. The obverse die for the Prince of Wales's marriage medal displayed in this exhibition (no. 269) appears in lot 215. Even the Mint's own collection of examples of all the coins it had struck since Matthew Boulton built the first Mint back in the late 1780s went under the hammer. It was a sad end to a great institution. DS

399 Door plate, c. 1848–95
Bronze, 27 × 32 cm.
BMAG, 1965 S 01946

After the death of James Watt junior in 1848, the engineering business at Soho Foundry continued as James Watt & Co., but the firm still used the strapline 'Late Boulton & Watt' to indicate its illustrious history. This plate was fixed at the Soho Foundry. The strapline was also used on maker's plates on steam engines. JK

400 A member of the Lines family (attributed), *Soho House, Handsworth*, c. 1850
watercolour, 27.4 × 39.7 cm.
BMAG, 1988 V 1538

401 Paul Braddon, *Soho House, Handsworth*, late nineteenth century
watercolour, 32.0 × 40.4 cm.
BMAG, 1959 V 196

Soho House is the only surviving feature of Matthew Boulton's Soho complex. In the mid-nineteenth century a member of the Lines family produced a detailed study of the house from the south-east, showing the main elevation, rounded entrance portico and part of the service wing adjoining the south wall. In the late nineteenth century Paul Braddon produced a similar view, in a more impressionistic style. JC

400

401

402

403

402–403 Two photographs of girls outside Soho House when it was a girls' school, 1893 and 1894

Birmingham Central Library, Stone Collection

After the Boultons left, Soho House was let to various tenants and then became a vicarage when the new St Michael's Church was built opposite. By 1893 the house was being used as a school, Birmingham Central High School for Girls. The photographs show the teachers, Mrs and Miss Taylor, and some of their pupils. LC

404 The wedding reception of Mr and Mrs Wilfred Spokes at Soho House, June 1922.

This photograph shows a wedding group seated in front of the portico at Soho House. By this time, the house was called Soho Hall Hotel and the building had undergone a series of alterations. It remained a hotel until the 1950s, when it became a hall of residence for GEC (General Electric Company) apprentices. Later, West Midlands Constabulary occupied it. In the 1990s Birmingham Museums & Art Gallery embarked on its restoration and development as a museum. It opened to the public in 1995. LC

404

EPILOGUE

In 1794 Matthew Boulton had finally bought the freehold of his whole estate, after setting out the arguments for and against such a move in a document headed 'Considerations upon the Propriety of Buying Soho'. He wanted to protect his investment for his descendants, and also to safeguard his beloved garden and views 'and all that contributes to render Soho a chearfull pleasant spot'. If he bought the estate, he reasoned:

I shall enjoy the Fruits of my own Labour;
Repose myself under my owne Vine and my own Fig tree;
or, I shall walk firmer upon my own Ground, and all my Paths will be peace.
I shall look back with more satisfaction on the days that are past,
and forward with the hopes of my Descendants being Opulent
and respectable Manufacturers, at Soho,
to the Third and Fourth Generation.

Notes

Abbreviations used in the Notes

BAH	Birmingham Archives & Heritage
BAO	Birmingham Assay Office
BRL	Birmingham Reference Library
B&F	Boulton & Fothergill
B&W	Boulton & Watt
JF	John Fothergill
JW	James Watt
JWJ	James Watt Junior
JWd	Josiah Wedgwood
MB	Matthew Boulton
MRB	Matthew Robinson Boulton
Mrs AB	Mrs Ann Boulton (wife)
Miss AB	Miss Anne Boulton (daughter)

Note: all primary source documents beginning with 'MS 3782', 'MS 3147', and 'MS 3219' are to be found in the Archives of Soho, Birmingham Archives & Heritage at Birmingham Reference Library.

Introduction

1 Samuel Smiles, *Lives of Boulton & Watt* (1865, republished Stroud, 2007).
2 H. W. Dickinson, *Matthew Boulton* (Cambridge 1936), pp. ix–xii.
3 For a detailed account of Boulton genealogy, see S. Mason: *The Hardware Man's Daughter* (Chichester, 2005), Appendix, pp. 197–212.
4 Smiles, *op. cit.*, p. 151. Smiles acknowledged the help of Boulton's grandson, Matthew Piers Watt Boulton, in researching his book, so some of the detail in it was possibly based on family tradition.
5 Parish Register, St Martin's Church (Birmingham Archives & Heritage).
6 Parish Register, St Philip's Church (Birmingham Archives & Heritage).
7 Eric Robinson, 'Matthew Boulton's Birthplace and his Home at Snow Hill: A Problem in Detection', in *Transactions of Birmingham Archaeological Society*, vol. 75, 1957.
8 Eric Robinson, 'Boulton and Fothergill, 1762–1782, and the Birmingham Export of Hardware', in *University of Birmingham Historical Journal*, vol. 6, no. 1 (1957), p. 63; see also Maxine Berg, *The Age of Manufactures, 1700–1820: Industry, Innovation and Work in Britain*, London and New York, 1994, pp. 265–6.
9 Smiles, *op. cit.*, 1865, p. 164; Jennifer Tann, *The Selected Papers of Boulton & Watt, vol. I, The Engine Partnership 1775–1825*, London 1981.
10 Smiles, *op. cit.*, p. 151.
11 Richard Schofield, *The Lunar Society of Birmingham* (1963); Eric Robinson, 'The Lunar Society, its Membership and Organisation', in *Transactions of the Newcomen Society*, 35 (1963), pp. 153–77; Jenny Uglow, *The Lunar Men* (London and New York 2002).
12 MS 3782/12/81/26, John Whitehurst (London)–MB, 31 Dec. 1785.
13 J. E. Cule, 'The Financial History of Matthew Boulton 1759–1800', unpubl. M.Comm. thesis, University of Birmingham (1935).
14 MS 3782/1/37, [Letterbook B], MB–John Lewis Baumgartner, 13 Jul. 1765.
15 Schofield, 1963, p. 83.
16 MS 3782/12/1/68 [Letterbook C], MB–J.H. Ebbinghaus, 2 Mar. 1768.
17 Robinson, 'Boulton and Fothergill', 1957, pp. 67–8 (note 8).
18 Ibid; see also Robinson, 'The Lunar Society', 1963 (note 11), p. 41.
19 Cule, 1935, pp. 4–6.
20 Ibid, 1935.
21 Cut-steel: goods such as buckles, buttons and jewellery, thickly ornamented with tiny steel studs, individually faceted in the manner of gemstones and highly polished to make them sparkle.
22 Nicholas Goodison, *Matthew Boulton: Ormolu* (London 2002), pp. 49–50.
23 Schofield, 1963, pp. 60–1.
24 H. B. Carter, *Sir Joseph Banks* (1988), p. 53.
25 H. W. Dickinson and R. Jenkins, *James Watt and the Steam Engine* (London 1927/1981), p. 244.
26 Laurence Ince, 'The Boulton and Watt Engine and the British Iron Industry', *Wilkinson Studies*, vol. 11, 1992, p. 81.
27 Dickinson and Jenkins, *op. cit.*, p. 244.
28 E. Robinson and A. E. Musson, *James Watt and the Steam Revolution: A Documentary History* (London 1969), p. 10; Tann, *op.cit.*, 1981, p. 43.
29 James Boswell, *Life of Dr Johnson* (1791; Oxford 1998).

Chapter 1: Matthew Boulton and the Lunar Society

1 MS 3782/13/37, James Keir–MRB, Memoir of Matthew Boulton, 3 Dec. 1809. See also R. E. Schofield, *The Lunar Society of Birmingham* (Oxford 1963) and Jenny Uglow, *The Lunar Men* (London 2002).
2 MS 3782/12/108/1, Notebook 1751–9.
3 MS 3782/12/108/2, Notebook 1757.
4 See note 2.
5 MS 3782/1/40, Letterbook Jan. 1757–May 1765, MB–Benjamin Huntsman, 19 Jan. 1757.
6 MS 3782/13/53/14, Erasmus Darwin–MB, [n.d., c. 1762]; see also Desmond King-Hele (ed.), *The Collected Letters of Erasmus Darwin* (Cambridge 2007), p. 48.
7 MS 3782/13/53/13, Erasmus Darwin–MB (undated, but from other evidence 1760s).
8 MS 3782/6/189/12, John Whitehurst–MB, 26 Jan. 1758.
9 MS 3782/13/53/1, John Michell (Cambridge)–MB, 5 Jul. 1758.

10 MS 3782/13/53/22, Erasmus Darwin–MB, 1 Jul. 1763; see also King-Hele, *op. cit.*, p. 49.
11 Schofield, R.: *op. cit.* (1963), p. 37.
12 Erasmus Darwin–JW, 18 Aug. 1767; King-Hele, *op. cit.*, p. 81.
13 Erasmus Darwin–Josiah Wedgwood, 14 Jun. 1768, King-Hele, *op.cit.*, p. 88.
14 MS 3782/12/65/3, James Keir–MB, 2 Oct. 1772.
15 MS 3782/1/10, B&F Letterbook 1774–77, MB–John Wyatt, 24 Feb. 1776.
16 MS 3782/13/53/87, Erasmus Darwin–MB, 5 Apr. 1778; King-Hele, *op. cit.*, p. 150.
17 MS3782/1/27/20, Alexander Aubert–MB, 25 Sep. 1778.
18 See note 1.
19 MS 3782/12/24/50, Joseph Priestley–MB, 6 Nov. 1775; see also Robert E. Schofield (ed.), *Scientific Autobiography of Joseph Priestley*, (Cambridge, Mass. 1966), p. 153.
20 MS 3782/12/5/106, MB–Logan Henderson, 6 Sep. 1781; see also, Samuel Smiles, *Lives of Boulton and Watt* (London 1865), p. 373.
21 MS 3219/4/123/10, JW–James Lind, 26 Dec. 1784. 'The history of Mr Boulton's explosive balloon'.
22 Desmond King-Hele, *Erasmus Darwin: A Life of Unequalled Achievement* (London 1999), p. 246.
23 Birmingham University Special Collections, 5/30, William Withering (Lisbon)–MB, 1 Feb. 1793: [ps] '7th Feb. we have just heard of the Execution of poor Louis the 16th. Surely M. Egalite will not escape for this useless piece of cruelty.'
24 MS 3782/12/36/156, MB–Charles Dumergue, 18 Aug. 1791.
25 William Withering–Dr Edward Ash, 12 May 1798, William Withering Papers, Special Collections, Birmingham University Library.
26 Christie's catalogue, *Fine Architectural Drawings and Watercolours*, 16 Dec. 1986, 32, 33.

Chapter 2: Soho House and the Boultons

1 Some early household bills are addressed to Boulton at Snow Hill. See for example MS 3782/6/189/ff. and MS 3782/6/190/ff.
2 MS 3782/12/112.
3 MS 3782/12/60/265, Memoranda by Matthew Boulton, respecting his partnership with John Fothergill. 3 Sep. 1765–20 Mar. 1766. Headed 'Case between B&F'.
4 MS 3782/12/5/180, MB–William Matthews, 20 Jun. 1782. Boulton & Fothergill was formally dissolved on 22 June 1782, twenty years to the day after it began, and three days after Fothergill's sudden death on 19 June 1782.
5 E. G. P. Fitzmaurice, *Life of William, Earle of Shelburne* (London 1912), p. 400.
6 Richard Morriss, Ron Shoesmith et al: *Soho House*, City of Hereford Archaeology Unit report, Hereford Archaeology Series no. 90, Nov. 1990.
7 MS 3782/6/189 (1752)–204 (1825), Soho House household bills over a long period.
8 MS 3219/4/66/1, MB–JW, 7 Feb. 1769.
9 MS 3782/14/76/49, MB–Miss Anne Boulton, 19 Jun. 1801.
10 MS 3782/12/16/1/80, MB–Mrs AB, 3 Oct. 1780.
11 MS 3782/13/53, Erasmus Darwin–MB, 19 Jun. 1769; also MS. 3782/12/23/199, John Hunter–William Small, 5 Apr. 1771. For further discussion on this matter see Shena Mason, *The Hardware Man's Daughter: Matthew Boulton and his 'Dear Girl'* (Chichester 2005), pp. 19–23.
12 MS 3782/14/76/28, MB–Miss AB, 22 Sep. 1792.
13 MS 3782/13/36/1, MB–MRB, 24 Dec. 1786.
14 MS 3782/12/108/17, Notebook 'Holland', 1779.

15 MS 3782/12/60/265, 'Case between B&F'.
16 For information on the Soho House gardens, see for example MS 3782/12/108/7, Notebook 1772; MS 3782/12/108/70, Notebook, 1795; MS 3782/6/192/100, Brunton & Forbes, 9 Apr. 1783; MS 3782/6/193/43, letter James Hunter–MB 18 Apr. 1788 plus bill Brunton & Forbes–MB, same date; Phillada Ballard: 'Soho House Gardens', Research Report for Birmingham Museum & Art Gallery (1992); drawings of John Phillp, Birmingham Museums & Art Gallery.
17 MS 3782/12/38/129, William Herschel–MB, Jul. 1793.
18 MS3782/12/68/106, MB–Charlotte Matthews, 17 Dec. 1794.
19 MS3782/12/85/131a, MB–James Wyatt, 7 Jul. 1796.
20 MS 3782/12/85/134, MB–James Wyatt, 18 Nov. 1796.
21 MS 3782/12/108/72, Notebook for 1796.
22 MS 3219/7/49/16, Gregory Watt–JW, 31 Aug. 1798.
23 MS 3782/12/47/194, Mrs Nicholson–MB, 30 Jun. 1802.
24 MS 3219/6/61/33, JWJ–John Furnell Tuffen, n.d.
25 MS 3219/4/49/88, John Furnell Tuffen–JW, 29 Aug. 1809.
26 Mary Anne Wilkinson (1795–1829) was the daughter of William Wilkinson, who had run the Wilkinsons' Bersham iron works in north Wales at the time James Watt junior was a trainee there in 1784. Her uncle was John 'Iron Mad' Wilkinson, and her aunt was Joseph Priestley's wife, Mary. Before his death in 1808 William Wilkinson appointed James Watt junior to be the guardian of Mary Anne and her sister, Elizabeth Stockdale Wilkinson. After Mary Anne's death Elizabeth Stockdale Wilkinson ('Aunt Bessie') helped to care for Matt Boulton's children.
27 The later history of the estate is taken from unpublished research by Val Loggie and Phillada Ballard.
28 MS 3782/6/200/3, James Newton–MB, 7 Aug. 1797–23 Oct. 1798.

Chapter 3: Picturing Soho

Author's acknowledgments:
The material on the estate draws on work undertaken by Dr Phillada Ballard in the form of a report for Birmingham Museum & Art Gallery, and other material. As these are not publicly accessible I have mainly cited references from the primary sources but it was Dr Ballard's research that uncovered many of them. See, however, Ballard, Loggie and Mason (forthcoming) in the Bibliography.

Thanks also to the descendants of John Phillp, particularly Joan Foden, and to Dr Richard Clay, George Demidowicz, Antony Griffiths, Professor Peter Jones, Shena Mason, Nicholas Molyneux, Victoria Osborne, Dr David Symons and Sue Tungate.

1 This chapter is based on a section of work for a forthcoming AHRC-funded collaborative Ph.D. co-supervised by the University of Birmingham and Birmingham Museums and Art Gallery.
2 A brand is defined by Nancy Koehn as 'a name, symbol or logo intended to distinguish a particular seller's offerings from those of competitors', Nancy F. Koehn, *Brand New: How Entrepreneurs Earned Consumers' Trust from Wedgwood to Dell* (Harvard Business School Press, 2001), p. 5.
3 Koehn, *op. cit.*, p. 6.
4 Eric Robinson, 'Eighteenth-Century Commerce and Fashion: Matthew Boulton's Marketing Techniques', *The Economic History Review* New Series, vol. 16, no. 1 (1963) pp. 39–60; Nicholas Goodison, *Matthew Boulton: Ormolu* (London 2002), pp. 157–80; Peter Jones in this volume, Chapter 9, pp. 71–79.
5 George Demidowicz, 'A Short History of the Soho Manufactory and Mint', unpublished MS, p. 5.
6 Demidowicz, *op. cit.*, p. 6; Jennifer Tann, *The Development of the Factory*

(London 1970), pp. 151–7.
7 MS 3782/12/60/32, John Fothergill–MB, 14 Dec. 1765, quoted in Demidowicz p. 5.
8 MS 3782/12/1/23, MB–J. H. Ebbinghaus, 28 Oct. 1767.
9 Demidowicz, *op. cit.*, p. 5; J. M. Robinson, *The Wyatts: An Architectural Dynasty* (Oxford 1979), p. 19.
10 Edgar Jones, *Industrial Architecture in Britain 1750–1939* (New York, 1985), pp. 14–23.
11 MS 3782/12/1/42, MB–J. H. Ebbinghaus. 2 Mar. 1768.
12 Demidowicz, *op. cit.*, p. 5.
13 Jones, *op. cit.*, p. 35; Demidowicz, *op. cit.*, p. 6.
14 Goodison, *op. cit.*, pp. 85–9; Robinson, *op. cit.*, p. 56.
15 Janet Wolfe, *The Social Production of Art* (New York, 1981). p. 51.
16 Koehn, *op. cit.*, pp. 5–6.
17 There is only one known copy in a scrapbook compiled by Samuel Timmins, Birmingham Archives and Heritage 82934, vol. 1, p. 59.
18 Demidowicz, *op. cit.*, p. 7.
19 MS 3782/12/23/128, William Jupp–MB, 25 May 1769, enclosing bill. MS 3782/1/18/7, Thomas Feilde–MB, 5 Jul. 1769; Stebbing Shaw, *The History and Antiquities of the Country of Stafford* (London 1798–1801).
20 Eric Robinson, 'Boulton and Fothergill 1762–1782, and the Birmingham Export of Hardware', *University of Birmingham Historical Journal*, vol. VII, no. 1 (1959–60), pp. 61–79; Eric Hopkins, 'Boulton before Watt: The Earlier Career Reconsidered', *Midland History*, vol. IX (1984), pp. 43–58; MS 3782/12/60/3,5,6,9, John Fothergill–MB, 1762, regarding Mr [probably Benjamin] Green engraving chapes; MS 3782/12/108/14, MB Notebook 1779 Holland, p. 26 regarding leaving books of plated goods.
21 Several published views and the sketches of John Phillp are taken at such an angle that Rolling Mill Row is visible but this shows only two sides. The later views published in the Penny Magazine, 5 Sep. 1835 and William West's *Picturesque Views […] in Staffordshire and Shropshire* (1830) both take a view of the Manufactory from the back. Another view of the back of the factory was part of a set of drawings produced following a robbery in 1801 and used to show the escape routes of the burglars, BCL MS 3069.
22 See for example, H. W. Dickinson, frontispiece. It is likely this date was arrived at because of the inclusion of Fothergill in the title and the fact that when he was writing aquatint had been considered rare in Britain until 1782.
23 MS 3782/12/72/5, John Scale–MB, 1 Feb. 1773; MS 3782/12/72/7, John Scale–MB, 7 Feb. 1773.
24 Antony Griffiths, 'Notes on Early Aquatint in England and France', *Print Quarterly* vol. IV, 1987, 3, pp. 255–270. I am grateful to Antony Griffiths for discussions on this aquatint. British Museum 1978,1216.3.1 is an example with a French caption.
25 Peter Jones in this volume, Chapter 9, pp. 71–79.
26 BMAG 2003.31.88 is an example from the later reworked plate.
27 *The New Birmingham Directory and Gentleman's Compleat Memorandum Book*, 1774, 1775 and 1777; Charles Pye's *Birmingham Directory*, 1788.
28 *The New Birmingham Directory*, 1774, preface.
29 Eric Roll, *An Early Experiment in Industrial Organisation: Being a History of the Firm of Boulton and Watt, 1775–1805* (New York, 1930), pp. 225–9; Shena Mason in www.revolutionaryplayers.org.uk. MS 3782/2/14, p. 48, John Hodges (Soho)–Dr Thomas Percival (Manchester), 16 Nov. 1782.
30 Science & Industry Collection of Birmingham Museums & Art Gallery 1951 S88.36; Roll, 1930, p. 229; the fact that the plate is intaglio while the text would have been relief meant that the poster version had to go through the printing press twice.
31 Richard Clay of the University of Birmingham argues that some of the workforce at Soho would have been able to interpret at least some of the iconography as they were working with elements of it on coins, tokens, medals and silverware. He also argues that the iconography was part of the common symbolic 'coinage' of tokens used widely in token-based commerce. Research seminar given at the University of Birmingham, 6 June 2008. Some of the iconography used is very straightforward, for instance the man with his arm in a sling, and is likely to have been widely understood.
32 P. Ballard, unpublished MS; MS 3782/12/37/54, William Withering–MB, 14 Mar. 1792; MS 3782/12/39/92, MB–Heneage Legge, 7 Apr. 1794; MS 3782/12/39/250, H. Whateley–MB, 13 Sep. 1794.
33 MS 3782/12/108/70, MB notebook 1795; MS 3782/12/111/150, 'Considerations upon the Propriety of buying Soho'.
34 Ballard, *op. cit.*
35 MS 3782/12/108/70, MB notebook 1795, p. 29.
36 Ballard, *op. cit.*
37 Ballard, *op. cit.*
38 This 'canal' formed part of the water system of the Manufactory, moving water between the pools rather than acting than as a navigable waterway; see George Demidowicz in this volume, Chapter 12, pp. 99–115.
39 Ballard, *op. cit.*
40 Ballard, *op. cit.*
41 MS 3782/12/108/75, MB notebook 1797.
42 MS 3782/12/111/150, 'Considerations upon the Propriety of buying Soho'.
43 *Bisset's Poetic Survey Round Birmingham* (1800), published with his *Magnificent Directory*; see also anonymous poem transcribed by John Phillp, BMAG 2003.31.81.
44 MS 3782/12/38/10, G. C. Fox–MB, 25 Jan. 1793; MS 3782/12/38/20, MB–G. C. Fox, 16 Feb. 1793. At the time there seem to have been rumours that Phillp was actually Boulton's son and this has been the tradition carried through his family but no conclusive evidence for this has been found. For further discussion see Richard Doty, *The Soho Mint & the Industrialization of Money* (Washington 1998), p. 49; Shena Mason, *The Hardware Man's Daughter: Matthew Boulton and his 'Dear Girl'* (Chichester 2005), p. 223 n.28; Brian Gould, 'John Phillp: Birmingham Artist (1778–1815),' unpublished typescript. No record of Phillp's birth has been found.
45 MS 3782/6/195/7, William Hollins–MB, 30 Jun. 1795; Michael Fisher, 'William Hollins' in ODNB accessed 24 June 2007; BMAG 2003.31.77 endorsed 'first under IB IP 1796'.
46 BMAG 2003.31.
47 MS 3782/12/108/70, MB notebook 1795, p. 36.
48 MS 3782/12/41/203, R. Riddle [*sic*]–MB, 15 Jun.1796. I am grateful to Professor Peter Jones for this reference. The current location of Riddell's views is not known.
49 MS3782/12/107/24, The Tablet, 1796; Victoria Osborne, 'Cox and Birmingham', in Scott Wilcox (ed.) *Sun, Wind, and Rain: The Art of David Cox* (New Haven and London 2008).
50 MS 3782/7/10/549, Joseph Barber's bill, 1792; BMAG 2003.31.77 endorsed 'first under IB IP 1796'.
51 *The Monthly Magazine*, No 17, 1797; G. Carnall, 'The Monthly Magazine', in *The Review of English Studies*, vol. 5 no. 18 (Apr. 1954), pp. 158–64; *Prospectus of a new Miscellany to be Entitled the Monthly Magazine or British Register* (London 1796).
52 MS 3147/3/25/7, JW–JWJ, 16 May 1796; MS 3147/3/418/42, R. Phillips–JW, 8 May 1796.
53 MS 3147/3/418/43 R. Phillips–JW, 1 May 1797.
54 *Copper-Plate Magazine*, no. 80, later reissued as John Walker, *The Itinerant: A Select Collection of Interesting and Picturesque Views in Great Britain and Ireland: Engraved from Original Paintings and Drawings. By Eminent Artists* (1799).
55 *New Copper-Plate Magazine*. Published by J. Walker, no. 44, Paternoster-Row. This day is published, number C. To be continued monthly, Eighteenth Century Collections Online.

56 Demidowicz in this volume, Chapter 12, pp. 99–115.
57 James Bisset, *A Poetic Survey round Birmingham* (1800), frontispiece.
58 James Bisset, 'To the Public', *Bisset's Magnificent Guide or Grand Copperplate Directory For the Town of Birmingham* (1808).
59 The artist of these watercolours is unknown but they may be by Francis Eginton junior, Joseph Barber, or could be some of the views produced by Robert Riddell as discussed. The link to Stebbing Shaw's volume is made stronger by the inclusion in the group of a view of Eginton senior's house which is also included in that volume. Copies of these views are British Library Ktop XLII 82n, o and p and William Salt Library, Stafford SV VII 24a, 25a and 28a.
60 Shaw, *op. cit.* (1798–1801), p. 117.
61 Shaw, *op. cit.* (1798–1801), p. 118.
62 MS 3782/12/43/257, Stebbing Shaw–MB, 4 Jun. 1798.
63 MS 3782/12/43/378, Stebbing Shaw–MB 12 Nov. 1800.
64 MS 3782/12/83/122, MB–John Woodward, 11 Apr. 1803.
65 MS 3782/12/43/257, Stebbing Shaw–MB, 4 Jun. 1798; MS 3782/12/90/103, MB–Joseph Franel, 30 Sep. 1799. I am grateful to Sue Tungate for this reference.
66 Edgar Jones, *op. cit.*, p. 13.

Chapter 4: The Context of Neo-Classicism

1 See Hugh Honour, *Neo-Classicism* (London 1968), p. 27, Svend Eriksen, *Early Neo-Classicism in France* (trans. by Peter Thornton, London 1974), pp. 48–51. There is no space here to dwell on the controversy that raged among the *cognoscenti* in the 1760s about the relative importance of Grecian architecture, lauded by J. J. Winckelmann as the fountainhead of all good taste, or Roman or Etruscan, terms which were also loosely adopted by designers and rather less loosely by architects. I don't think the practical makers of objects minded too much about this controversy, although there is a good chance that they will have caught up with it in the pages of Piranesi's polemical works, particularly his *Diverse maniere d'adornare i cammini ed ogni altra parte degli edifizi desunte dell' architettura egizia, etrusca, e graeca* (Rome 1769). In modern parlance *le goût grec* was a brand.
2 Eriksen, *op. cit.*, p. 30.
3 Ibid., p. 34. Eriksen argues that de Marigny was hardly an avid convert to classical taste, but that the journey certainly made him an opponent of the rococo, if a less rabid opponent than his companions.
4 Abbé Le Blanc, *Letter to le Comte de C*** (assumed to be the Comte de Caylus), 1737–44, quoted in full in Eriksen, pp. 226–32. For criticism of the rococo see Wolfgang Hermann, *Laugier and Eighteenth-Century French Theory* (London 1962).
5 For Le Lorrain's radical furniture for Lalive de Jully in 1758–9, see Eriksen, *op. cit.*, p. 46.
6 Pierre Patte, *Cours d'architecture* (Paris 1777), vol. V, p. 86 (a continuation of J-F. Blondel's work of the same title). The translation is quoted from Eriksen, *op. cit.*, p. 22.
7 Eileen Harris, *The Furniture of Robert Adam* (London 1963), plate 1.
8 John Harris, 'Early Neo-Classical Furniture', *Furniture History*, vol. II (1966), pp. 1–6.
9 G. B. Piranesi, *Diverse Maniere*, p. 33. The German art historian, Johann Joachim Winckelmann, whom Piranesi was in essence attacking when he criticised the fervent promoters of all things Greek, also believed in this creative imitation. Piranesi himself came in for some criticism from his contemporaries for his somewhat over-lavish and idiosyncratic use of antique ornament.
10 Joshua Reynolds, *Discourses on Painting and the Fine Arts* (London 1837), Sixth Discourse, pp. 91, 96.
11 MS 3782/1/10, pp. 514–5, MB–Earl of Findlater, 20 Jan. 1776.
12 Kenneth Quickenden, 'Boulton and Fothergill Silver' (unpublished thesis, Westfield College, London 1989), p. 162.
13 Ibid., pp. 162–3, plates 5, 33, 34.
14 The half-hearted survival of rococo motifs was not peculiar to Boulton and Fothergill. There are many objects of decorative art of this period embodying this transitional taste.
15 MS 3782/12/23/218, MB–Mrs Montagu, 16 Jan. 1772.
16 Baron Grimm, *Correspondence Littéraire* (Paris 1813), part I, vol. III, p. 362, quoted in Harris, *Furniture of Robert Adam*, p. 9.
17 See note 11.
18 MS 3782/1/21/4, Lord Cathcart–B&F, 21 Feb. 1772.
19 Pierre Verlet, *Les Bronzes dorés français du XVIIIe siècle* (Paris 1987), p. 198, fig. 227.
20 See Nicholas Goodison, *Matthew Boulton: Ormolu* (London 2002), pp. 66–7.
21 François Vion, *Livre de Dessins*, no.13 (Bibliothèque Doucet), folio 20/1.
22 Robert Sayer, *The Compleat Drawing Book* (3rd edition, London 1762), plate 61. See Goodison, *op. cit.*, p. 96.
23 Interestingly this was one of the features of the new taste that Cochin criticized severely in his *Mémoires*. He did not admire Lalive de Jully's furniture and blamed it for causing all sorts of excesses in the work of subsequent makers such as 'garlands like well ropes, and clocks with rotating bands instead of dials and shaped like vessels that the Ancients had made to hold cordials; all very pleasing inventions in themselves but, now that they are imitated by all kinds of ignorant persons, we find Paris flooded with all sorts of stuff in the Greek manner.' Quoted in Eriksen, *op. cit.*, p. 270.
24 Christie's, 12 Dec. 1986, 'Books from the Library of Matthew Boulton and his Family', lots 63, 69.
25 MS 3782/6/190/167, James Sketchley–MB, 17 Dec. 1764.
26 See Goodison, *op. cit.*, p. 91.
27 Christie's, 12 Dec. 1986, *op. cit.*, lot 43. It is possible that some of the books in this sale came from Boulton's descendants, but in the case of the eighteenth-century books of design this seems unlikely.
28 See Goodison, *op. cit.*, pp. 94–5.
29 MS 3782/12/107/2, MB's Diary 1767: 'Les Antiquite Des Godetz'.
30 Christie's, 12 Dec. 1986, *op. cit.*, lot 115.
31 MS 3782/1/9, pp. 203, MB–Peter Elmsley, 28 Sep. 1771.
32 See Goodison, *op. cit.*, pp. 91–5.
33 See Goodison, *op. cit.*, p. 88.
34 MS 3782, Early Accounts 1768, subscription paid to Mainwaring; B&F–William Matthews, 2 Feb. 1771 and 4 May 1773.
35 See Goodison, *op. cit.*, pp. 89–90.
36 MS 3782/1/9, pp. 207–8, MB–Peter Elmsley, 2 Oct. 1771.
37 William Chambers, *Treatise on Civil Architecture* (London 1759), p. 21.
38 Ibid., pp. 22–3.
39 There are many examples in the pattern books, particularly of tea urns, with these grooved feet.
40 MS 3782/12/16/1/38, MB–Mrs AB, 6 Mar. 1770.
41 For a discussion of the design and its origins, see Goodison, *op. cit.*, pp. 79–80.
42 Exhibition at the Royal Academy 1770, catalogue, p. 6, no. 39 (Royal Academy archives).
43 See Goodison, *op. cit.*, pp. 80–1.
44 The second and third volumes were not published until after Stuart's death in 1788 – volume II in 1789 and volume III in 1795.
45 James Stuart–Thomas Anson, 23 Sep. 1769 (Staffordshire County Record Office).
46 First day, lot 87; second day, lots 69, 92; third day, lot 83.
47 Nicholas Goodison, 'Mr Stuart's Tripod', *Burlington Magazine*, Oct. 1972, and *Matthew Boulton: Ormolu*, pp. 72–6.
48 See Goodison, *Matthew Boulton: Ormolu*, pp. 271–5.

49 Goodison, *op. cit.*, p. 76.
50 MS 3782/1/9, pp. 161–2, B&F–James Stuart, 1 Aug. 1771; MS 3782/1/9, pp. 212–3, B&F–James Stuart, 5 Oct. 1771. The dolphins that support the tureen (cast from a clay model supplied by Wedgwood) echo the dolphins on Stuart's design for the tripod on the Lanthorn of Demosthenes at Shugborough. See Quickenden, *Boulton and Fothergill Silver*, p. 170 and note 181. The design for this tureen, dated 1781, survives in Boulton & Fothergill's Pattern Book I, p. 111. Neither this tureen nor the copy of it supplied to the Admiralty in 1781 has survived in the Ministry of Defence collections.
51 Sir John Soane, *Lectures on Architecture*, A. T. Bolton ed. (London 1929), vol. XI, 'Decoration and Composition', pp. 172, 180.
52 See Goodison, *op. cit.*, pp. 167–8. The property was in Durham Yard, part of the Adam brothers' ambitious and speculative Adelphi scheme. See Alistair Rowan, *Vaulting Ambition: The Adam Brothers, Contractors to the Metropolis in the reign of George III* (London 2008).
53 See Goodison, *op. cit.*, pp. 240–3 and 249–54.
54 Frances Fergusson, 'Wyatt Silver', *Burlington Magazine*, Dec. 1974, pp. 751–5.
55 Victoria and Albert Museum. See Fergusson, *art. cit.*
56 Goodison, *op. cit.*, plates 251, 252.
57 The same figures appear in another Wyatt drawing in the Noailles album for a candelabrum and stand. See Fergusson, *art. cit.*, fig. 55.
58 Goodison, *op. cit.*, plates 94, 304.
59 Kenneth Quickenden, 'Boulton and Fothergill Silver: an Epergne Designed by James Wyatt', *Burlington Magazine*, June 1986.
60 Fergusson, *art. cit.*, p. 752, Quickenden, *art. cit.*, pp. 174–7.
61 MS 3782/12/85/142, John Wyatt [James Wyatt's cousin and B&F's agent in London]–B&F, 14 Feb.; MS 3782/12/85/153/4, 27 Feb. 1776.
62 Fergusson, *art. cit.*, p. 752, Quickenden, *art. cit.*, p. 176 and note 272.
63 MS 3782/12/1/5, MB–P. J. Wendler, n.d. (July 1767).
64 See note 11.

Chapter 5: Matthew Boulton's Silver and Sheffield Plate

1 MS 3782/12/60, JF–MB, 11 Feb. 1764.
2 MS 3782/13/37, James Keir Memorandum, 3 Dec. 1809.
3 H. W. Dickinson, *Matthew Boulton* (Cambridge 1937), pp. 25–43.
4 R. A. Pelham, 'The Water Power Crisis in Birmingham in the Eighteenth Century', in *University of Birmingham Historical Journal*, vol. IX, no. 1 (1963), p. 75.
5 MS 3782/12/108/1, p. 53, MB Notebook 1, 1751–9.
6 *Sketchley and Adams's Tradesman's True Guide* (Birmingham 1770), p. 54.
7 G. Crosskey 'The Early Development of the Plated Trade', in *Silver Society Journal* (hereafter *SSJ*) 12 (2000), pp. 27–38 (pp. 29–31).
8 MS 3782/1/1, p. 136, 1 Aug. 1776.
9 MS 3782/1/3, p. 1, 1 Jan. 1776.
10 MS 3782/1/3, p. 101, Feb.–Mar. 1776.
11 MS 3782/1/1, p. 170, 9 Oct. 1776.
12 K. Quickenden, 'Boulton and Fothergill Silver: Business plans and miscalculations', in *Art History*, vol. 3, 3, 1980, pp. 274–94 (p. 274).
13 Shena Mason, *The Hardware Man's Daughter* (Chichester 2005), pp. 7–8.
14 Quickenden, *op. cit.*, p. 277.
15 Crosskey, *op. cit.*, p. 32.
16 Quickenden, *op. cit.*, p. 275.
17 Frederick Bradbury, *History of Old Sheffield Plate* (London 1912, reprinted 1968), pp. 47–8.
18 MS 3782/1/37, p. 50, B&F–Thomas Jeffries, 12 Jan. 1765.
19 Bradbury, *op. cit.*, p. 33.
20 Quickenden, *op. cit.*, pp. 280–1.
21 MS 3782/1/9, p. 107, B&F–John Porzelius, 17 May 1771.
22 MS 3782/1/9, p. 2, MB–Earl of Shelburne, 7 Jan. 1771.
23 J. Paul de Castro, *The Law and Practice of Hall-marking Gold and Silver Wares* (London, 2nd ed. 1935), pp. 71–2.
24 Quickenden, *op. cit.*, pp. 275–6.
25 A. Westwood, *The Assay Office at Birmingham Part 1: its Foundation* (Birmingham 1936), pp. 14–15.
26 MS 3782/1/9, pp. 212–13, B&F–James Stuart, 5 Oct. 1771.
27 MS 3782/12/2, p. 72, MB–Duke of Richmond, 4 Dec. 1772.
28 MS 3782/21/1, pp. 154–5, B&F–Mr Udney, 12 Jun. 1773
29 P. Glanville, *Silver in England* (London 1987), p. 179.
30 MS 3782/1/9, p. 281, B&F–Kelly Lot & Co., 23 Nov. 1771.
31 MS 3782/1/9, p. 500, A. J. Cabrit [clerk at Soho, hereafter AJC]–William Matthews, 12 Jul. 1772.
32 K. Quickenden, 'Lyon-faced candlesticks and candelabra', in *SSJ*, vol. 11 (1999), pp. 196–209, pp. 197–206.
33 MS 3782/1/9 p. 395, B&F–B. Pullan, 11 Mar. 1772.
34 MS 3782/1/9, p. 451, AJC–B. Pullan, 15 May, 1772.
35 MS 3782/1/10, pp. 204–5, B&F–William Matthews, 19 Dec. 1774.
36 BAO, The Register of Plate and Silver Wares Assayed and Marked or Broke at the Birmingham Assay Office [hereafter RP]. In 1773–4 thirty-eight pairs of silver candlesticks, plus one single candlestick.
37 MS 3782/1/9, p. 62, John Wyatt–William Matthews, 6 Mar. 1771.
38 MS 3782/1/10 p. 63, B&F–Humphrey Palmer, 12 Jul. 1774.
39 MS 3782/1/10 p. 121, B&F–William Matthews (for Mr Pickett), 10 Sep. 1774.
40 K. Quickenden, 'Did Boulton Sell Silver Plate to the Middle Class? A Quantitative Study of Luxury Marketing in Late Eighteenth-Century Britain', *Journal of Macromarketing*, 1, 1 (2007), pp. 51–64 (p. 56).
41 K. Quickenden, 'Elizabeth Montagu's service of plate, Part 1', in *SSJ*, 16 (2004), pp. 131–41 (pp. 134–8).
42 K. Quickenden, 'Elizabeth Montagu's service of plate, Part 2' in *SSJ*, 19 (2005), pp. 19–37 (pp. 28–34).
43 Quickenden, *op. cit.* (note 41), pp. 134–5.
44 Quickenden, *op. cit.* (note 12), pp. 283–4.
45 Quickenden, *op. cit.* (note 42), pp. 34–5.
46 Quickenden, *op. cit.* (note 12), pp. 284–7.
47 Westwood, *op. cit.* (note 25), p. 15.
48 Weighing between 80 ozs and 90 ozs, MS 3782/1/10, p. 544, B&F–John Wyatt, 24 Feb. 1776.
49 Supplied by Ansill & Gilbert, weighing 94oz. 9dwts., costing £26. 0s. 0d. in silver, and they charged £14. 2s. 0d. for labour, John Parker and Edward Wakelin, Workman's ledger No.2, 1) 1766–72, vol. 3 (Victoria and Albert Museum [VAM] 8) p. 70, 8 Dec. 1768).
50 Parker and Wakelin charged £48. 5s. 10d. (John Parker and Edward Wakelin, Gentlemen's Ledger 1765–75, vol. 7 (VAM 7) p. 83, 3 Jan. 1769).
51 MS 3782/1/9, p. 8, MB–V. Green, 6 Aug. 1774.
52 Quickenden, *op. cit.* (note 12), pp. 285–6.
53 MS 3782/1/9 p. 379, B&F–London silversmiths, 29 Feb. 1772.
54 William Webb was invited but later only bought Sheffield Plate (MS 3782/1/1, p. 181, 21 Sep. 1776).
55 Quickenden, *op. cit.* (note 12), pp. 279–80.
56 MS 3782/21/1 p. 121, AJC–Andrew Vezian, merchant, London, 8 May 1773.
57 MS 3782/12/26/116, J. S. Clais–MB, 10 Dec. 1781.
58 MS 3782/1/10, p. 23, B&F–Lord Craven, 10 Jun. 1774.
59 MS 3782/12/87/75, Resolutions of the Birmingham Commercial Committee, 19 Sep. 1783.
60 MS 3782/1/38, p. 511, B&F–William Aitkin, 22 Jun. 1774.
61 MS 3782/1/11 p. 16, B&F–J.C. Preidel, 7 Jun., 1777.
62 MS 3782/1/1 p. 263, 23 Aug. 1777.

63 Quickenden, *op. cit.* (note 40), pp. 58–9.
64 Quickenden, *op. cit.* (note 12), pp. 288–89.
65 K. Quickenden, 'Silver, "plated" and silvered products from the Soho Manufactory, 1780', in *SSJ*, 10 (1998), pp. 77–95 (pp. 89–90).
66 K. Quickenden, 'Richard Chippindall and the Boultons', in *SSJ*, 22 (2007), pp. 51–66 (pp. 52–3).
67 Quickenden, *op. cit.* (note 66) pp. 57–8.
68 MS 3782/2/14 p. 101 John Hodges [JH]–Samuel Silver, 15 Apr. 1783.
69 Dickinson, *op. cit.* (note 3) p. 14.
70 MS 3782/2/15 p. 314, JH–Mr Sobakin, 2 Sep. 1799.
71 MS 3782/2/15 p. 144, JH–J. Johnes, 8 Jun. 1797.
72 MS 3782/2/14 p. 506, MB–Richard Chippindall [RC], 3 Feb. 1786.
73 Dickinson, *op. cit.* (note 3), p. 124.
74 MS 3782/12/8 p. 163, MB–RC, 25 Nov. 1789.
75 MS 3782/2/15 p. 171, JH–Smith & Son, 13 Oct. 1797.
76 MS 3782/2/15 p. 248, JH–RC, 20 Nov. 1800.
77 MS 3782/1/11 p. 886, JH–Wakelin and Tayler, 5 Dec. 1781.
78 T. S. Ashton, *Economic Fluctuations in England 1700–1800* (Oxford 1959), pp. 162–4.
79 J. S. Forbes, *Hallmark: A History of the London Assay Office* (London 1998), p. 228.
80 Bradbury, *op. cit.* (note 17), p. 49.
81 Quickenden, *op. cit.* (note 66), p. 54.
82 BAO, RP, In 1798–9 sixty-four pairs of candlesticks were assayed and only seventy other items (apart from mounts).
83 MS 3782/2/15, p. 422, MB Plate Co.–Dr. H. Edgar, n.d. Jul. 1800.
84 MS 3782/2/15, p. 404, MB Plate Co.–James Acheson Co., 7 Apr. 1800.
85 MS 3782/2/15, p. 325, MB Plate Co.–Robert Jones, 25 Sep. 1799.
86 Quickenden, *op. cit.* (note 66), p. 57.
87 David S. Shure, *Hester Bateman: Queen of English Silversmiths* (Garden City, N.Y. 1959), plate XXXVIII and p. 18.
88 Quickenden, *op. cit.* (note 66), pp. 54–6.
89 MS 3782/12/59, p. 230, RC–MB, 1 Jan. 1807.
90 E. Cule, 'The Financial History of Matthew Boulton 1759–1800' (unublished Master of Commerce thesis, University of Birmingham, 1935), pp. 294–5.
91 Quickenden, *op. cit.* (note 66), p. 60.
92 Bradbury, *op. cit.* (note 17), pp. 42–3.
93 Glanville, *op. cit.* (note 29) p. 106. Crespell made at least 20,000 ozs in 1778–9.
94 BAO, RP. In 1805–6 MB Plate Co. sent one quarter of all the plate assayed at the Birmingham Assay Office.
95 John Bentley left Soho in 1776 to work in Birmingham (MS 3782/1/10 p. 785, B&F–Adams & Sons, 23 Dec. 1776).
96 MS 3782/13/136, Proposals for Purchasing the Plate Trade at Soho, 7 Nov. 1833.

Chapter 6: 'I Am Very Desirous of Becoming a Great Silversmith': Matthew Boulton and The Birmingham Assay Office

1 MS 3782/1/9. MB–Earl of Shelburne, 7 Jan. 1771.
2 Plate Assay (Sheffield and Birmingham) Act (13 Geo.3.c.52). The use of the term 'plate' requires some clarification; 'plate' denotes objects made of solid precious metal, as opposed to articles that are made of base metal 'plated' with a covering of silver or gold. Sheffield Plate is an example of the latter, made by rolling sheets of copper and silver together by a method developed by Thomas Bolsover of Sheffield, c. 1742. This method was used to produce a cheaper product which closely resembled solid silver.
3 H. W. Dickinson, *Matthew Boulton* (Cambridge 1937) p. 53. Dickinson arrives at this date from evidence given before the Parliamentary Committee of 1773 by Samuel Garbett, a manufacturer and refiner of precious metals in Birmingham who supplied Boulton and Fothergill with silver.
4 Arthur Westwood, *The Assay Office at Birmingham. Part 1: Its Foundation* (Birmingham 1936); Dickinson, *op. cit.*, pp. 63 70; Eric Robinson, 'Matthew Boulton and the Art of Parliamentary Lobbying' in *The Historical Journal*, VII, 2 (1964), pp. 217–20; Jennifer Tann, *Birmingham Assay Office 1773–1993* (Birmingham 1993) pp. 13–26.
5 Lord Edmund Fitzmaurice, *Life of William, Earl of Shelburne*, vol. 1, 1737–66 (London 1875) quoted in Westwood, *op. cit.*, p. 4.
6 Kenneth Quickenden, 'Boulton and Fothergill Silver', unpublished PhD thesis, Westfield College, University of London, 1989, p. 26.
7 MS 3782/1/9, MB–Earl of Shelburne, 7 Jan. 1771.
8 Quickenden, *op. cit.* (1989), p. 27.
9 On Boulton's original intention, see Dickinson, *op. cit.*, p. 66. On Boulton's activities at this time, see Quickenden, *op. cit.* (1989), p. 29.
10 Westwood, *op. cit.*, p. 12.
11 MS 3782/12/89/23, *Memorial relative to Assaying and marking wrought plate at Birmingham.*
12 Ibid. Bearing in mind that some of Boulton's designs were the first of their kind to be produced in this country, on this point he was clearly justified. On the originality of Boulton's designs, see Kenneth Quickenden, 'Lyon-faced Candlesticks and Candelabra' in *The Silver Society Journal*, 11, Autumn 1999, pp. 196–210.
13 MS 3782/12/89/23.
14 Ibid.
15 Westwood, *op. cit.*, p. 10.
16 Dickinson, *op. cit.*, p. 66.
17 MS 3782/12/89/23.
18 MS 3782/12/88/10. Petition of the Wardens and Assistants of the Company or Mystery of Goldsmiths of the City of London.
19 Ibid.
20 Ibid.
21 MS 3782/12/88/11. Petition of the Goldsmiths, Silversmiths and Plateworkers, of the City of London, and Places Adjacent.
22 Ibid.
23 Ibid.
24 Ibid.
25 MS 3782/12/88/13. Garbett supplied Boulton and Fothergill with silver and later served with Boulton as a Guardian of the Birmingham Assay Office. Skipwith was also well-acquainted with Boulton. Dickinson, *op. cit.*, p. 67; Quickenden, *op. cit.* (1989), p. 29.
26 Ibid.
27 Westwood, *op. cit.*, p. 16.
28 Westwood, *op. cit.*, p. 17.
29 MS 3782/12/88/16, 36 and 36b. Joseph Wilkinson wrote first to Boulton on this subject on 27 Feb. 1773, defending May's ingenuity rather than decrying his attempt at fraud. May wrote to Boulton in April apologising for any embarrassment or trouble caused by his manufacture of these candle-snuffers, and stating that silver buckles often contained more iron than the offending articles due to the chape being made of steel. The chape is the part of the buckle to which the strap or ribbon is attached.
30 MS 3782/12/88/26. Precious metal was, and often still is, measured in Troy ounces and pounds. A Troy ounce was made up of 20 pennyweights, abbreviated to 'dwts'.
31 MS 3782/12/88. See accounts of the Committee's meetings in the documents relating to the establishment of the Birmingham Assay Office.
32 MS 3782/12/88/15. I do not mean to imply here that the manufacture of silver at Soho was an entirely industrialised process, using die-stamping as the only technique. Undoubtedly, this was the method favoured at Soho for the manufacture of candlesticks, but many

examples of the articles Boulton produced were made by a combination of more industrial and older craft techniques. For example, the sauce tureens (Fig. 45) are an example of Boulton using a mixture of craft and industrial techniques. See Kenneth Quickenden, 'Boulton and Fothergill Silver: Business Plans and Miscalculations' in *Art History*, vol. 3, no. 3, Sep. 1980, p. 283.

33 Westwood, *op. cit.*, p. 18.
34 MS 3782/12/89/13 and 6.
35 MS 3782/12/89/13.
36 Ibid.
37 Ibid.
38 Ibid.
39 Ibid.
40 MS 3782/12/89/6.
41 Ibid.
42 Ibid.
43 Ibid.
44 MS 3782/12/88/40.
45 MS 3782/12/88. Documents relating to the establishment of the Birmingham Assay Office where the proceedings of the Special Parliamentary Committee are detailed.
46 Westwood, *op. cit.*, p. 24.
47 MS 3782/12/88/19.
48 Ibid.
49 MS 3782/12/88/20.
50 J. S. Forbes, *Hallmark: A History of the London Assay Office* (London 1999) p. 223.
51 Birmingham Assay Office, Plate Register, 1773–92.
52 Westwood, *op.cit.*, pp. 30–31.
53 Dickinson, *op. cit.*, p. 70.
54 Birmingham Assay Office, Register of Sponsors' Marks, Book A (1773–1858), p. 1.
55 Ibid., p. 21.
56 Birmingham Assay Office Act (2 Edw.7.c.27).
57 Birmingham Assay Office, Register of Sponsors' Marks, Book A (1773–1858), p. 1; Plate Register, 1773–92.
58 Shirley Bury, 'Assay Silver at Birmingham – I: The Soho Designers and Their Successors' in *Country Life*, 13 June 1968, pp. 1610–15.
59 Ibid., p. 1615.
60 Ibid., p. 1615.
61 Maurice H. Ridgway, *Chester Silver 1727–1837* (Chichester 1985).
62 Quickenden, *op. cit.* (1989), p. 175.
63 Ibid., p. 176.
64 Ibid., p. 175.
65 Ibid., pp. 175–6.
66 Ibid., p. 176.
67 Ibid., p. 177.
68 Ibid., p. 177.
69 Kenneth Quickenden, 'Elizabeth Montagu's Service of Plate, Part 1' in *Silver Studies: The Journal of the Silver Society*, no. 16, 2004, p. 135.
70 Quickenden, *op. cit.* (1980), p. 283.
71 Quickenden, *op. cit.* (2004) and 'Elizabeth Montagu's Service of Plate, Part 2' in *Silver Studies: The Journal of the Silver Society*, no. 19, 2005, pp. 19–37.
72 *Matthew Boulton and the Toymakers: Silver from the Birmingham Assay Office* (London 1982). Catalogue of an Exhibition at Goldsmiths' Hall, 15–26 Nov., 1982, p. 38. On the economic failure of the Boulton & Fothergill silver business, see Quickenden, 'Boulton and Fothergill Silver: Business Plans and Miscalculations' in *Art History*, vol. 3, no. 3, Sept. 1980, pp. 274–94.
73 Dickinson, *op. cit.*, p. 86.

Chapter 7: Ormolu Ornaments

1 M. Swinney, *The New Birmingham Directory* (Birmingham 1773). According to this Directory, the ornaments were 'admired by the nobility and gentry, not only of this kingdom, but of all Europe; and are allowed to surpass anything of the kind made abroad'.
2 MS 3782/1/27, Matthew Boulton to various patrons, 18 May 1778.
3 Josiah Wedgwood (hereafter JWd)–Thomas Bentley (TB), Jan. 1769 (Wedgwood Museum Archives, hereafter WMA, 18216–25).
4 JWd–TB, 1 May 1769 (WMA, 18240–25).
5 JWd–TB, 2 Aug. 1770 (WMA, 18314–25).
6 JWd–TB, 15 Mar.1768 (WMA, 18193–25).
7 JWd–TB, 15 Mar.1768 (WMA, 18193–25).
8 JWd–TB, 15 Mar. (WMA, 18193–25) and 21 Nov. 1768 (WMA, 18215–25).
9 MS 3782/12/2, p. 4, MB–JWd, 28 Dec. 1768.
10 MS 3782/12/2, p. 7, MB–JWd, 4 Jan. 1769.
11 JWd–TB, 27 Sep. 1769 (WMA, 18261–25). For a fuller account of this relationship see Nicholas Goodison, *Matthew Boulton: Ormolu* (London 2002), pp. 45–7.
12 MS 3782/12/2/6, MB–John Whitehurst, 28 Dec. 1768.
13 JWd–TB, 6 Jun.1772 (WMA, 18376–25).
14 MS 3782/6/191/2, receipt from John Platt, 2 Mar. 1769.
15 MS 3782/MB/Correspondence 12/23/177, Robert How–MB, 31 December 1769.
16 MS 3782/12/2, pp. 4, 7, MB–JWd, 28 Dec. 1768 and 4 Jan. 1769.
17 MS 3782/12/108/5, Notebook, 1768–75, p. 2.
18 For a full discussion of Boulton's designs, see Goodison, *op. cit.*, Chapter 3.
19 MS 3782/12/2/13, MB–Solomon Hyman, 23 Jan. 1769.
20 MS 3782/12/23/149, Thomas Pownall–MB, 8 Sep. 1769.
21 MS 3782/12/16/1/37, MB–Mrs AB, n.d. (3 Mar. 1770).
22 Dorothy Richardson, 'Tours', vol. 2, pp. 208–17 (John Rylands University of Manchester Library).
23 JWd–TB, 1 Oct. 1769 (WMA, 18264–25).
24 MS 3782/1/19/10, MB–JF, n.d. (4 Mar. 1770).
25 *The Public Advertiser*, 13 March 1770 and several later dates in March and April (Burney Collection 571B, British Library).
26 Horace Walpole to Horace Mann, 6 May 1770, W. S. Lewis, Warren Hunting Smith and George L. Lamb (eds), *Horace Walpole's Correspondence with Horace Mann* (New Haven 1967), p. 211.
27 MS 3782/12/23/184, James Adam–MB, 14 Aug. 1770.
28 The property was in Durham Yard, part of the Adam brothers' ambitious and speculative Adelphi scheme. See Alistair Rowan, *Vaulting Ambition: The Adam Brothers, Contractors to the Metropolis in the reign of George III* (London 2008).
29 MS 3782/1/19/1, MB to JF, n.d (end Feb. 1770), and MS 3782/1/19/10, n.d. (4 March 1770).
30 MS 3782/1/10/11, MB–Earl of Warwick, 30 Dec. 1770.
31 JWd–TB, 24–6 Dec. 1770 (WMA, 18334–25).
32 MS 3782/12/2, p. 47, draft letter to patrons (in this case to Dukes), 25 Mar. 1771.
33 The catalogue survives in the archives of Christie's.
34 *The Gazetteer and New Daily Advertiser*, 26 Mar. 1771 and several later dates in March and April (Burney Collection 581B, British Library).
35 MS 3782/1/10/77, B&F–Nathaniel Jefferys, 16 Feb. 1771.
36 The ornaments mentioned here that are not shown in this bicentenary exhibition are all described and illustrated in Goodison, *op. cit.*, 2002, Survey of Ornaments.
37 See Goodison, *op. cit.*, Chapter 4.
38 Mrs Delany to the Viscountess Andover, 11 Apr. 1771, Lady Augusta Llanover (ed.), *Autobiography and Correspondence of Mary Granville, Mrs Delany* (1862), vol. I, p. 335.

39 MS 3782/1/9, p. 58, MB–James Christie, 2 Mar. 1772.
40 MS 3782/1/9, p. 254, B&F–Duchess of Portland, 6 Nov. 1771.
41 MS 3782/12/23/259, MB–Lady Dashwood, n.d. (Apr. 1772).
42 MS 3782/1/10/89, MB–John Whitehurst, 23 Feb. 1771.
43 MS 3782/12/16/1/43, MB–Mrs AB, n.d. (11 Apr. 1772).
44 MS 3782/12/60/66, JF–MB, 28 Mar. 1772.
45 MS 3782/12/60/70, JF–MB, 20 May 1772.
46 MS 3782/12/2/39, MB–Earl of Warwick, Nov. 1772; MS 3782/1/11, p. 143, MB–Earl of Dartmouth, 10 Nov. 1772.
47 MS 3782/12/60/91, JF–MB, 2 Mar. 1773.
48 MS 3782/1/10, pp. 514–5, B&F–William Lewis, 6 Mar. 1775.
49 MS 3782/1/10/114, B&F–Robert Bradbury jun., Bakewell, 7 Mar. 1775.
50 MS 3782/12/24/10, Sir Robert Gunning (St Petersburg)–MB, 13 Sep. 1774; MS 3782/1/10, pp. 179–80, B&F–Marquess of Granby, 19 Nov. 1774.
51 MS 3782/1/1, Ledger 1776–8, pp. 120, 138, 174, etc.
52 JWd–TB, 14 Jul. 1776 (WMA, 18684–25).
53 MS 3782/12/63/3, John Hodges–MB, 31 Jan. 1778; MS 3782/12/63/4, John Hodges–MB, 6 Feb. 1778.
54 MS 3/82/12/60/126, JF–MB, 14 Aug. 1778 [incorrectly docketed 14 Feb. 1778].
55 The catalogue survives in the archives of Christie's.
56 The ornaments mentioned here that are not shown in this bicentenary exhibition are all described and illustrated in Goodison, *op. cit.* (2002), Survey of Ornaments.
57 MS 3782/2/13, Inventory 1782, pp. 133–4, 156, 181.
58 MS 3782/12/59/124, R. Chippindall (Chipindale on outside) (London)–MB, 12 Oct. 1790. I am grateful to Kenneth Quickenden for alerting me to this reference. Benjamin Vulliamy was Clockmaker to George III and made, largely through subcontracting, and sold a large number of clocks, watches and metalwork ornaments. Much of his ormolu and bronze was finely crafted.
59 MS 3782/13/37, James Keir, 'Memoir of Mr Boulton', p. 5.
60 Ibid., p. 4.

Chapter 8: The Soho Steam-Engine Business

1 J. Kanefsky and J. Robey, 'Steam Engines in Eighteenth Century Britain: A Quantitative Assessment', *Technology & Culture* (1980), pp. 161–86. Also J. Tann, 'Makers of Improved Newcomen Engines in the late 18th Century', in *Transactions of the Newcomen Society* [TNS], 50 (1979–80), pp. 181–92.
2 MS 3147/4/1, Calculation Blotter pp. 7, 8 and 9, tests on 7 and 24 June 1779.
3 MS 3147/3/80, Letter Book no. 1, p. 1. Letter JW–Ralph Lodge, 12 June 1775.
4 J. Andrew, 'The Costs of Eighteenth Century Steam Engines', *TNS* 66 (1994–5), pp. 83, 84.
5 MS 3147/3/80, Letter Book No. 1, p. 44. JW–Charles Cook, 26 Jul. 1776; MS 3147/3/80, Letter Book no. 1 p. 106, Watt–N. Henshall, 27 Apr. 1777; MS 3147/4/1, Calculating Blotter, pp. 1–3, calculations of the premium for the Hawkesbury Engine, 25 Apr. 1779.
6 H. W. Dickinson and R. Jenkins, *James Watt and the Steam Engine* (Oxford 1927), pp. 51 and 117.
7 MS 3147/3/80, Letter Book no. 1, p. 27, JW–John Blight, 22 Oct. 1775, and MS 3147/3/80, Letter Book no. 1, p. 30, JW–John Scott, 14 Dec. 1775.
8 MS 3147/3/431/142B, Houghton of Birmingham Canal Company–B&W, 27 Jul. 1778; MS 3147/3/81; Letter Book 2, p. 53; B&W–Houghton, 23 Oct. 1778. The National Archives, Public Record Office, Kew: Rail 810/247 30 May 1778 and Rail 810/248 18 September 1779.
9 R. L. Hills, *James Watt*, 3 vols (Ashbourne, 2006), III, p. 267.
10 Dickinson and Jenkins, *op. cit.* (note 6), p. 345.
11 Ibid.
12 Hills, *op. cit.* (note 9), II, p. 140.
13 J. Tann, 'Boulton and Watt's Organisation of Steam Engine Production before the opening of Soho Foundry', *TNS* 49 (1977–8), p. 44.
14 Numbers taken from listings in given in L. Ince, 'The Soho Engine Works', *The Journal of the International Stationary Steam Engine Society*, 16 (2000), pp. 103–31.
15 Andrew, *op. cit.* (note 4), p. 84.
16 J. Andrew, 'The Copying of Engineering Drawings and Documents', TNS 53 (1981–2), pp. 1–13; Hills, *op. cit.* (note 9), II, pp. 190–211.
17 J. Tann, 'Mr Hornblower and his Crew: Steam Engine Pirates in the late 18th Century', *TNS* 51 (1979–80), pp. 95–107 (pp. 103–5).
18 Kanefsky and Robey, *op. cit.* (note 1), p. 169.
19 L. T. C. Rolt and J. S. Allen, *The Steam Engine of Thomas Newcomen* (Hartington 1977, reprinted Ashbourne 1997), p. 106.
20 Ibid., pp. 110–18.
21 J. Andrew, 'Steam Power Patents in the Nineteenth Century – Innovations and Ineptitudes', *TNS* 72.1 (2000–1), p. 17.
22 See note 14.

Chapter 9: 'I had L[or]ds and Ladys to wait on yesterday …': Visitors to the Soho Maunfactory

1 MS 3782/12/1/8, MB–Mr and Mrs Jeffrys, 26 Jul. 1767.
2 MS 3782/12/60, J. Fothergill–MB, 14 Dec. 1765.
3 K. Morgan (ed.), *An American Quaker in the British Isles: the Travel Journal of Jabez Maud Fisher, 1775–1779*. Records of Social and Economic History, new series XVI. Published for the British Academy by Oxford University Press (Oxford 1992), p. 253.
4 M. B. Rowlands, *A History of Industrial Birmingham* (Birmingham 1977), p. 18.
5 Morgan, *op. cit.*, p. 253.
6 *Letters of Josiah Wedgwood*, 3 vols (Manchester and Stoke-on-Trent, n. d.), II, p. 83.
7 MS 3782/12/2, MB–J. H. Ebbinghaus, 24 Oct. 1772.
8 Cornwall Record Office, AD/1583/2/77, MB–T. Wilson, 6 Nov. 1787.
9 MS 3219/4/12, MB–JW, 7 Oct. 1804.
10 MS 3219/6/2/B, MRB–JWJ, 10 Sep. 1801.
11 A. G. Cross, *'By the Banks of the Thames': Russians in Eighteenth-Century Britain* (Newtonville, MA 1980), p. 196.
12 MS 3219/4/123, JW–JWJ, 13 Aug. 1786.
13 MS 3782/12/32, MB–C. F. Greville, 9 Feb. 1787.
14 See J. Tann (ed.), *The Selected Papers of Boulton & Watt*, vol. 1: *the Engine Partnership, 1775–1825* (Cambridge, MA 1981), p. 163.
15 W. von Kroker, *Wege zur Verbreitung technologischer Kentnisse zwischen England und Deutschland in der zweiten Hälfte des 18 Jahrhunderts* (Berlin, 1971), p. 93.
16 MS 3782/12/57/37, MB–MRB, 26 Oct. 1789.
17 W. von Dyck, *Georg von Reichenbach. Deutsches Museum Lebenbeschreibungen und Urkunden* (Munich 1912), p. 3; for a translation, see F. Klemm, *A History of Western Technology* (London 1959), pp. 259–60.
18 L. Dutens, *L'Ami des étrangers qui voyagent en Angleterre* (London 1789), p. 184.
19 C. A. G. Goede, *The Stranger in England or Travels in Great Britain*, 3 vols (London 1807), III, pp. 106–7.
20 MS 3782/12/47, MB–Sir W. Hamilton, n. p., 27 Aug. 1802.
21 MS 3147/3/5, MB–JW, 26 Jun. 1781.

22 MS 3782/12/63, J. Hodges–MB, 12 Sep. 1780.
23 MS 3782/12/45, F. De Luc–MB, 7 Sep. 1800. Miss De Luc was the daughter of Boulton's friend Jean André De Luc, a Swiss geologist and meteorologist who was reader to Queen Charlotte at Windsor.

Chapter 10: Matthew Boulton's Mints: Copper to Customer

Author's acknowledgements:
Thanks to the staff of the Birmingham Libraries & Archives Services and to Dr David Symons, Dr Richard Clay, Professor Peter Jones, and Val Loggie.

1 This chapter is based on a section of work for a forthcoming Ph.D. at the University of Birmingham and Birmingham Museums and Art Gallery.
2 James Watt: 'Memorandum concerning Mr Boulton, commencing with my first acquaintance with him', 17 Sep. 1809; in H. W. Dickinson *Matthew Boulton* (Cambridge 1936) p. 205.
3 MS 3782/12/23/198, Francis Cobb–MB, 3 Apr. 1771.
4 *Monthly Review*, September 1771, stated that twenty tons of copper coin had been counterfeited; the *Gentleman's Magazine*, 1789, talked about counterfeiting in Scotland.
5 MS 3782/21/1, MB–Joseph Banks, 10 Sep. 1789.
6 Eric Robinson, 'Eighteenth-Century Commerce and Fashion: Matthew Boulton's Marketing Techniques' in *Economic History Review*, vol. XVI, no. 1 (1963), pp. 39–60.
7 MS 3782/13/36/37, MB–MRB, 12 Nov. 1789.
8 MS 3782/12/61/44, Samuel Garbett–MB, 2 Sep 1782; /45, 22 Oct. 1782; /47, 5 Dec. 1782.
9 Walter Breen, *The Minting Process* (American Institute of Professional Numismatists, 1970), p. 13.
10 Stebbing Shaw, *History and Antiquities of the County of Stafford* (London 1798). Shaw noted 'all the operations are concentrated on the same spot; such as rolling the cakes of copper hot into sheets; 2ndly fine rolling the same cold in steel polished rollers; 3rdly cutting out the blank pieces of coins….; 4thly the steam engine also performs other operations, such as shaking the coin in bags; and 5thly it works a number of coining machines with greater rapidity and exactness…. as the machine itself lays the blanks up in the dies perfectly concentral with it, and when struck displaces one piece and replaces another.'
11 MS 3782/12/74/128, Zaccheus Walker–MB, 7 Dec. 1786; also MS 3782/12/74/152, 20 Oct. 1787; MS 3782/12/74/168, 23 Jun. 1788. There is discussion about the need for a new warehouse in letters starting in December 1786. Plans were drawn up by October 1787 and the warehouse was in use by June 1788.
12 MS 3782/13/36/19, MB–MRB, 8 Feb. 1788.
13 For example: MS 3782/13/120, Folder 6, Inventory of Property belonging to Coinage Account taken 31 Dec 1790.
14 MS 3782/13/36/54, MB–MRB, 26 May 1791. An example is a list of eighteen resolutions sent by MB.
15 Shaw, *op. cit.*, from information received from MB.
16 MS 3782/21/1, MB–Joseph Banks, 10 Sep. 1789.
17 MS 3782/13/36/55, MB–MRB, 17 May 1791.
18 MS 3782/12/108/53, Mint Book 1788, pp. 56, 57 and 62.
19 MS 3782/13/49/90, MB–John Rennie, 26 Sep. 1791. 'This Steel is for fine Medal Dies & must be the best possible or it will be worth nothing to me I will not limit you in Price, charge it what you please so that it be as good as ever you made.'
20 MS 3782/12/42/165, MB–Benjamin Huntsman, 6 Jul. 1797.
21 MS 3782/12/108/53, Mint Book 1788, p. 16. MB notes that 'The strips cast by Ryley are about 7lb each and are before rolling about 3/4 inch thick. Rolled 1st time from 17 inches to 24 inches long, and 2nd time to 30 inches long, then annealed, then rolled third time to 39 inches, annealed 2nd time, rolled 4th time to 50 inches, 4th time to 58 inches. Annealed 3rd time, rolled 6th time to 72 inches and 7th time to 82 inches, then annealed 4th time, scoured.'
22 MS 3782/12/108/53, Mint Book 1788, p. 32.
23 MS 3782/12/108/53, Mint Day Book 1788. Reference is made for example on 29 Sep. 1791: 'For Richard Skeldon for shaking bags £4 2s'; and 30 June 1792 '20 bags of sawdust from T Lucas £2 10'. Boulton's friend Samuel Garbett provided both 'Silver and Vitriol £41 15 8d' on 16 Feb. 1792.
24 MS 3782/12/66/2, James Lawson–MB, 27 Jun. 1789.
25 *Conder Token Newspaper*, vol. 1, no. 3, 15 Feb. 1997 and vol. 1, no. 4, 15 May 1997. Discussion between David Vice and Richard Doty as to when the layer-in was introduced.
26 MS 3782/13/36/64, MB–MRB, 21 Mar. 1792. Boulton gave very detailed instructions for coining an order for the Monneron Frères in 1792: 'When these Medals are bronzed they must be taken into the packing room, wiped Clean, for they will spoil the Dies, with 2 Cloths & put into the Copper tubes … All Medals must be cut out of pickled Clear Metal double polished & milled either blank plain or some may be done with a new Roulet … They must be anneald in Tubes before they are Bronzed & as fast as they are struck they must be taken up by the Edge one by one & placed on boards 10 by 10= 100 & Each piece kept in fine Cambrich paper & then soft brownish paper & then packed in small parcels about the usual size & Each of those parcels packed or wrapt in London Brown paper & those into Casks.'
27 MS 3782/12/108/53, Mint Day Book 1788. Orders were given on various dates for tubs and casks to Richard Bicknall, for example on 2 Aug. 1791 'for making casks to pack up the Et Ia Co's Coin'; on 7 Feb. 1792 'for Tubs of different sizes in 1791 £4 3 6d'.
28 *Conder Token Newspaper*, vol. 1, no. 3, 15 Feb. 1997.
29 MS 3782/12/46/109, MB–Charles Hatchett, 26 Mar. 1801.
30 Nicholas Goodison, *Matthew Boulton: Ormolu* (London 2002) p. 128.
31 Phoenix Art Museum: *Copper as Canvas* (Oxford 1999), p. 129. Production rose from 6,000 tons per year in 1725 to 28,750 tons in 1770, and prices of copper ore fell from £7 15s per ton to £6 15s per ton respectively.
32 MS 3782/12/73/62, Thomas Williams–MB, 20 Jun. 1781.
33 MS 3782/12/61/49, Samuel Garbett–MB, 17 May 1783.
34 MS 3782/12/73/66, Thomas Williams–MB, 12 Jul. 1786. 'It would be folly in us to give you any opposition therein, and you may be assured I shall wish no further concern in the business than furnishing the Sheet Copper that may be wanted to which I dare say you will have no objection, but all this to yourself only.'
35 MS 3782/12/73/69, Thomas Williams–MB, 12 Oct. 1787.
36 MS 3782/13/36/9, MB–MRB, 30 Jul. 1787.
37 MS 3782/17/4, Coinage Licence, 9 Jun. 1797; MS 3782/17/5, Coinage Licence 4 Nov. 1799; MS 3782/17/6, Licence to coin, 18 Apr. 1805.
38 MS 3782/12/90/71–80, prices for battery copper in England and in Calcutta sold by public auction; and copper in cakes from Bussorah and Europe from 1787 to 1794. The complex price comparison required converting from the local currency of sicca rupees and the local weight of mamodies.
39 MS 3782/12/59/169, William Cheshire–MB, 26 Apr. 1798.
40 MS 3782/12/59/186, William Cheshire–MB, 4 Mar. 1799. 'We shall write to Bristol for Information respecting any Vessels that may be going soon from thence to Philadelphia, but if you have no particular Motive for wishing them ship'd at that Port, Mr Brown suggests that Liverpool would be preferable as the Port charges are much higher at the former than at the latter place.'

41 MS 3782/3/13, Mint Day Book, 1791–5. 26 Dec. 1792: For example, the French Monneron orders in 1792 were sent via the Soho wagon or by Thomas Toye's cart, first to Hugh Henshall's wharf in Birmingham, then by canal and along the River Trent to Gainsborough, Lincolnshire, to Charles Bradley, who shipped it via Hull to London, Calais, or Rouen.

42 MS 3782/12/23/173, Samuel Glover–MB, 23 Dec. 1769.

43 MS 3782/13/36/174, MRB–MB, Feb. 1795.

44 MS 3782/12/66/86, John Southern–MB, 16 Jul. 1797.

45 MS 3782/17/4, Coinage Licence 9 Jun. 1797; MS 3782/17/5, Coinage Licence 4 Nov. 1799.

46 MS 3782/12/108/53, Mint Notebook 1788, pp. 10, 52.

47 MS 3782/13/36/64, MB–MRB, 21 Mar. 1792. 'Some Medals may be given to Nelson to Gild as he may fill up his Gilders time & it may be a Convenience to both parties'.

48 MS 3782/13/36/64, MB–MRB, 21 Mar. 1792.

49 MS 3782/3/13, Mint day Book 1791–5. 1 Dec. 1791, 'Thomas Wyon for sinking the Negro die and other things in the months of Aug & Sept last (Vide his account) £4 1s.'

50 MS 3782/13/36/48, MB–MRB, 30 Aug. 1790.

51 MS 3782/12/66/9, James Lawson–MB, 23 May 1791. 'I have received the Letter Punches from Phillips, and shall get the "nut" finished as fast as possible for Southampton…… Mr. Wyon has not yet finished the Guinea die but promised to have it finished tomorrow.'

52 BMAG 2003.0031.129. Designs for halfpennies, 1802. Phillp album.

53 MS 3782/3/13, Mint day Book 1791–5. 24 Dec. 1792. 'Monnerons advise they have paid Dupre for an engraving of Hercules £10 2 2d'; MS 3782/12/66, Item 29, Letters of Lawson 12 April 1792, 'I have got Dies of the Serment du Roi & J. J. Rousseau, & can get some made from La Fayette without waiting for the Punch which was sent to Dumarest'; MS 3782/12/66, Item 30, 14 Apr. 1792, 'I have got Medal Dies ready both for Serment du Roi and for J. J. Rousseau, also the Reverse for La Fayette I have got from Ponthon this day and am making a punch for it'.

54 J. G. Pollard, 'Matthew Boulton and Conrad Heinrich Küchler' in *Numismatic Chronicle* (1970), vol. X, p. 271.

55 Eric Robinson, 'Matthew Boulton: Patron of the Arts', in *Annals of Science*, 9 (1953) p. 375.

56 MS 3782/13/36/126, MB–MRB, 13 May 1795.

57 David Vice, 'The Indian Token Coinage of Major-General Claude Martin', in *Format*, 37, Sep. 1988.

58 Eric Roll, *An Early Experiment in Industrial Organisation* (London 1930/1968), to p. 23.

59 Robinson, *art. cit.*, p. 369.

60 MS 3782/12/23/189, John Flaxman Sr.–MB, 12 Dec. 1770. Bill for 'rams head, deer's head, lion's head, bas relief of lion etc; figure of a sleeping Bacchus etc £3 plus also Hercules and M'angelo's anatomy'.

61 Shaw, *op. cit.* Boulton sent some of his products in coin to Paul, Emperor of Russia, and received a gold medal with his likeness in April 1797. He was also sent some Siberian minerals, Russian money and 200 medals.

62 F. Pridmore, *The Coins of the British Commonwealth of Nations Part 4 India* (London 1975), p. 220.

63 Pollard, *art. cit.*, pp. 277–9; pp. 314–16.

64 MS 3782/13/39/46, JWj–MB, Jun. 1795: 'I am sorry to inform you that the word TheSsaliarcha [sic] does not exist either in Greek or Latin…I think you should not publish the Medal as it now stands, or your reputation as a classical scholar will stand a bad chance & may influence upon your future performances.'

65 BRL Reference 82934, Timmins album, vol. 1.

66 MS 3782/12/42/142, Patrick Colquhoun–MB, 15 Jun. 1797.

67 MS 3782/12/42/139, Brook Watson–MB, 14 Jun. 1797.

Chapter 11: 'Bringing to Perfection the Art of Coining': what did they make at the Soho Mint?

Author's acknowledgements:
I am grateful to David Vice for generously providing me with a typescript of his paper in advance of publication and for commenting on an early draft of this paper. My debt to him will be obvious from these pages. I would also like to thank Pamela Magrill and Sue Tungate for their help and advice.

1 D. Vice, 'A numismatic history of Soho Manufactory and Mint 1772–1850', to be published by the British Numismatic Society. The other main source for the story of the Soho Mint is R. Doty, *The Soho Mint and the Industrialization of Money* (London 1998). Readers who wish to investigate these areas, or simply to find out in more detail about Soho's products, should consult David Vice's forthcoming paper, which provides a complete account of Soho's numismatic products from the 1770s down to 1850.

2 Usually described as bronze, these medals were in fact struck from platina, a soft brass alloy (Vice, art.cit.). See also L. Brown, *British Historical Medals 1760–1960. Vol. I, The Accession of George III to the Death of William IV* (London 1980) (hereafter *BHM*), p. 38, no. 165. John Westwood's career is conveniently summarised in D. Dykes, 'John Gregory Hancock and the Westwood brothers: an eighteenth-century token consortium', *BNJ* 69 (1999), pp. 173–86.

3 Banks presented one of the gold medals to King George III and the other to his sister, Sarah Sophia Banks, the great coin, medal and token collector (Vice, *op. cit.*). Banks had sailed as the expedition's botanist on Cook's first Pacific voyage (1768–71) and was supposed to go on the second voyage, but withdrew before the ships sailed. He remained a good friend of Boulton for the next 37 years.

4 Vice, *op. cit.* Vice suggests that the St Eustatius medal is probably *BHM*, p. 55, no. 235.

5 Boulton had upgraded Soho Mill in 1785 and the new Rolling Mill was more than adequate for the needs of the Manufactory. The free capacity could usefully be employed on this order (Vice, *op. cit.*).

6 Vice, *op. cit.*; Doty, *op. cit.*, pp. 299–301. Some original documents relating to this coinage were published in B. M. Gould, 'Matthew Boulton's East India Company Mint in London, 1786–88', *Seaby Coin and Medal Bulletin*, no. 612 (August 1969), pp. 270–7. It is not certain how many of each denomination were struck, but, even if all the copper had been made into the largest, three keping coins, this amounts to at least five million coins. The correct figure must be considerably higher than this.

7 James Watt, 'Memorandum concerning Mr. Boulton', 17 September 1809 (MS 3782/13/37/111). See in general Vice, *op. cit.*; Doty, *op. cit.*, pp. 24–7.

8 At this period there were thirteen Irish pennies to a shilling rather than twelve as in Britain, so an Irish halfpenny weighed slightly less than its British counterpart. 'Evasive' coins got their name because they evaded the laws on forgery by differing in some way from genuine regal coins, while still looking very similar to them. It was usually their legends that were changed, for example substituting GEORGE RULES for the normal GEORGIUS III REX around the king's head.

9 Sir John Craig, *The Mint. A History of the London Mint from A.D. 287 to 1948* (Cambridge 1953), pp. 252–3.

10 For this paragraph in general, see G. P. Dyer and P. P. Gaspar, 'Reform, the new technology and Tower Hill, 1700–1966', in *A New History of the Royal Mint*, ed. C. E. Challis (Cambridge 1992), pp. 398–606 (at pp. 431–46).

11 Vice, *op. cit.*

12 For the full story of the relationship between Boulton and Droz see

J. G. Pollard, 'Matthew Boulton and J.-P. Droz', *Numismatic Chronicle* (hereafter NC) 7th series vol. 8 (1968), pp. 241–65; Doty, *op. cit.*, pp. 26–45; Vice, *art. cit.* In a letter to his son of 30 July 1790, Boulton, describes Droz as 'the most ungratefull [*sic*], most ungenerous, & basest man I ever had any concern with' (MS 3782/13/36/48).

13 Doty, *op. cit.*, pp. 302–3; Vice, *op. cit.* (There is some dispute whether Soho also struck tokens at the same time for the related firm of Roe and Co. of Macclesfield: Dr Doty thinks that it did, Mr Vice does not.)

14 Here and in later instances these are rounded figures, based on the estimates given in Vice, *op. cit.* In some instances they may vary slightly from those given by Doty, *op. cit.*, pp. 302–3.

15 Vice, *op. cit.*; Doty, *op. cit.*, pp. 306–10.

16 Vice, *op. cit.*; Doty, *op. cit.*, pp. 322–3.

17 Vice, *op. cit.*; Doty, *op. cit.*, pp. 312–13.

18 For a detailed account of Boulton and the Monnerons, see R. Margolis, 'Matthew Boulton's French ventures of 1791 and 1792; tokens for the Monneron Frères of Paris and the Isle de France', *BNJ* 58 (1988), pp. 102–9. See also Vice, *op. cit.*; Doty, *op. cit.*, p. 308.

19 For details of the case, see R. Margolis, 'Those pests of canals: a theft of Monneron tokens intended for France', *BNJ* 75 (2005), pp. 121–31. As Mr Margolis points out (p. 128), it is an interesting reflection on the state of the British coinage that English shopkeepers were prepared to accept payment in tokens inscribed in French.

20 Vice, *op. cit.*; Doty, *op. cit.*, pp. 305–6, 311–12, 321, 324–6, 331–32; F. Pridmore, *The Coins of the British Commonwealth of Nations to the End of the Reign of George VI 1952. Part 4, India. Volume 1: East India Company Presidency Series c.1642–1835* (London 1975), pp. 34–5, 50, 124–5; F. Pridmore, *The Coins of the British Commonwealth of Nations to the End of the Reign of George VI 1952. Part 2, Asian Territories* (London 1962), p. 33.

21 On the coins for Sierra Leone, see especially D. Vice, *The Coinage of British West Africa and St Helena 1684–1958* (Birmingham 1983), pp. 19–35. See also Vice, *op. cit.*; Doty, *op. cit.*, pp. 308–9.

22 Vice, *op. cit.*; Doty, *op. cit.*, pp. 321–2.

23 For an account of Boulton's dealings with the United States, see particularly Doty, *op. cit.*, pp. 275–86. Vice, *op. cit.*, supplies details of the shipments of blanks.

24 G. P. Dyer, 'The currency crisis of 1797', *BNJ* 72 (2002), pp. 135–42.

25 Although 'dollar' was the normal English term for these coins, they were actually worth 8 reales in the Spanish currency system.

26 Dyer, *art. cit.*, pp. 136–40.

27 MS 3782/12/42/35, Lord Liverpool–MB, 3 March 1797.

28 Dyer, *art. cit.*, pp. 140–1; Vice, *op. cit.*; Doty, *op. cit.*, pp. 315–18. 'Cartwheels' remained in use in substantial numbers for many years – as late as 1857 up to a quarter of the pennies in circulation were 'ring pence', very worn Cartwheel pennies still immediately recognisable by their raised rims (G. P. Dyer, 'Thomas Graham's copper survey of 1857', *BNJ* 66 (1996), pp. 60–66).

29 K. Eustace, 'Britannia: some high points in the history of the iconography on British coinage', *BNJ* 76 (2006), pp. 323–36 (p. 330).

30 Vice, *op. cit.*; Doty, *op. cit.*, pp. 319–20, 330–31.

31 Vice, *op. cit.*; Doty, *op. cit.*, pp. 326–7. Because these were issued by the Bank of England and not by the Royal Mint, these dollars were technically tokens rather than coins.

32 Vice, *op. cit.*; Doty, *op. cit.*, pp. 327–8.

33 Vice, *op. cit.*; Doty, *op. cit.*, pp. 328–9.

34 Vice, *op. cit.*; Doty, *op. cit.*, p. 36; *BHM*, p.73, no. 311.

35 J. G. Pollard, 'Matthew Boulton and Conrad Heinrich Küchler', NC 7th series vol. 10 (1970), pp. 259–318; the text of the agreement between Boulton and Küchler about the medals appears on pp. 263–4.

36 Vice, *op. cit.*; Pollard, 'Boulton and Küchler', pp. 269–75. See *BHM*, pp. 84–85, no. 363 for the Cornwallis medal.

37 Vice, *op. cit.*; Pollard, 'Boulton and Küchler', pp. 277–9, 288–9, 293; *BHM* pp. 90–91, no. 383, p. 117 no. 479, and p. 128 no. 523. Soho Mint records show that 524 medals were sent to Naples for sale while 611 were sold to 'toyshops' in London or Birmingham or were held in stock at Soho; Boulton and Küchler split a profit of £55 on this venture (Vice, *op. cit.*).

38 Vice, *op. cit.*

39 MB-MRB, 12 Feb. 1795 (MS 3782/13/36/118).

40 Vice, *op. cit.*; Pollard, 'Boulton and Küchler', pp. 279–81; *BHM* p. 93, no. 392.

41 Vice, *op. cit.*; Pollard, 'Boulton and Küchler', pp. 282–4.

42 Vice, *op. cit.*; Pollard, 'Boulton and Küchler', pp. 284–6; *BHM*, pp. 106–8, no. 447.

43 The background to the Trafalgar medal is fully discussed in N. Goodison, *Matthew Boulton's Trafalgar Medal* (Birmingham 2007). See also Vice, *op. cit.*; Pollard, 'Boulton and Küchler', pp. 303–9; *BHM*, p. 44, no. 584.

44 Vice, *op. cit.*; W. J. Davis and A. Waters, *Tickets, Checks and Passes of Great Britain and Ireland* (Leamington Spa 1922), p. 6, nos. 55–7, p. 111 no. 89, p. 128 no. 212, p. 178 no. 597.

45 Vice, *op. cit.*; Davis and Waters, Tickets, p. 132, no. 247.

46 Vice, *op. cit.*

47 Matthew Boulton to Samuel Garbett, quoted in H. W. Dickinson, *Matthew Boulton* (Cambridge 1937), p. 162.

Chapter 12: A Walking Tour of the Three Sohos

1 There is no room here to provide a systematic list of published and unpublished descriptions and Peter Jones's article in this volume (Chapter 9) covers the subject in more depth. It is worth mentioning one for each site: (Manufactory) *An American Quaker in the British Isles, The Travel Journals of Jabez Maud Fisher, 1775–1779*, ed. K. Morgan (New York 1992), pp. 253–5; (Mint) H. B. Hancock and N. B. Wilkinson, 'Joshua Gilpin, An American manufacturer in England and Wales, 1795–1801 – part II', in *Transactions of the Newcomen Society*, 33 (1962), pp. 57–9; (Foundry) *J. C. Fisher and his Diary of Industrial England 1814–51*, ed. W. O. Henderson (London 1966), pp. 68–9, 133–4.

2 MS 3782/12/80/94, Ambrose Weston–MB, 30 Apr. 1800; MS 3782/12/80/99, Ambrose Weston–MB, 24 May 1800; MS 3782/12/63/74, John Hodges–MB, 11 Aug. 1800.

3 G. Demidowicz: *The Soho Industrial Buildings: Manufactory Mint and Foundry* (forthcoming); G. Demidowicz: *The Soho Foundry Smethwick West Midlands A Documentary and Archaeological Study*, report for Sandwell Borough Council and the Heritage Lottery Fund (2002).

4 Jabez Maud Fisher, *op. cit.*, p. 253.

5 The term 'great bank' was not used at the time, but has been coined as a convenient description of this important topographical feature of the site.

6 Originally the pool immediately below Soho was Hockley Pool alone. Boulton then constructed another dam immediately above Hockley Pool and created Soho Pool.

7 The normal designation of horsepower at this time was a capital 'H' and will be used during tours.

8 The machinery in this part of the Mint on the axonometric view is shown incorrectly.

9 MS 1682/8/1 (BRL Archives & Heritage). The design was by William Hollins.

10 The axonometric view (see Fig. 1) was drawn before this information came to light.

11 Views of the Manufactory, however, show the scheme completed, a useful reminder that drawings commissioned by building owners,

particularly of proposed constructions, are not always reliable. The crescent was finally completed in 1825–6.

12 This is the first building visited that survives today. Part of the pattern stores and the 6H shop also survive. There is no doubt that the robust and deep archaeology of much of the original Foundry complex lies undisturbed below the ground.

13 This same tunnel can be walked down today; it no longer has exits at either end into the open air but enters later buildings. Both sets of small vaults have also survived.

Chapter 13: How do we know what we know? The Archives of Soho

1 MS 3782/12/8 Letter Book, 1789–91.
2 MS 3782/2/13.
3 MS 3782/13/37, MB's draft of speech at the Soho Foundry feast; also reported in *Arisi's Birmingham Gazette*, 1 Feb. 1796.
4 MS 3782/6/137/137, James Weale junior–William Cheshire, 19 Nov. 1808.

Select Bibliography

Archive material

Matthew Boulton Papers (MS 3782), Boulton & Watt Papers (MS 3147), James Watt Papers (MS 3219), all in Birmingham Archives & Heritage, Birmingham Reference Library.

Articles and papers

Cule, J. E., 'Finance and Industry in the 18th Century: The firm of Boulton & Watt', in *Economic History Supplement to Economic Journal* (1940), 4, 319–25.

—— 'The Bill Account: Industrial Banking for Matthew Boulton', in *The Three Banks* (1964), 63, 42–51.

—— 'The Financial History of Matthew Boulton 1759–1800', unpubl. M.Comm. thesis (University of Birmingham, 1935).

Robinson, Eric, 'Boulton and Fothergill 1762–82 and the Birmingham Export of Hardware', in *University of Birmingham Historical Journal* (1959), vol. VII, no. 1.

—— 'Matthew Boulton and the Art of Parliamentary Lobbying', in *Historical Journal*, 7 (1964), 209–29.

—— 'Matthew Boulton's Birthplace and his Home at Snow Hill: A Problem in Detection', in *Transactions of Birmingham & Warwickshire Archaeological Society* (1957), vol. 75, pp. 85–9.

—— 'The Lunar Society and the Improvement of Scientific Instruments', in *Annals of Science* (1956, 1957), 12, 296–304, and 13, 1–8.

—— 'The Lunar Society, its Membership and Organisation', in *Trans. Newcomen Society* (1963), 35, 153–77.

—— 'The Origins and Life-span of the Lunar Society', in *University of Birmingham Historical Journal* (1967), vol. XI, 1, 5–16.

—— 'Training Captains of Industry: The Education of Matthew Robinson Boulton [1770–1842] and The Younger James Watt [1769–1848]', in *Annals of Science* (1954), 10, 301–13.

Quickenden, Kenneth, 'Boulton & Fothergill Silver: Business Plans and Miscalculations', in *Art History* (September 1980), vol. 3, no. 3.

—— 'The Planning of Boulton & Fothergill's Silver Business', in *Silver & Jewellery Production & Consumption since 1750* (UCE, Birmingham, 1995).

Quickenden, Kenneth and Arthur J. Kover, 'Did Boulton Sell Silver Plate to the Middle Class? A Quantitative Study of Luxury Marketing in late Eighteenth-century Britain', in *Journal of Macromarketing* (August 2006), 27.1.

Books

Ballard, Phillada, Loggie, Val and Mason, Shena, *Lost Landscape: Matthew Boulton's Gardens at Soho* (forthcoming).

Berg, M., *The Age of Manufactures 1700–1820* (London, 1985).

Clay, Richard and Tungate, Sue (eds), *Matthew Boulton and the Art of Making Money* (Studley, 2009).

Delieb, Eric, *The Great Silver Manufactory*, Studio Vista (London, 1971).

Demidowicz, G., *The Soho Industrial Buildings: Manufactory, Mint and Foundry* (forthcoming).

Dick, Malcolm (ed.), *Joseph Priestley and Birmingham* (Birmingham, 2005).

Dick, Malcolm (ed.), *Matthew Boulton: A Revolutionary Player* (Birmingham, 2009).

Dickinson, H., *Matthew Boulton* (Cambridge, 1936, republished in paperback, Leamington Spa, 1999).

Doty, R., *The Soho Mint and the Industrialization of Money* (Smithsonian, Washington, DC, 1998).

Gale, W. K. V., *Boulton, Watt and the Soho Undertakings*, Birmingham Museums and Art Gallery (Birmingham, 1970).

Goodison, Nicholas, *Matthew Boulton: Ormolu*, Christie's (London, 2002).

Goodison, Nicholas: *Matthew Boulton's Trafalgar Medal* (Birmingham, 2007).

Hills, Richard, *James Watt*, 3 vols (Ashbourne, 2002).

Jones, Peter, *Industrial Enlightenment: Science, Technology and Culture in the West Midlands, 1760–1820* (Manchester University Press, 2008).

Mason, Shena, *Jewellery Making in Birmingham, 1750–1995* (Chichester, 1998).

—— *The Hardware Man's Daughter* (Chichester, 2005).

Ransome-Wallis, Rosemary, *Matthew Boulton and the Toymakers, Silver from the Birmingham Assay Office*, exhibition catalogue, The Worshipful Company of Goldsmiths (London, 1982).

Schofield, R., *The Lunar Society of Birmingham* (Oxford, 1963).

Selgin, George, *Good Money: Birmingham Button Makers, the Royal Mint, and the Beginnings of Modern Coinage, 1775–1821* (Michigan, 2008).

Smiles, Samuel, *Lives of Boulton & Watt* (1865, republished Stroud, 2007).

Tann, Jennifer, *Birmingham Assay Office 1733–1993* (Birmingham, 1993).

Timmins, S. (ed.), 'The Plated Wares and Electro-plating Trades', in *Birmingham and the Midland Hardware District* (London, 1866).

Uglow, Jenny, *The Lunar Men* (London, 2002).

Picture credits

Illustrations in the book appear by kind permission of the following copyright © holders:

Figures in the essays:

Birmingham Archives & Heritage: 1, 3, 6, 7, 9, 16, 19, 21, 22, 30, 32, 36, 51, 53, 58, 59, 60, 61, 81, 85, 86, 87, 88

Birmingham Assay Office: 39, 40, 41, 45, 52, 122

Birmingham Assay Office objects, photographs © Birmingham Museums & Art Gallery: 33, 37, 38, 42, 43, 44

Birmingham Museums & Art Gallery: pp. i, ii, x; figs 8, 11, 13, 14, 15, 17, 23, 24, 25, 49, 52, 65, 66, 67, 68, 69, 70, 71, 72, 73, 74, 75, 76, 77, 78, 80, 82

Bremner & Orr Design Ltd: 79
George Demidowicz: 83, 84
Derby Museum & Art Gallery: 10
Deutsches Museum, Munich: 57
Sir Nicholas Goodison: 29, 31, 46, 47, 48, 50
Professor Peter Jones: 54, 55
National Archives, Stockholm: 56
Trustees of the William Salt Library, Stafford: 18, 20
The Speed Art Museum, Louisville, Kentucky: 34, 35
Sue Tungate: 64
Lapworth Museum of Geology, University of Birmingham: 63
Victoria & Albert Museum: 26, 27, 28
Private collectors: 3, 4, 5, 12

Illustrations in the exhibition pages:

Birmingham Archives & Heritage: 43, 56, 60, 74, 92, 120, 129, 144, 227, 231, 241, 243, 250, 382, 397, 302, 403

Birmingham Museums & Art Gallery: 15, 8, 17, 22, 23, 24, 26, 29, 30, 33, 34, 35, 36, 47, 50, 51, 52, 53, 55, 57, 58, 63, 67, 68, 70, 71, 72, 73, 76, 78, 78a, 79, 85, 89, 91, 96, 97, 101, 102, 104, 105, 106, 107, 110, 111, 113, 114, 131, 133, 134, 143, 146, 149, 150, 152, 155, 156, 157, 158, 159, 161, 168, 170, 172, 173, 176, 177, 178, 183, 184, 182, 193, 194, 197, 198, 200, 201, 202, 207, 208, 209, 210, 211, 212, 213, 223, 228, 229, 236, 237, 238, 240, 247, 248, 249, 254, 255, 256, 257, 266, 277, 284, 288, 289, 291, 292, 298, 299, 305, 314, 319, 320, 324, 326, 335, 336, 347, 351, 352, 353, 355, 361, 369, 370, 373, 385, 386, 388, 391, 392, 394, 396, 400, 401, 404

The British Museum: 124, 230
Christie's: 166
Gordon Crosskey: 181, 185, 186, 187, 188, 189, 190, 191, 192
Sir Nicholas Goodison: 162, 169, 170, 171, 174, 175
The National Gallery: 39
The National Maritime Museum, London: 236a, 236b
The National Portrait Gallery, London: 37, 65
The Royal Alpha Lodge No. 16: 140–2
The Royal Collection © 2008, Her Majesty Queen Elizabeth II: 160, 163, 164, 165, 167
The Science Museum/Science and Society Picture Library: 251a, 251b
Sotheby's Picture Library: 81–4
Tate Britain: 2
Thinktank Trust: 3, 64, 253
University College London: 46
Victoria & Albert Museum/V&A Images: 86, 87, 88, 112
Wolverhampton Art Gallery: 38

Index

Page numbers in *italic* refer to illustrations.
MB refers to Matthew Boulton.

Abbott, Lemuel Francis (c.1760–1802): *Portrait of Matthew Boulton* frontispiece, 121
Adam, James (1732–1794) 39, 59, 109
Adam, Robert (1728–1792) 23, 34, 35, 39, 40, 59, 181
   design for door escutcheon and knob *38*, 39
   Ruins of the Palace of the Emperor Diocletian at Spalatro in Dalmatia 34
Alexander, William (1767–1816): *The Emperor of China approaching his Tent in Tartary* ... 225, *225*
Albion Mill, London 39, 113
Alston, Amelia 205
American Revolution 198–200
Anderson, Diederich Nicolaus (d. 1767) 38
   candelabra vase (attrib.) *168*, 169
Andras, Catherine: *Horatio Nelson* 202, *202*
Anson, Thomas (1767–1818) 18, 35
Antichità di Ercolano, Le 34
'Archives of Soho Project' (1998–2003) 109–15
Argand, Aimé (1755–1803) 109, 189
   lamps 9, *44*, 45, *188*, 189
Aris, Samuel 194
Aris, Thomas (d.1761) 124
Arrowsmith, John: *Plan of Soho* 19
Ash, Dr John (1723–1798) 191
   portrait (Reynolds) *191*, 192
Aubert, Alexander (1730–1805) 10
Avery, W. &. T. 70, 219

Baader, Joseph von (1763–1835) 77
Banks, Sir Joseph (1743–1820) 6, 21, 74, 81, 83, 109, 113, 224
   MB's medals for 89
   portrait (Russell) *5*
Barber, Joseph (1757–1811) 27, 28
Baskerville, John (1706–1775) 124
   Boulton family bible 150, *150*
Baskerville, Mrs 71
Bate, Robert: orrery *131*, 131–2
Beddoes, Thomas (1760–1808) 109
Beechey, Sir William (1753–1839) 87
   Portrait of Matthew Boulton 206, 208

Bentley, Thomas (1731–1780) 8
Billington, Elizabeth (1768–1818) 19, 21
Birch, George (1739–1807) 26
Birmingham, growth of 71–2, 122–5, 189, 196–7
   Plan of Birmingham (Westley) opp. *1*, 122, 123, *123*
Birmingham Assay Office 4, 41–2, 47–54, 161
   King's Head inn sign *51*, 192–3
   registers *49*, *50*, 193
Birmingham General Dispensary 191, 192, 230
Birmingham General Hospital 16, 191, 192, 230
Birmingham Jacobin Society 190, *196*, 197
Birmingham Loyal Association (est. 1797) 203, *203*
   medal to mark Peace of Amiens 204, *204*
Bisset, James (1762–1832) 190, 196
   Magnificent Directory of Birmingham 29, *29*, 30, 190, 219
   A Poetic Survey Around Birmingham 189
   promotional medals 190, *190*
Blondel, J.-F. (1705–1774) 31, 34
blue john 9, 57–8, *57*, 161, *169*–79
Bombelles, Marquis de (1744–1822) 71, 72
Booth, Joseph 220
Boswell, James (1740–1795) 6
Boulton & Fothergill 4–5, 23, 52, 55
   archives 110, 115
   pattern books *38*, *39*, 39, 39–40, 40, *44*, 163, *163*
Boulton & Plate Company *see* M. Boulton & Plate Company
Boulton & Scale 4
   archives 111, 115
   pattern book 4, *4*, 143, *143*
Boulton & Smith 110–11
Boulton & Watt 6, 63–70, *64*, *68*
Boulton, Ann (née Robinson) (MB's 2nd wife) (1733–1783) 2, 15, 16, 17, 41, 71, 114, 148, 149, 156
Boulton, Anne (MB's daughter) (1768–1829) 18, 19, 20, 130, 159, 227, 233
   childhood 15, 16–17, 158
   health 157
   letter from MB 228, *228*
   in MB's will 230
   papers 113
   portrait (attrib. Kettle) *16*, 17
Boulton, Christiana (née Piers) 1, 15, 118

Boulton, Hugh William (1821–1847) 113, 114
Boulton, John 148
Boulton, M, & Plate Company *see* M. Boulton & Plate Company
Boulton, M. (firm) *see* M. Boulton
Boulton, M. R. (Button Company) *See* M. R. Boulton (Button Company)
Boulton, Mary Anne (née Wilkinson) (1829–1912) 20, 113–14, 233
Boulton, Mary (née Robinson) (MB's 1st wife) 2, 15
    MB's poem for 15, 149, *149*
Boulton, Matthew, senior (MB's father) (1700–1759) 1, 2, 15, 118
Boulton, Matthew Piers Watt (MB's grandson) (1820–1894) 114, 140, 233
Boulton, Matthew Robinson (Matt) (MB's son) (1770–1842)
    childhood 15, 16–17, *17*, 18, 19, 157, 189
    marriage and family 20, 233
    buys Tew Park, Oxfordshire 20, 112, 233
    and MB's businesses 46, 52, 54, 65, 70, 75, 77–9, 85–6, 97, 140, 210
    and Loyal Birmingham Volunteers 205
    in MB's will 230
    records in archives 103, 111, 112, 113, 114, 115
    portrait (Lawrence and Shee) 232, *232*
    portrait (Liotard) *17*, 149
Bown, John 111
Boyd, Robert 51
Braddon, Paul: *Soho House, Handsworth* 234, *234*
Bradford, Samuel: *Plan of Birmingham* 122, 123
Breda, Carl Frederic von (1759–1818) 87
    Portrait of Matthew Boulton 226, 227
Brown, William D. 110, 111, 113
Brunton, William 103
Buck, Samuel and Nathaniel: *Prospect of Birmingham* 123–4
buckles 122, *122*, 126, 144, *144*
    MB with Wedgwood 147, *147*
Bückling, Carl Friedrich (1750–1812) 75
Buffon, Comte de (1807–1888) 9
Burke, Edmund 159
Büsch, Johann Georg (1728–1800) 75
Busch, John (d. 1830) 86
buttons 41, 111, 121, 122, *122*
    MB with Wedgwood 147, *147*
    Phillp's designs for 148, *148*
Butty, Francis (fl. 1757–1776) 53

Calvert, F.: *Soho from the Nineveh Road . . .* (eng by Radclyffe) 142, *142*
candelabra
    Persian candelabra 35, *36*, 61
    tripod perfume burner with candelabra *37*, 38, 61
    two-light candelabra 185, *185*
candle/candelabra vases 55, *56*, *58*, 169, 170, *170*, 178, 178–9, 181, *181*
    Anderson design (attrib.) *168*, 169
    Cleopatra 170, *170*
    dismantled vase *60*, 61
    goat's head *169*, 169–70
    King's candle vase 59, *172*, 173
    'lion faced' 33, *57*, 58
candlesticks 41, *42*, 125, *125*, 164, 166, *166*, 184

chambersticks 185, *185*
    James Wyatt design 40, 165, *165*
    'lion faced' 33, *42*, *43*, 164, *164*
    Masonic 162, *162*
    rococo style *52*, 53
Carburi, Jean-Baptiste (1722–1804) 71
Carless, Thomas 156
Carmichael, Dr John 228, 233
Carpenter, Charlotte (later Mrs Walter Scott) 19
Catherine the Great, Empress (1729–1796; r. 1762–1796) 33, 45, 55, 62, 75, 175
Caylus, Comte de (1692–1765) 31
    Recueil d'Antiquités Égyptienne, Étrusques, Greques et Romains 33
Chamberlain, Joseph (1836–1914) 20
Chambers, William (1723–1796) 23, 34–5, 38, 39, 58, 59, 159, 170, 173, 175
    'Franco-Italian Album' *34*, 35
    Treatise on the Decorative Part of Civil Architecture 34, *35*
chambersticks 185, *185*
Chantrey, Sir Francis 136, *136*
Charlotte, Queen (1744–1818; r. 1761–1818) 5, 55, 59, 87, 168, 170
chatelaines 144, *145*
Cheshire, William 111, 113, 115, 227, 228
Chester Assay Office 41, 47, 48, 49, 53, 161
Child, Robert (1739–1782) 61, 178
Chippindall, Richard (1751–1826) 45, 46, 62, 111
Christie's, London 59–61, 62, 168, 173, 175, 182
    catalogue 174, *174*
Clay, Henry 71–2
Cliff, Thomas 51
clocks and timepieces
    clock mechanism 127, *127*
    geographical clock 61, 175, *176*
    King's Clock 59, 170–3, *171*
    Narcissus clock 62, 183, *183*
    sidereal clock 35, 61, 175, *175*, *177*
    Titus clock 33, 61, 182, *182*
    watch chains 122, 143, *143*, 144, *144*
    watch key 144
    watch stands 61
    watchcocks 127, *127*
coasters 187, *187*
Cochin, C.-N. 31
coffee pot *186*, 187
    James Wyatt design (attrib.) 39–40, *192*, 193
coins
    Bahamas 213, *213*
    Bermuda 213, *213*
    Bombay *93*, 213
    British copper coinage 89–90, *91*, 94–5, *95*, *214*, 214–16, *215*, 217
    cartwheel 95, *95*, 210, 215, 216, *216*, 217
    Denmark 214, *214*
    'evasive' halfpennies 214, *214*
    Gold Coast 213
    Isle of Man *93*, 94, 213
    Russia 214, *214*

Sierra Leone 93, 94, 213, *213*
Spanish counterfeit/countermarked dollars 216, *217*
Sri Lanka 213, *213*
Sumatra 89, *90*, 93, 213
see also Soho Mint
Cook, Captain James (1728–1779) 89, 224, 225
Cooke, Joseph 156
Copper-Plate Magazine 28–9, *103*, 224
Cotton, George 93
Cottrell, Sir Stephen (1738–1818) 109
cups and covers
    for Cornelius O'Callaghan 44, 166, *167*
    silver gilt 46, *46*
    for Sir George Shuckburgh *186*, 187

Darwin, Erasmus (1731–1802) 2, 7, 8, 9, 10, 11, 12, 30, 109, 133, 198
    letter to MB *130*, 130–1, 157
Davenport & Farrant 156
Davison, Alexander (1750–1829) 97
Dawson, John 197
Day, Thomas (1748–1789) 9, 10, 109, 113, 133
Delafosse, J.-C.: *Nouvelle Iconologie Historique* (1734–1789) 32, 33
Delaney, Mrs (1700–1788) 61
Desgodetz, Antoine (1653–1728): *Les Édifices Antiques de Rome* 33
Dickinson, H. W. 1
dish rings 164, *164*
Divier, John 197
Dixon, Cornelius (d. 1826) 18
Dixon, Gilbert 48
Donisthorpe, George 131
door furniture
    escutcheon and knob (Adam) *38*, 39
    lock 127, *127*
Droz, Jean-Pierre (1746–1823) 86, 90, 95, 113, 211, 215
    copper pattern halfpennies *91*
    sexpartite collar 83, *83*
Duesbury, William (1725–1786) 34
Dumarest, Rambert (1760–1806) 86, 87, 211
Dumée, Nicholas (fl. 1758–1776) 53
Dumergue family 19, 74
Dumergue, Charles (1739–1814) 19, 109, 157, 227
Duncomb, John 125, *126*
Duplessis, Joseph Siffred: *Portrait of Benjamin Franklin 128*, 129
Dutens, Louis 77

East India Company 82, 85, 87, 89, 94, 210, 213, 225
Eckstein, Johannes: *John Freeth and His Circle 196*, 197
Edgar, Dr H. 46
Edgeworth, Maria (1767–1849) 2, 11
Edgeworth, Richard Lovell (1755–1817) 9, 11, 133
egg frame and egg cups 167, *167*
Eginton, Francis (1737–1805) 44, 220, 224
    portrait (Millar) *220*, 220
    Soho Park 14
    Vue des Magasins &c &c appartenants à la manufacture de Boulton & Fothergill . . . 24, *25*

Winter and Summer (polygraphs after de Loutherberg) 221, *221*
Eginton, Francis, junior (?1775–1723) 30, 141, 219
Elkington & Co. 115
Elmsley, Peter (fl. 1774–1781) 87
Enlightenment, Age of 128
entrée dish *186*, 187
epergnes *168*, 169
Equiano, Olaudah (?1745–1797) 11, 197
Ewart, Peter (1767–1842) 84, 227
ewers *see* jugs and ewers

Feilde, Reverend Thomas (1768–1781) 24, 30
Fitzherbert, Mrs Maria (1756–1837) 19, 227
Flaxman, John (1755–1826) 87, 175
    Portrait Bust of Matthew Boulton 136, 136
Fothergill, John (1730–1782) 3–4, 15, 45, 52, 55, 59, 61, 62, 71, 194
    portrait (Schaak) *3*
    see also Boulton & Fothergill
Fox, George C. (1752–1807) 27
Franel, Joseph (d. 1801) 30
Franklin, Benjamin (1706–1790) 2, 5–6, 7, 8, 71, 109, 124
    letter to MB 137, *137*
    portrait (Duplessis) *128*, 129
Freeth, John 197
French Revolution 201–2
    medals *96*, 97, 212

Galton, John 197
Galton, Samuel, junior (1753–1832) 131, 133, 137, 198
Garbett, Samuel (1717–1803) 49, 50–1, 66, 81, 109, 110, 113, 124, 225
George III (1738–1820; r. 1760–1820) 5, 35, 55, 84, 86, 87, 168, 170, 182
    medal commemorating recovery from illness 95, *96*
George, Prince of Wales (later George IV) (1762–1830; r. 1820–1830) 19, 61, 184, 227
    medal commemorating marriage 97, *97*, 212
Gilpin, William 159
Goede, Christian 79
Goldsmith's Hall 49, 51
Gorden, Reverend William (d. 1837) 115
Gori, Antonio Francesco: *Museum Florentinum* 33
Gouthière, Pierre (1732–1813) 33
Graham-Gilbert, John (1794–1866): *William Murdock* 67
Greaves, Mary 156
Grimm, Baron Friedrich von (1723–1807) 32
Gule, J. E. 3
guns/gunmaking 125
    flintlock muskets 202
    flintlock pistol 126, *126*
Gustavus III of Sweden (1746–1792), medals 87, 95

Hamilton, Lady Emma (c.1761–1815) 79
Hamilton, Sir William (1730–1803) 34, 79, 109
Hancarville, Baron d' (1719–1805) 34
Hancock, John Gregory (b. c.1749) 86
Hancock, Joseph (d.1791) 184

Hanson, Thomas: *Plan of Birmingham* 189
Harbord, Sir Harbord (1734–1810) 53, 179
Harrison, Joseph (d. c.1790) 86
Hausted, Reverend John 2
Hector, Edmund 124, 157
Heming, Thomas 184
Henderson, Logan 11
Herschel, Sir William (1738–1822) 17, 21, 109
Hertford, Lady (1726–1782) 43
Hodges, John (d. 1808) 45, 46
Hollins, William (1763–1843) 27, 159
Holstein-Gottorp, Duke of (1755–1829) 45
Hornblower, Jonathan (1753–1815) 109
Howe, Admiral Richard (1726–1799): medal to commemorate victory 87, 88, 97, 212
Huntsman, Benjamin (1704–1776) 8, 83
Huquier, Gabriel (1672–1742) 33
Hutton, Catherine 196
Hutton, William (1723–1815) 124, 189

ice pails 58
illumination of Soho Manufactory, 1802 204

Jackson, Sir George (1725–1822) 93
James Watt & Co 70, 115
    copying process 68, 103, 222, *222*
Jefferson, Thomas (1743–1826) 8
    letter to William Small *198*, 199, *199*
Johnes, Colonel Thomas 98
Johnson, Dr Robert Augustus (1745–1799) 9, 135
Johnson, Samuel (1709-1784) 124
Jones, Inigo (1573–1652) 31
jugs and ewers
    ewers 179, *179*
    flagon from St Bartholomew's Chapel, Birmingham 165, *165*
    James Wyatt design *39*, 40
    water and wine jug 166, *166*
    wine jug 53, *53*
Jupp, William (d. 1788) 23–4

Kauffman, Angelica (1741–1807) 72, 220
Kedleston Hall, Derbyshire 32, 38, 39
Keir, James (1745–1820) 7, 9, 10, 12, 19, 62, 81, 109, 133, 136, 137, 178
Kellet, John 86
Kellet, Thomas (d. 1823) 101
Kempson, Peter, tokens issued by 190, *190*
Kettle, Joseph 51
Kettle, Tilly (1735–1786): (attrib.) *Miss Anne Boulton 16*
King, Rufus 200
Knight, Richard Payne 159
Küchler, Conrad Heinrich (d. 1810) 86, 87, 93, 95, 97, 212, 231

latchets 45, 105, 110–11
Lavoisier, Antonine-Laurent 109
Lawrence, Sir Thomas (1769–1830)
    Portrait of James Watt 207, 208
    Portrait of Matthew Robinson Boulton (completed by Shee) 232, *232*
Lawson, James (1760–1818) 83, 84, 85, 113
Le Blanc, the Abbé (1707–1781) 31
Le Geay, Jean-Laurent (c. 1710–1786) 31, 33, 35
Le Lorrain, L.-J. (1715–1759) 31
Lee, Mr 74
Lettice, John (1737–1832) (with Martyn): *Antiquities of Herculaneum* 34
Lines family: (attrib.) *Soho House, Handsworth 234*
Liotard, Jean Etienne (1702–1789): *Miniature Portrait of Matthew Robinson Boulton 17*, 149
Liverpool, Robert Jenkinson, 2nd Earl of 94
Ljungberg, Jøns Matthias (b. 1748) 77
    drawing of rolling mill *76*
Lloyd, Samson, II (1699–1779) 124
London Assay Office 49, 51, 161
London Banking Agency, archives of 110
Loutherberg, Philippe de (1740–1812) 220
    Winter and Summer (polygraphs by Eginton) 221, *221*
Loyal Birmingham Light Horse Volunteers (est. 1797) 203
Loyal Birmingham Volunteers (est. 1803) 115, 205
Luc, Fanny de (d. 1824) 79
Lunar Society 2, 7–13, 16, 133–5, 137, 191, 197, 198

M. & R. Boulton, J and G. Watt, & Company 110
M. Boulton & Plate Company 45–6, 52, 111, 115, 162
M. Boulton (est. 1782) 110, 115
M. R. Boulton (Button Company) 111
Macartney, George, 1st Earl (1737–1806) 109
MacLeish, Robert, junior
    Plans of the Principle Building (Soho Manufactory) *101*
Magnificent Directory of Birmingham 29, *29*, 30
Major, Thomas (1714–1799): *The Ruins of Paestum* 34
Manchester, Duchess of: ormolu cabinet for 62, *180*, 181
Marigny, Marquis de (1727–1781) 31
Marsden, John 88
Martin, Major-General Claude (1735–1800) 87
Martyn, Thomas (1735–1825) (with Lettice): *Antiquities of Herculaneum* 34
Mason, William 204
Matthew Boulton Plate Company *see* M. Boulton & Plate Company
Matthews, Charlotte (d. 1801) 17, 109, 110, 113
Matthews, William (d. 1792) 17, 109, 110, 113
May, Benjamin 49
Mayhew and Ince: cabinet for Duchess of Manchester 62, *180*, 181
mazerine or strainer dish *52*, 53
medals 86–8, 89, 95–8, 212–13
    Birmingham Loyal Association 204, *204*
    Birmingham volunteer regiments 203
    Bisset's commission 190, *190*
    Cornwallis medal *87*, 88, 97, 212
    George, Prince of Wales, marriage *97*, 212
    George III, recovery from illness 95, 212
    George Washington 200, *200*
    Hafod Friendly Society 98, *98*, 213
    Howe's victory, Glorious First of June *87*, 88, 97, 212
    King of Naples, restoration 96

Louis XIV (1638–1715; r. 1643–1715) 97, 212
Marie Antoinette, execution *96*, 97, 212
medallic scale medal 211, *211*
memorial medals for MB 230, *230*, *231*, 231–2
Monneron Frères 87, *92*, *93*, 217
'Otaheite' ('Resolution and Adventure') 89, *90*, 224
St Albans Female Friendly Society 98
Trafalgar medal *96*, 97, 98, 202, *202*
Union of Britain and Ireland 212, *213*
see also Soho Mint
Michell, John (1724–1793) 8
Middlehurst, John 86
Millar, James (1735–1805)
  (attrib.) *An Allegory of Wisdom and Science 128*, 129
  Portrait of Francis Eginton 220, *220*
  (attrib.) *Portrait of Matthew Boulton* 139, *139*
Monneron Frères: tokens and medals 87, *92*, *93*, 217
Montagu, Mrs Elizabeth (1720–1800) 32, 43–4, 109, 158, 163, 220, 224
  dinner services for *43*, 43–4, 53–4, *54*, *162*, 163, *163*, 184
Montfauçon, Bernard de (1655–1741): *L'Antiquité Expliquée et Représentée en Figures* 33, 183
Monthly Magazine, The 28, *28*, 141
Moore, Joseph 191
Mosley, John 110
Motteux, John 74
Murdock, William (1754–1839) 11, 66, 70, 103, 105, 107, 135, 204
  collections of *84*, 219
  model of locomotive 138, *138*
  portrait (Graham-Gilbert) *67*
Mynd, Ann (Nancy) (1763–c.1807) 112, 148, 158
Mynd, George 148
Mynd, Thomas 148

nails, handmade 125, *125*
Napoleonic Wars 74, 202–5
Nelson, Horatio (1758–1805) 21, 79, 227
  portrait plaque (Andras) 202, *202*
  Trafalgar medal *96*, 97, 98, 202, *202*
neo-classicism 31–40, 55, 58
Neufforge, J.-F. de (1714–1791) 31
  Recueil Élémentaire d'Architecture 32, 33
New Birmingham Directory and Gentleman's Compleat Memorandum Book 24–5, *25*
New Street Theatre (later Theatre Royal), Birmingham 191, 194–5, *195*
  passes 194, *194*
  playbill 194, *194*
Newcomen, Thomas (1663–1729): steam engines 63, 65, 66, 69
Newton, James (1773–1829) 18, 21
  japanned armchair 154, *154*
  Kismos chair 154, *154*
Newton, John 197
Nichols, Mr (diemaker) 83
Nicholson, Sarah (Sally) (d. 1839) 19, 227
Nicholson, William (1753–1815) 130
Noailles, Vicomte de (1756–1804) 40

Nollet, Abbé (1700–1770) 7
Norton, Charles: *Proposals . . . for Building the Crescent in Birmingham* 196

O'Callaghan, Cornelius, cup and cover for 44, 166, *167*
ormolu 5, 32–3, 41, 55–62, 161, 168–83
orrery *131*, 131–2

Palladio, Andrea (1508–1580) 31
Parlby, Reverend Samuel 189
Parrott, Richard 66
Patte, Pierre (1723–1814) 31–2
Paul, Emperor of Russia (1754–1801; r. 1796–1801) 87
Pearson, James (d. 1804) 111
Pearson, Thomas 28, 132
Pedley, J. D.: *A Bird's Eye View of Vauxhall Gardens 195*, 195–6
Peploe, John (d. c.1805) 86
perfume burners 55, 61
  tripod perfume burner with candelabra *37*, 38, 61
perfume burner vases 174, *175*
  sphinx vase perfume burners 59, 173, *173*
  Venus vase perfume burner 33, *56*, 61
Pether, William (1713–1819): *A Philosopher Giving a Lecture on the Orrery* (after Wright) *13*, 131
Petit, Dr 8
Petitot, E.-A. (1727–1801) 31, 35
pewter 125, *125*, *126*
Peyre, Marie-Joseph (1730–1785) 34
Phillips, Richard (1767–1840) 28, 86
Phillp, John (d. 1815) 27, 86, 98, 224
  portrait (Rouw) 150, *151*
  designs
    buttons 148, *148*
    cane head terminal 205
    Hafod medal 213
    halfpennies 211
    MB's coffin plate 229, *229*
    metalware 150
  drawing of a Gothic building 132, *132*
  drawing of a sphinx *158*, 159
  drawings of machinery at the Soho Mint 210, *210*, *211*
  drawings of objects from Otaheite 225
  painting of a spaniel's head 150, *151*
  The Handsworth Troop of Horse drilling 205, *205*
  Lawn and Park at Soho House 152, *152*
  Soho House Stables 152, *152*
  Soho Pool with the Boat and Boathouse, and *Soho Pool with the Boathouse and a Side View of the Manufactory* 153, *153*
  Temple of Flora, and *Soho Pool and Garden Buildings* 152–3
  Temple of Flora, Soho 18
  View across Birmingham Heath 141
  View across Soho Pool 142, 143
  View across the roof of the Manufactory 27, *27*
  View of Handsworth Church taken on the Spot 230, *230*
  View of Soho Manufactory 141
  View of Soho Manufactory taken on the Spot 141, *141*
  View of the Soho Foundry 218, 219

View over the Soho Manufactory 152
Views of Soho House from Birmingham Heath 151, *151*
Views of the Hermitage 152, *153*
Phipson, Thomas 71
Pidgeon, G. F. 231
Pingo, Lewis (1743–1830) 215
  copper pattern halfpennies *91*
Piranesi, Giambattista (1720–1778) 32, 33–4
Pitt, William, the Younger (1759–1806; PM 1783–1801, 1804–1806) 81, 83
Pocock, Nicholas: *The Battle of Trafalgar* . . . 203, *203*
polygraphs 220–1, *221*
Ponthon, Noel-Alexandre (1769/70–1835) 86, 87
Price, Uvedale 159
Priestley, Joseph (1733–1804) 7, 8, 9, 10, 11, 12, 13, 109, 135, 198, 201, 202
Prieur, Jean-Louis (c.1725–c.1785) 33
Pryce, William: *Mineralogia Cornubiensis* 132
purse 144
purse rings 122

Radclyffe, T. : *Soho from the Nineveh Road* . . . (eng. after Calvert) 142, *142*
Raspe, Rudolf (1736–1794) 109
Ravensworth, Lord 43
Rawst(h)orne, John (fl. 1780s)
  Perspective View of the Crescent . . . (eng. by Tukes) 196
  Proposed plan and elevation for alterations to Soho House 12, 13, *130*
Reichenback, Georg Friedrich von (1771–1826) 77
  sketch of lap engine *76*
Rennie, John (1761–1821) 83, 109
Revett, Nicholas (1720–1804) (with James Stuart): *The Antiquities of Athens* 34
Reynolds, Sir Joshua (1723–1792) 32, 220
  A Portrait of Dr Ash 191, 192
Richardson, Dorothy (1748–1819) 59
Richardson, George (1737/8–1813): *Iconology* 33
Richmond, Duke of 42
Riddell, Robert Andrew 28
Ripa, Cavaliere Cesare (1560?–1625?): *Iconolgia* 33
Roberts, John 111
Robinson, Luke (1683–1749) 2
rococo style 31, 32, *52*, 53, 55
Roebuck, John (1718–1795) 63, 109, 113
Rooker, Michael Angelo (1746–1801) 24
Rouw, Peter (1770–1852) 87, 231
  Portrait of John Phillp 150, *151*
Rowlandson, Thomas
  Reform advised. Reform begun. Reform compleat. 201, *201*
  Transplanting of teeth 157, *157*
  The Wonderful Charms of a Red Coat & Cockade 203, *203*
Royal Society, London 2, 6, 74
Rudge, Edward: three soldiers of the Birmingham Loyal Association (eng. by Forbes) 203, *203*
Russell, John (1745–1806)
  The Face of the Moon 134, 135
  Sir Joseph Banks 5
Ruston, Edward 156
Ryland, Samuel 71

Sabakin, Lev Fedorovich (1746–1813) 75
St Bartholomew's Chapel, Birmingham: silverware 165, *165*
St Fond, Faujas de: *Descriptions des expériences de la Machine Aérostatique de MM de Montgolfier 11*, 132
Saly, J.-F. (1717–1776) 31, 33, 35
Saunders, E. Gray: sphinxes 159, *159*
Saunders, George 27
Scale, John (1736/7–1793) 4, 111, 113, 114, 225
  see also Boulton & Scale
Scasebrick, John (d. 1785) 49
Schaak, J. S. C.
  John Fothergill 3
  Portrait of Matthew Boulton 160, 161
Schweppe, Jacob: bill for 'alkaline water' 156
Scott, Charlotte (née Carpenter) 19
Shakespear & Johnson 204
Shaw, Reverend Stebbing 30, 83
  History and Antiquities of the County of Staffordshire 14, 24, 30, 141
Shee, Sir Martin Archer (1769–1850): (with Lawrence) *Portrait of Matthew Robinson Boulton* 232, *232*
Sheffield Assay Office 42, 48, 49, 50, 51
Sheffield plate 4, 41–6, 184–9
Shelburne, Lady (1745–1771) 15, 71, 223
Shelburne, Lord (1737–1805) 15, 47, 59, 71, 223
Shuckburgh, Sir George (1742–1804), cup and cover for *186*, 187
Siddons, Sarah (1755–1831) 19, 21
silver 4, 32–3, 41–6, 162–7
  see also Birmingham Assay Office; Sheffield plate
Simcox, George 71
Skipwith, Thomas 49
slavery 11, 197–8
Small, Dr. William (1734–1775) 6, 8, 9, 10, 63, 109, 135
  Jefferson's letter to *198*, 199, *199*
  portraits *5*, 156, *156*
Smeaton, John (1724–1792) 63, 69
Smiles, Samuel 1, 2
Smith, Benjamin 105
Smith, James 105, 110
snuffboxes 124, *124*
snuffer tray 187, *187*
Soane, Sir John 39
Soho Estate 20, 139
  axonometric projection *100*
  plans of *19*, *112*, 151, *151*
  Soho Park (Eginton) *14*
  Soho Pool with the Boat and Boathouse, and Soho Pool with the Boathouse and a Side View of the Manufactory (Phillp) 153, *153*
  'Sphinx Walk' 159, *159*
  Temple of Flora, and Soho Pool and Garden Buildings (Phillp) 152–3
  Temple of Flora, Soho (Phillp) *18*
  View across Birmingham Heath (Phillp) 141
  View across Soho Pool (Phillp) 143, *143*

Views of the Hermitage (Phillp) 152, *153*
　walking tour 99–107
Soho Foundry 65, 67, 70, 104, 105–7, *106*, 205, *218*, 219
　View of the Soho Foundry (Phillp) *218*, 219
Soho House 9, 15–21, *21*, 71, *111*, 159, 235, *235*
　archives 111, 112
　James Wyatt's rear elevation for the extended Soho House, 1796 20
　Lawn and Park at Soho House (Phillp) 152, *152*
　Proposed plan and elevation for alterations to Soho House (Rawstorne) *12*, *13*, 130
　Soho House, Handsworth (attrib. Lions family) 234, *234*
　Soho House, Handsworth (Braddon) 234, *234*
　Soho House Stables (Phillp) 152, *152*
　Views of Soho House from Birmingham Heath (Phillp) 151, *151*
Soho Insurance Society 25–6
　Soho Manufactory (from poster) *26*, 222
Soho Manufactory 3, 20, 23–4, 41, 99–104, 124, 139, 140, 233
　plans of *101*, *102*, 140
　visitors to 71–9, *72*, *73*, *78*, 99, 161, 223, 224
　visual images 2, 22, 23–30, *24*, *25*, *26*, *28*, *29*, *103*, 117–18, 223, 233
　　illuminations for the Peace of Amiens 204, *204*
　　Soho from the Nineveh Road . . . (Radclyffe aft. Calvert) 142, *142*
　　View across the roof of the Manufactory (Phillp) 27, *27*
　　View of Soho Manufactory (Phillp) 141
　　View of Soho Manufactory taken on the Spot (Phillp) 141, *141*
　　View over the Soho Manufactory (Phillp) 152
　　Vue des Magasins &c &c appartenans à la manufacture de Boulton & Fothergill . . . (Eginton) 24, *25*
Soho Mint 6, 74, 79, 81–8, 89–98, 104, *104*–5, 210–17, 234
　archives 110
　coining presses *82*, *83*
　Phillp's drawings of machinery 210, *210*, *211*
　see also coins; medals; tokens
Soho Plate Company 140
Soho Rolling Mill archives 111
Soufflot, J.-G. (1713–1780) 31
Southern, John (1761/2–1815) 85, 103, 104
Spencer, Dowager Lady 98
sphinxes
　Phillp *158*, 159
　Saunders 159, *159*
　sphinx vase perfume burners 59, 173, *173*
　sphinx vases 61
steam engines 5–6, 63–70, *64*, *68*, 205–9
　fragment of cylinder 209
　model of lap engine 209, *209*
　model of locomotive (Murdock) 138, *138*
　model of rotative beam engine *208*, 209
　sketch of lap engine (Reichenbach) *76*
Stein, Heinrich Friedrich vom 75
Stieglitz, Henry 110
Stokes, Dr Jonathan (1755–1831) 10, 135
Stretch, L. M. 157
Stuart, James 'Athenian' (1713–1788) 32, 34, 35–9, *39*, 42, 169, 174, 220
　(with Nicholas Revett) *The Antiquities of Athens* 34
sweetmeat dish 167, *167*

Swinney, Myles 24
swords 146, *146*, *147*
　patterns for sword hilts *4*, 143, *143*

Tablet, The (almanac) 28
Tangye, George (1835–1920) 109
tankard 186, *187*
Taylor, Edward Richard: The Crescent, Cambridge Street 196
Taylor, John (1711–1775) 71, 121, 122, 124
tea urns/tea vases 174, *174*, 184, *184*, 193, *193*
　Boulton family tea vase 154, *154*, *155*
Tew Park, Oxfordshire 20, 112, 233
Theatre Royal, Birmingham *see* New Street Theatre, Birmingham
Thomason, Edward (1769–1848) 86
Thompson, Sir Benjamin (1753–1814) 77
timepieces *see* clocks and timepieces
tokens 90–3
　Birmingham workhouse 190, *190*
　Bishops Stortford *92*, 93, 212
　Cronebane halfpennies (Associated Irish Mine Co.) 91, *92*, 212
　Glasgow 91, *92*, 212
　Inverness 91, 212
　Monneron Frères *92*, 93, 217
　Parys Mine Co. 91, 212
　Pen-y-Darran Ironworks truck tickets 212, *212*
　Southampton 91, 212
　see also Soho Mint
Tongue, William: apprenticeship indenture 209, *209*
tooth transplants 157
toy-making 2, 3–4, 41, 121–2
trade tokens *see* tokens
treadle lathe 121, *121*
Tuffen, John Furnell (d. 1820) 20
Tukes, Francis: *Perspective View of the Crescent . . .* (eng. after Rawstorne) 196
tureens
　designs incl. Admiralty tureen *38*, 39
　for Mrs Montagu *43*, *43*, 53–4, *54*

vases 55–8, 61
　Bacchanalian vase 62, 183, *183*
　black basalt vase and cover (Wedgwood & Bentley) 137
　see also candle/candelabra vases; perfume burner vases
Vion, François (1764–1800) 33
Virly, Grossart de (1754–1805) 72
Vivian, John 113
Vulliamy, Benjamin (1747–1811) 62, 170, 182

Wailly, Charles de (1730–1798) 34, 35
Wakelin, Inigo 49
Walker, J.: engraving of Soho Manufactory *103*, 224
Walker, John (fl. 1784–1802) 29
Walker, Zaccheus, junior ('Zack') (1768–1822) 148, 225
Walpole, Horace (1717–1797) 59
Warwick, Earl of 59
watch chains 122, 143, *143*, 144, *144*

watch key 144
watch stands 61
watchcocks 127, *127*
Watt, Gregory (1777–1804) 19, 110
Watt, James (1736–1819)
    and Beechey's portrait of MB 208
    burial at Handsworth Church 231
    copying machine 68, 103, 222, *222*
    steam engines 6, 16, 17, 63, 65, 66, 69, 75, 205, 209
    and Lunar Society 2, 7, 8, 9, 11, 12, 135
    and MB's death 20, 81, 227, 229
    and the *Monthly Magazine* 28
    papers and letters 109, 113
    portrait (Lawrence) *207*, 208
    see also Boulton & Watt; James Watt & Co
Watt, James, junior (1769–1848) 17, 19, 28, 208, 229
    and Anne Boulton 233
    and the London Banking Agency 110
    and Lunar Society 137
    and the Soho businesses 65, 70, 77–9, 88, 89, 104, 111, 219
Weale, James, junior 115
Wedgwood, Josiah (1730–1795) 1, 34, 55, 62, 72, 135
    Bust of George Washington 200, *200*
    joint ventures with MB 56, 59, 135, 147, *147*
    and Lunar Society 7, 8, 9, 11, 12, 197, 198
    on MB 3, 161
    MB's correspondence with 109, 113
    Wedgwood & Bentley black basalt vase and cover 137
Wesley, John (1703–1791) 124
West, Benjamin 220
Westley, William
    Plan of Birmingham opp. *1*, 122, 123, *123*
    Prospect of Birmingham 123
Westwood, Arthur (1865–1957) 47, 51, 52
Westwood, Henry 52
Westwood, John, senior (d. 1792) 86, 89
Wheeler, Robert 202
Whitehurst, John (1713–1788) 2, 8, 10, 57, 109, 113, 135, 175, 198

angle barometer 138, *138*
wig powderer box 188, *188*
Wilkes, Elizabeth 158
Wilkinson, John (1728–1808) 6, 63, 65, 91, 109, 113, 209, 212
Wilkinson, Joseph 50
Wilkinson, William 219
Williams, Thomas (1730–1802) 85, 90, 91, 212
Wilson, Benjamin (1721–1788) 7
Wilson, James 20
wine coolers 185, *185*
wirework 166–7
Withering, Dr William (1741–1799) 2, 9, 10, 11, 12, 13, 17, 26, 110, 135
    An Account of the Foxglove 10, *10*
Wood, Robert (1717–1771)
    Ruins of Balbec 34
    Ruins of Palmyra 34
Woodcock's Bank, Enniscorthy 93
Woodward, John 110, 113
Woronzow, Count (1744–1832) 21, 110
Woulfe, Peter (1727–1803) 9
Wright, Joseph, of Derby (1734–1797)
    An Experiment on a bird in the Air Pump 129, *129*
    An Iron Forge 120, 121
    A Philosopher Giving a Lecture on the Orrery (Pether after Wright) *13*, 131
Wright, Thomas 170
Wyatt, Benjamin (1755–1813) 21, 110, 112
Wyatt, James (1746–1813) 23, 34, 39–40, 44, 53, 110
    candlesticks 40, 165, *165*
    coffee pot, burner and stand (attrib.) 39–40, *192*, 193
    James Wyatt's rear elevation for the extended Soho House, 1796 20
    jug design from pattern book *39*, 40
Wyatt, Samuel (1737–1807) 13, 18, 23, 39, 110
Wyatt, William (1734–1780) 23
Wynn, Sir Watkin Williams (1742–1789) 61, 220
Wyon, Thomas (1767–1830) 86

Yenn, John (1750–1821) 35

Considerations upon the propriety of *Buying Soho* and the [...]
Mr. Scale's, Mr. Ford's pool; The Mill pool with the Land in the Slade; N[...]
Reasons, *For*, and *Against*, such purchase considered in two p[...]
First, As it relates to my *Health* and *Happiness*; and [...]

If I do *not* buy Soho, the Consequences will be, That Houses, Workshops & other Nuisances may be built upon the Lawn facing the Parlour Windows —

I *shall* be deprived of a View of the Birmingham Churches, and all that contributes to render Soho a chearfull pleasant Spot —

I shall have the Mortification to see the Plantations & Shruberies which I have raised, Cut down & applied to other mens Uses, And the Soil I have collected Spread on other mens Grounds —

I shall, with pain, see the Flower Garden I have formed, & its Temple, destroyed; My pools drained; My Cascade & Water falls no more please the Eye or the Ear —

My Mill pool will be taken from me & the Spirit of the Manufactury thereby abated —

I shall be driven from my Habitation (which alone can bring to my remembrance the agreeable circumstances of early days) and in my Old Age be obliged to seek another Dwelling, I know not where. I shall, probably, be plunged into disputes and Law-Suits, about Roads, Water courses, boundries &c: to the destruction of my peace and Health —

The Monument I have raised to myself, with so much pains & Expence, I shall see Gliding upon the Wings of Time into the possession of other Families —

The Value of my Buildings will daily diminish; insomuch, that if they were to be offered to Sale, after my decease, they would not bring half the Value, upon a Sixty Years Lease, which they would do, provided the Land, Water, and Buildings, were *Freehold* —

After living so many Years in the Country, I can not live in a Great Town nor far from the various conveniences which Great Towns afford —

Nor should I like to live in any mans House that will not let me drive a Nail, or hang up a picture, or suffer me to have a Library, or other conveniences which [...]